PLANETARY SYSTEMS AND THE ORIGINS OF LIFE

Several major breakthroughs in the last decade have helped to contribute to the emerging field of astrobiology. These have ranged from the study of micro-organisms, which have adapted to living in extreme environments on Earth, to the discovery of over 200 planets orbiting around other stars, and the ambitious programs for the robotic exploration of Mars and other bodies in our Solar System. Focusing on these developments, this book explores some of the most exciting and important problems in this field.

Beginning with how planetary systems are discovered, the text examines how these systems formed, and how water and the biomolecules necessary for life were produced. It then focuses on how life may have originated and evolved on Earth. Building on these two themes, the final section takes the reader on an exploration for life elsewhere in the Solar System. It presents the latest results of missions to Mars and Titan, and explores the possibilities for life in the ice-covered ocean of Europa. Colour versions of some of the figures are available at www.cambridge.org/9780521875486.

This interdisciplinary book is a fascinating resource for students and researchers in subjects in astrophysics, planetary science, geosciences, biochemistry, and evolutionary biology. It will provide any scientifically literate reader with an enjoyable overview of this exciting field.

RALPH PUDRITZ is Director of the Origins Institute and a Professor in the Department of Physics and Astronomy at McMaster University.

PAUL HIGGS is Canada Research Chair in Biophysics and a Professor in the Department of Physics and Astronomy at McMaster University.

JONATHON STONE is Associate Director of the Origins Institute and SHARCNet Chair in Computational Biology in the Department of Biology at McMaster University.

Cambridge Astrobiology

Series Editors

Books in the series:

1. Planet Formation: Theory, Observations, and Experiments
   Edited by Hubert Klahr and Wolfgang Brandner
   ISBN 978-0-521-86015-4
2. Fitness of the Cosmos for Life: Biochemistry and Fine-Tuning
   Edited by John D. Barrow, Simon Conway Morris, Stephen J. Freeland, and Charles L. Harper, Jr.
   ISBN 978-0-521-87102-0
3. Planetary Systems and the Origins of Life
   Edited by Ralph Pudritz, Paul Higgs, and Jonathon Stone
   ISBN 978-0-521-87548-6

PLANETARY SYSTEMS AND THE ORIGINS OF LIFE

RALPH PUDRITZ, PAUL HIGGS, JONATHON STONE
McMaster University, Canada

CAMBRIDGE UNIVERSITY PRESS
Cambridge, New York, Melbourne, Madrid, Cape Town, Singapore, São Paulo

Cambridge University Press
The Edinburgh Building, Cambridge CB2 8RU, UK

Published in the United States of America by Cambridge University Press, New York

www.cambridge.org
Information on this title: www.cambridge.org/9780521875486

First published 2007

Printed in the United Kingdom at the University Press, Cambridge

A catalogue record for this publication is available from the British Library

ISBN 978-0-521-87548-6 hardback

Contents

Contributors

Denis Baurain
Department de Biochimie
Université de Montréal
Montréal, Quebec
Canada

Szaniszló Bérczi
Department of General Physics
Cosmic Material Research Group
Eötvös University
Budapest
Hungary

Henner Brinkmann
Department de Biochimie
Université de Montréal
Montréal, Quebec
Canada

David Deamer
Department of Chemistry & Biochemistry
University of California, Santa Cruz
Santa Cruz, CA
USA

Tibor Gánti
Collegium Budapest
Institute for Advanced Study
Budapest
Hungary

J. Peter Gogarten
Department of Molecular & Cell Biology
University of Connecticut
Storrs, CT
USA

Richard Greenberg
Lunar and Planetary Observatory
University of Arizona
Tucson, AZ
USA

Paul G. Higgs
Department of Physics & Astronomy
McMaster University
Hamilton, Ontario
Canada

András Horváth
Collegium Budapest
Institute for Advanced Study
Budapest
Hungary

Ákos Kereszturi
Collegium Budapest
Institute for Advanced Study
Budapest
Hungary

Chris P. McKay
NASA Ames Research Center
Moffett Field, CA
USA

Hervé Philippe
Department de Biochimie
Université de Montréal
Montréal, Quebec
Canada

Tamás Pócs
Department of Botany
Eszterházy Károly College
Eger
Hungary

Alexandra Pontefract
Department of Biology
McMaster University
Hamilton, Ontario
Canada

Ralph E. Pudritz
Department of Physics &
Astronomy
McMaster University
Hamilton, Ontario
Canada

François Raulin
Universités Paris 7 et Paris 12
Creteil
France

L. Jeremy Richardson
NASA Goddard Space Flight Center
Greenbelt, MD
USA

Lynn Rothschild
NASA Ames Research Center
Moffett Field, CA
USA

Sara Seager
Department of Earth,
Atmospheric and Planetary
Sciences
Massachusetts Institute of Technology
Cambridge, MA
USA

Adolf Seilacher
University of Tubingen
Tubingen
Germany

András Sik
Department of Physical Geography
Eötvös University
Budapest
Hungary

Karl O. Stetter
Lehrstuhl fur Mikrobiologie
Universiät Regensburg
Regensburg
Germany

Jonathon Stone
Department of Biology
McMaster University
Hamilton, Ontario
Canada

Eörs Szathmáry
Collegium Budapest
Institute for Advanced Study
Budapest
Hungary

Edward W. Thommes
Canadian Institute for Theoretical
Astrophysics
University of Toronto
Toronto, Ontario
Canada

Olga Zhaxybayeva
Department of Biochemistry &
Molecular Biology
Dalhousie University
Halifax, Nova Scotia
Canada

Shay Zucker
Geophysics & Planetary Science
Department
Faculty of Exact Sciences
Tel Aviv University
Tel Aviv
Israel

Preface

The inspiration for this book arises from the creation of the Origins Institute (OI) at McMaster University, which formally started operating in July 2004. Many of the greatest questions that face twenty-first century scientists are interrelated in fundamental ways. The OI was established to address several of these major interdisciplinary questions from within a broad framework of 'origins' themes: space-time, elements, structure in the cosmos, life, species, and humanity.

The origin of life has a privileged position in this great sweep of scientific endeavour and ideas. It addresses, arguably, the most surprising and most fundamental transition to have arisen during the entire evolution of the universe, namely, the transformation of collections of molecules from the inanimate to animate realm. Substantial progress in solving this great problem has been achieved relatively recently but may be traced back to ideas first proposed by Darwin. The great excitement in our era is the realization that physical properties of planetary systems play an important role in setting the stage for life, and that microbial life, on Earth at least, is incredibly robust and has adapted itself to surprisingly 'extreme' conditions. Progress can be traced to four scientific revolutions that have occurred over the last two decades:

(i) the discovery, since 1995, of over 200 extrasolar planets (one which is only 7.5 times more massive than the Earth) around other stars and the possibility that at least a few of these systems may harbour life-sustaining planets;
(ii) the discovery of extremophile microorganisms on Earth that have adapted to conditions of extreme temperatures, acidity, salinity, etc., which considerably broadens the range of habitats where we might hope to find life on other planets in our solar system and other planetary systems;
(iii) the rapid advances in genome sequencing that enable comparative analysis of large numbers of organisms at the whole genome level, thereby enabling the study of evolutionary relationships on the earliest branches of the tree of life; and

(iv) the enormous efforts being made by National Aeronautics and Space Agency (NASA) and European Space Agency (ESA) (and, more recently, the Canadian Space Agency (CSA)) to send robotic probes to search for water, biomolecules, and life on Mars and Titan and possibly the ice-covered, oceanic moon of Jupiter – Europa.

These are some of the major drivers of the emergent science of astrobiology and were the central themes explored during the two-week conference and workshop sponsored by the OI and held at McMaster University in Hamilton, Ontario, Canada, on 24 May–4 June 2005. Our conference featured invited review lectures as well as invited and contributed talks from many of the international leaders in the field (for a full list, please consult the conference internet site at http://origins.mcmaster.ca/astrobiology/).

How to use this book – a user's manual

The chapters of this book are derived from invited, one-hour review talks, as well as a few invited shorter talks, given at the conference. These constitute an outstanding set of lectures delivered by masters of fields such as planetary science, evolutionary biology, and the interdisciplinary links between the two. One of our major goals was to create a volume that would be useful for teaching an interdisciplinary audience at the level of senior undergraduate or junior graduate students. Our intent was to capture the exciting interdisciplinary research atmosphere that attendees experienced at the conference and, thereby, create a volume that is an excellent resource for research. The OI plans to use this volume for a third year undergraduate course about the origins of life, which will be offered for the first time in 2006. The authors were all aware of these two aspects of the book as they prepared their manuscripts. To accommodate and educate a broad interdisciplinary audience, we have tried to ensure that technical jargon is kept to a minimum without compromising scientific accuracy and a clear analysis of the important principles and latest results at an advanced scientific level.

The editors made every effort to keep the authors of individual chapters informed of the content of related chapters. All of the chapters in this book were peer reviewed by arm's-length experts in relevant fields. In addition to receiving useful referee reports, the authors also received comprehensive comments from the editors designed to help integrate their chapters with other related chapters. We hoped by these means to create an integrated book of the highest scientific standard and not just a collection of unrelated review talks that are typical of many conference proceedings. The users of this book will be the ultimate judges of how well we succeeded in attaining this goal.

There are three parts of this book. The first takes the reader from the domain of planetary systems and how they are formed, through the origins of biomolecules and water and their delivery to terrestrial planets. It then focuses on general questions about how the genetic code may have appeared and how the first cells were assembled. These chapters marshal general arguments about the possible universality of basic processes that lead to the appearance of life, perhaps on planets around most stars in our Galaxy and others.

The second part – life on Earth – begins with an exploration of microbial life on our planet and how it has adapted to extreme environments. These are analogous to environments that will be explored on Mars and other worlds in the Solar System. The part then moves on to the results of genomics – as exploited by phylogenetic methods. This allows us to explore the interrelationships of organisms to try to create a tree of life. This is central to efforts designed to address what the earliest organisms might have been like, and two chapters are devoted to such issues. This part then moves on to explore ideas on how metazoans originated approximately 560 million years ago.

The topic of the final part of the book – the search for life in the Solar System – constitutes a synthesis of those from the first two parts and lies at the heart of modern 'astrobiology'. Its four chapters review the latest results on the physical environments and the search for life in the Solar System, specifically on Mars, Titan, and Europa.

Acknowledgements

There are many people to thank for helping to put together the conference out of which this book originated. Financial support for sponsoring research conferences and workshops run by the OI comes ultimately from the Office of the Vice President Research at McMaster University – Professor Mamdouh Shoukri. We are indebted to him for his keen interest and support in helping us launch the OI and this first conference.

The scientific organizing committee for the conference consists of OI members – who are also faculty members in departments across the Faculty of Science. The list of orgnanizers is:

- Professors Paul Higgs and James Wadsley of the Dept. of Physics and Astronomy
- Professors Brian Golding and Jonathon Stone (also the Associate Director of the OI) from the Dept. of Biology
- Professor Ralph Pudritz, the chair of the organizing committee, Director of the OI, and member of the Dept. of Physics and Astronomy

The hard work and scientific insights of these committee members were essential in driving the very successful conference programme, discussions, workshop, and, ultimately, the foundations of this book.

The enormous amount of work in actually organizing and running this international conference and workshop was carried out with great skill and dedication by two outstanding individuals:

- Ms Mara Esposto, administrator for the Dept. of Physics and Astronomy and part time administrator support for the OI and
- Ms Rosemary McNeice, the OI secretary and also secretary in the Dept. of Physics and Astronomy.

The design of the posters and website was carried out by:

- Mr Steve Janzen, graphic designer and media production services at McMaster, as well as by
- Mr Dan O'Donnell, an undergraduate physics and astronomy student. Dan also performed all of the many tasks needed to keep the conference website updated, and ran all of the audiovisual equipment at the conference.

Ultimately, the value of this book rests with the outstanding efforts and insights of our chapter authors, all of whom wrote admirable contributions and did a lot of extra work in addressing referee and editorial reports. The editors could not have finished this book without the outstanding services of Dan O'Donnell who performed all the LaTeX formatting required for this volume.

Finally, we wish to thank our excellent editors and assistants at Cambridge University Press for their interest in this volume, for their patience and many helpful suggestions, and for their quick responses to the many issues that came up in producing this volume. We thank in particular Miss Jacqueline Garget, who was the commissioning editor for astronomy and space science in charge of the Astrobiology series and whose early interest in our proposal helped to launch this volume. We also thank Vince Higgs, editor, astronomy and astrophysics who followed Jacqueline as well as his assistant, Ms Helen Morris (publishing assistant, physical sciences), for all of their help. Their continued support and help has been most welcome.

We close this preface with the hope that the reader of this volume will find much fascination, inspiration, and enjoyment in its pages. The scope and promise of this vibrant new area of science is extraordinary. We, as editors as well as authors, enjoyed our task and feel privileged to have worked with so many outstanding individuals during this project.

Part I

Planetary systems and the origins of life

1

Observations of extrasolar planetary systems

Shay Zucker
Tel Aviv University

1.1 Introduction

A decade has passed since the first discovery of an extrasolar planet by Mayor and Queloz (1995) and its confirmation by Marcy *et al.* (1997). Since this groundbreaking discovery, about 190 planets have been found around nearby stars, as of May 2006 (Schneider, 2006). Here we shall review the main methods astronomers use to detect extrasolar planets, and the data we can derive from those observations.

Aitken (1938) examined the observational problem of detecting extrasolar planets. He showed that their detection, either directly or indirectly, lay beyond the technical horizon of his era. The basic difficulty in directly detecting planets is the brightness ratio between a typical planet, which shines mainly by reflecting the light of its host star, and the star itself. In the case of Jupiter and the Sun, this ratio is 2.5×10^{-9}. If we keep using Jupiter as a typical example, we expect a planet to orbit at a distance of the order of 5 astronomical units (AU, where 1 AU $=$ 1.5×10^{11} m $=$ distance from Earth to Sun) from its host star. At a relatively small distance of 5 parsecs from the Solar System, this translates into a mean angular separation of the sources on the sky of 1 arcsecond. Therefore, with present technology, it is extremely demanding to directly image any extrasolar planet inside the overpowering glare of its host star, particularly from the ground, where the Earth's atmosphere seriously affects the observations. Nevertheless, significant advances have been made in the fields of coronagraphy and adaptive optics and positive results are very likely in the coming few years.

The vast majority of the known extrasolar planets were detected by *indirect* means, which matured in 1995 to allow the detection of the giant planet around the star 51 Peg through the spectroscopic *radial-velocity* (RV) technique. RV monitoring is responsible for most of the known extrasolar planets. *Transit* detection, another indirect method, has gained importance in the last few years,

Planetary Systems and the Origins of Life, eds. Ralph E. Pudritz, Paul G. Higgs, and Jonathon R. Stone.
Published by Cambridge University Press.

already yielding a few detections. The next two sections will review those two methods, their advantages and their drawbacks. Section 1.4 will provide a brief account of the emerging properties of the extrasolar planet population. In Section 1.5 we will briefly review other possible methods of detection. Section 1.6 will conclude the chapter with some predictions about future observations from space.

1.2 RV detections

Consider a star–planet system, where the planet's orbit is circular, for simplicity. By a simple application of Newton's laws, we can see that the star performs a reflex circular motion about the common centre of mass of the star and planet, with the same period (P) as the planet. The radius of the star's orbit is then given by:

$$a_\star = a\left(\frac{M_\mathrm{p}}{M_\star}\right), \tag{1.1}$$

where a is the radius of the planet orbit, and M_p and $M_\star$ are the planet mass and the star mass, respectively. The motion of the star results in the periodic perturbation of various observables that can be used to detect this motion. The RV technique focuses on the periodic perturbation of the line-of-sight component of the star's velocity.

Astronomers routinely measure RVs of objects ranging from Solar System minor planets to distant quasars. The basic tool to measure RV is the spectrograph, which disperses the light into its constituent wavelengths, yielding the stellar *spectrum*. Stars like our Sun, the so-called *main-sequence* stars, have well-known spectra. Small shifts in the wavelengths of the observed spectrum can tell us about the star's RV through the *Doppler effect*. Thus, a Doppler shift of $\Delta\lambda$ in a feature of rest wavelength λ in the stellar spectrum corresponds to an RV of:

$$v_\mathrm{r} = \frac{\Delta\lambda}{\lambda}c, \tag{1.2}$$

where c is the speed of light.

The most obvious parameters which characterize the periodic modulation of the RV are the period, P, and the semi-amplitude, K (Figure 1.1(a)). These two parameters are related to the planet mass via the general formula (e.g., Cumming *et al.* (1999)):

$$K = \left(\frac{2\pi G}{P}\right)^{\frac{1}{3}} \frac{M_\mathrm{p}\sin i}{(M_\star + M_\mathrm{p})^{\frac{2}{3}}} \frac{1}{\sqrt{1-e^2}}. \tag{1.3}$$

In this formula G is the universal gravitational constant, and e is the orbital eccentricity. The inclination of the orbital axis relative to the line of sight is denoted

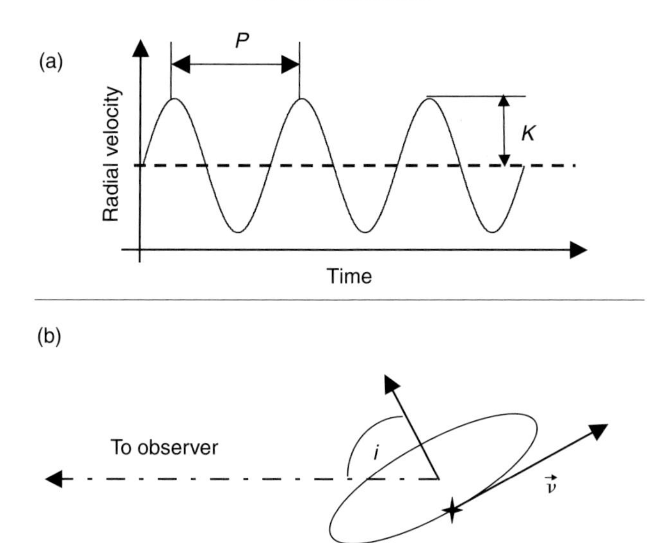

Fig. 1.1. (a) A schematic illustration of a periodic RV curve of a planetary orbit, showing the two quantities P (period) and K (semi-amplitude). (b) Visualization of the inclination angle (i), the angle between the orbital axis and the line of sight.

by i (Figure 1.1(b)). In a circular orbit we can neglect e and, assuming that the planet mass is much smaller than the stellar mass, we can derive the empirical formula:

$$K = \left(\frac{P}{1\ \mathrm{day}}\right)^{-\frac{1}{3}} \left(\frac{M_\star}{M_\odot}\right)^{-\frac{2}{3}} \left(\frac{M_\mathrm{p} \sin i}{M_\mathrm{J}}\right) 203\ \mathrm{m\,s^{-1}}. \qquad (1.4)$$

M_J denotes Jovian mass, and $M_\odot$ stands for the Solar mass. (Extrasolar planets are typically the mass of Jupiter, hence we normalize our formulae using 'Jovian mass'.)

Close examination of Equation (1.4) reveals several important points. First, K has a weak inverse dependence on P, which means that the RV technique is biased towards detecting short-period planets. Second, the planet mass and the inclination appear only in the product $M_\mathrm{p} \sin i$, and therefore they cannot be derived separately using RV data alone. In principle, a planetary orbit observed edge-on (i close to 90°) will have exactly the same RV signature as a stellar orbit observed face-on (i close to 0). Statistics help to partly solve the conundrum, since values of $\sin i$ which are close to unity are much more probable than smaller values (e.g., Marcy and Butler (1998)). In fact, for a randomly oriented set of orbits, the mean value of $\sin i$ is easily shown to be $4/\pi \approx 0.785$. Obviously, a better solution would be to seek independent information about the inclination.

Equation (1.4) shows the order of magnitude of the desired effect – tens or hundreds of metres per second. Detecting effects of this magnitude requires a precision of the order of metres per second. Such a precision was almost impossible to achieve before the 1990s. Before that time, the only claim of a very low-mass companion detected via RV was of the companion of the star HD 114762. The semi-amplitude of the RV variation was about 600 m s^{-1}, and the companion mass was found to be around 10 M_{J} (Latham *et al.*, 1989; Mazeh *et al.*, 1996). Although the existence of this object is well established, the question of its planetary nature is still debated. Alternatively, it could be a *brown dwarf* – an intermediary object between a planet and a star. The detection of smaller planet candidates had to await the development of instruments that could measure *precise* RVs.

Campbell and Walker (1979) were the first to obtain RVs of the required precision. They introduced an absorption cell containing hydrogen fluoride gas in the optical path of the stellar light in order to overcome systematic errors in the RVs, using the known spectrum of the gas for calibration. They carried out a pioneering survey of 16 stars over a period of 6 years, which yielded no detections, probably because of the small sample size (Campbell *et al.*, 1988).

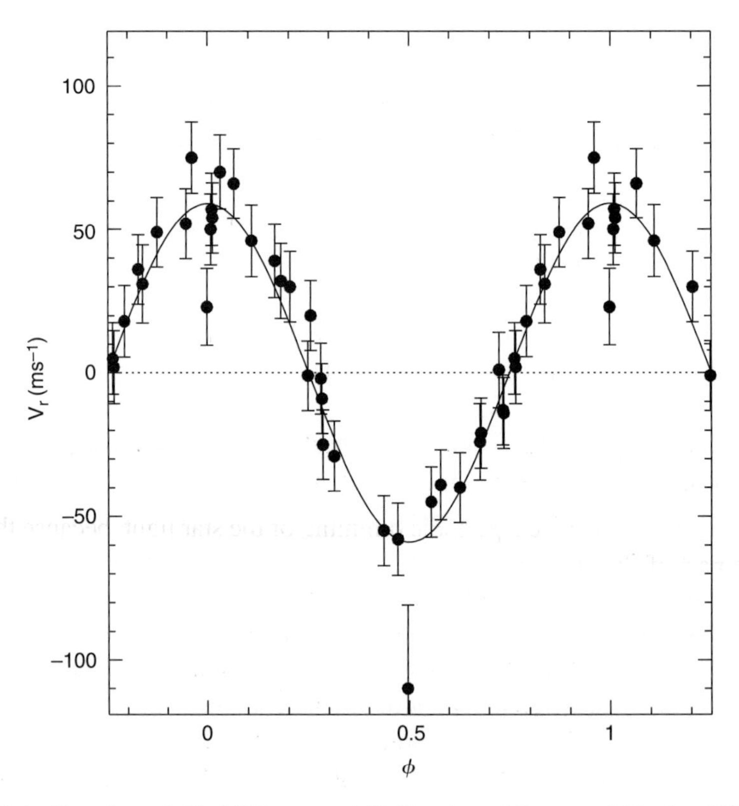

Fig. 1.2. The phase-folded RV curve of 51 Peg, from Mayor and Queloz (1995).

The first planet candidate detected using precise RV measurements was 51 Peg b. Mayor and Queloz (1995) used the fibre-fed ELODIE spectrograph in the Haute-Provence Observatory (Baranne *et al.*, 1996), and obtained an RV curve of 51 Peg corresponding to a planet with a mass of $0.44\,M_{\mathrm{J}}$ and an orbital period of 4.23 days (Figure 1.2). This short period means an orbital distance of 0.05 AU from the host star. The discovery was soon confirmed by Marcy *et al.* (1997), using the Hamilton echelle spectrograph at the Lick Observatory, with the iodine absorption cell technique (Butler *et al.*, 1996). This proximity to the host star was a major surprise and it actually contradicted the previous theories about planetary system formation. This discovery, and those of many similar planets that followed (now nicknamed 'Hot Jupiters'), led to a revision of those theories, and to the development of the planetary migration paradigm (e.g., Lin *et al.* (1996)). The current state of the formation and evolution theories is reviewed in Chapter 3.

Since that first detection, several groups have routinely performed RV measurements. The most prominent groups are the Geneva group, using fibre-fed spectrographs (Baranne *et al.*, 1996), and the Berkeley group, using the iodine-cell technique (Butler *et al.*, 1996).

1.3 Transit detections

We have seen in Section 1.2 that the basic drawback of the RV method is the lack of independent information about the orbital inclination, which leads to a fundamental uncertainty in the planet mass. Currently, the most successful means of obtaining this information is via the detection of planetary *transits*. In the Solar System, transits are a well-known rare phenomenon in which one of the inner planets (Mercury or Venus) passes in front of the solar disk. The most recent Venus transit occurred on 8 June 2004, and attracted considerable public attention and media coverage. Extrasolar transits occur when an extrasolar planet passes in front of its host-star disk. Obviously, we cannot observe extrasolar planetary transits in the same detail as transits in the Solar System. With current technology, the only observable effect would be a periodic dimming of the star light, because the planet obscures part of the star's surface. Thus, transits can be detected by *photometry*, i.e., monitoring the stellar light intensity.

The probability that a planetary orbit would be situated in such a geometric configuration to allow transits is not very high. For a circular orbit, simple geometrical considerations show that this probability is:

$$\mathcal{P} = \frac{R_\star + R_{\mathrm{p}}}{a}, \tag{1.5}$$

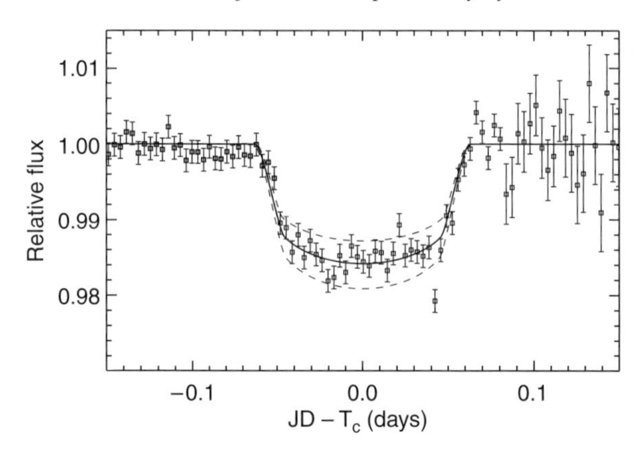

Fig. 1.3. The transit light curve of HD209458, from Charbonneau *et al.* (2000).

where $R_\star$ and R_p are the radii of the star and the planet, respectively (e.g., Sackett (1999)). For a typical hot Jupiter, this probability is about 10%.

The idea of using transits to detect extrasolar planets was first raised by Struve (1952), but the first extrasolar transit was observed only in 1999. Mazeh *et al.* (2000) detected a planet orbiting the star HD 209458, using 'traditional' RV methods. Soon after the RV detection, Charbonneau *et al.* (2000) and Henry *et al.* (2000) detected a periodical dimming of the light coming from HD 209458, at exactly the predicted orbital phase and with the same period as the RV variation, of 3.52 days. The light dimmed by about 1.5% for about 1.5 hours (Figure 1.3). The two teams detected the transits using small and relatively cheap telescopes, demonstrating that it was realistic to achieve the required photometric precision with ground-based observations.

The depth of the transit (i.e., the amount by which the light intensity drops) depends on the fraction of the stellar disk obscured by the planet. Thus, assuming there is a reasonable estimate of the star's radius, we can use the depth to derive the planet radius. This is the first direct estimate we have of a physical property of the planet itself. Obviously, the detection of transits immediately constrains the orbital inclination (i) to values close to 90° (transits occur only when we observe the orbit edge-on or almost edge-on). Furthermore, the transit duration depends strongly on the orbital inclination, and we can use it to explicitly derive i (e.g., Sackett (1999)). Thus, in combination with RV data, we can finally obtain a measurement of the planet mass, M_p.

Brown *et al.* (2001) used the Hubble Space Telescope (HST) to obtain a very precise light curve of HD 209458. This light curve led to a very precise estimate of the planet radius: 1.347 Jupiter radii (R_J). Using the inclination and the mass estimate from the RV orbit, the planet mean density could be derived: 0.35 g cm^{-3}.

The special circumstances of a transiting extrasolar planet were exploited by many more observations of HD 209458, with new clues about its atmosphere. Those observations are reviewed in Chapter 2.

The successful observations of HD 209458 encouraged many teams to try to detect more transiting extrasolar planets. Currently about 25 surveys are being conducted by teams around the world (Horne, 2006). Some of these surveys use small dedicated telescopes to monitor nearby stars (which are relatively bright) in large fields of view, like TrES (Alonso *et al.*, 2004) or HATnet (Bakos *et al.*, 2004). Other surveys, such as OGLE-III (Udalski *et al.*, 2002) or STEPSS (Burke *et al.*, 2004), focus on crowded fields like the Galactic Centre or globular clusters, and monitor tens of thousands of stars.

So far, the only successful surveys have been OGLE, with five confirmed planets, and TrES and XO, with one planet each. Their success highlights the difficulties such surveys face, and the problems in interpreting the observational data. The basic technical challenge is transit detection itself. The transits last only a small fraction of the planet's orbital time around its central star, and the drop in the stellar brightness is usually of the order of 1–2% at most. The first obvious challenge is to reach a sufficient photometric precision. The next challenge is to obtain sufficient phase coverage, on the observational front, and efficient signal analysis algorithms, on the computational front.

The OGLE project has yielded so far 177 transiting planet candidates (Udalski *et al.*, 2004). However, only five thus far have been confirmed as planets. This is due to the fundamental problem in using photometry to detect giant planets. Since the transit light curve alone does not provide any information regarding the mass of the eclipsing companion, we have to rely on its inferred radius to deduce its nature. However, it is known (e.g., Chabrier and Baraffe (2000)) that in the substellar mass regime, down to Jupiter mass, the radius depends extremely weakly on the mass. Therefore, even if we detect what seems to be a genuine transit light curve, the eclipsing object may still be a very low-mass star or a brown dwarf. The only way to determine its nature conclusively is through RV follow-up that would derive its mass. Thus, while only five of the OGLE candidates have been shown to be planetary companions, many others have been identified as stellar companions.

The proven non-planetary OGLE candidates demonstrate the diversity of events that can be mistaken for planetary transits. OGLE-TR-122 is a perfect example of a low-mass star which eclipses its larger companion, like a planet would. Only RV follow-up determined its stellar nature (Pont *et al.*, 2005). Other confusing configurations are 'grazing' eclipsing binary stars, where one star obscures only a tiny part of its companion, and 'blends', where the light of an eclipsing binary star is added to the light of a background star, effectively reducing the measured eclipse depth. In principle, these cases can be identified by close scrutiny of the light curve

(Drake, 2003; Seager and Mallén-Ornellas, 2003), or by using colour information in addition to the light-curve shape (Tingley, 2004). However, the most decisive identification is still through RVs.

Although transit detection alone is not sufficient to count as planet detection, transit surveys still offer two important advantages over RV surveys. First, they allow the simultaneous study of many more stars – the crowded fields monitored by transit surveys contain thousands of stars. Second, they broaden the range of stellar types which we examine for the existence of planets. While obtaining the required precision of RV measurements puts somewhat stringent constraints on the stellar spectrum, transit detection relies much less on the stellar type. The star simply has to be bright enough and maintain a stable enough brightness so that we can spot the minute periodical dimming caused by the transiting planet. Since stellar radii depend only weakly on the stellar mass, we should be able to detect planetary transits even around stars much hotter (and therefore more massive) than the Sun.

1.4 Properties of the extrasolar planets

As of May 2006, 193 extrasolar planets are known (Schneider, 2006). This number, although not overwhelming, is already enough to make some preliminary statistical observations. Obviously, these findings have a significant effect on the development of theories concerning the formation and evolution of planets in general, and the Solar System in particular. We shall now review the most prominent features of the growing population of extrasolar planets.

1.4.1 Mass distribution

The definition of planets, especially a definition that would distinguish them from stars, has been a central research theme since the very first detections. The most obvious criterion, which remains the most commonly used one, is simply the object mass. The boundary between stars and substellar objects, at $0.08\,M_\odot$, is already well known and physically understood (the mass above which hydrogen burning is ignited). A similar boundary that would apply for planets was sought. This was set at the so-called 'deuterium burning limit', at $13\,M_\mathrm{J}$ (Burrows *et al.*, 1997). This arbitrary limit is not related to the hypothesized formation mechanisms. The tail of the mass distribution of very low-mass companions suggests that objects with masses as large as $20\,M_\mathrm{J}$ exist. The question of whether the mass distribution of the detected planets indeed exhibits two distinct populations – planets and stars – remains. The evidence is mounting that this is indeed the case. It seems that the mass regime between $20\,M_\mathrm{J}$ and $0.08\,M_\odot$ is underpopulated (Jorissen *et al.*, 2001;

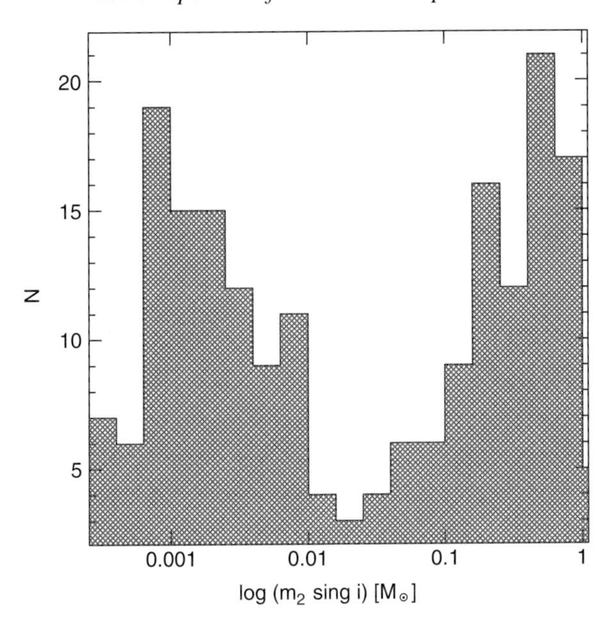

Fig. 1.4. The mass distribution of companions to solar-type stars. Note the dearth of planets at masses between 0.02 and 0.08 solar masses – 'the Brown-Dwarf Desert'. From Udry *et al.* (2004).

Zucker and Mazeh, 2001b), as can be seen in Figure 1.4. This depletion is nicknamed *the Brown-Dwarf Desert* (Marcy and Butler, 2000; Halbwachs *et al.*, 2000).

1.4.2 Mass–period distribution

The existence of 'Hot Jupiters' implied the possible presence of massive planets close to their host stars. Since the RV technique is mostly sensitive to short periods and massive planets, we expect to find a dense population in this part of the mass–period diagram. However, this is not the case, as was shown by Zucker and Mazeh (2002), Udry *et al.* (2002), and Pätzold and Rauer (2002). This may hint that the planetary migration process is less effective for very massive planets (Nelson *et al.*, 2000; Trilling *et al.*, 2002). Alternatively, it could mean that at very short distances the planet 'spills over' part of its mass onto the host star (Trilling *et al.*, 1998).

1.4.3 Orbital eccentricities

The second extrasolar planet detected, 70 Vir b, turned out to have a considerably eccentric orbit, with an eccentricity of 0.4 (Marcy and Butler, 1996). HD 80606 b currently holds the eccentricity record, with an eccentricity of 0.93 (Naef *et al.*, 2001). Such high orbital eccentricities were also a surprise. The matter in protoplanetary disks was assumed to orbit the central star in circular Keplerian orbits, and the planetary orbits were supposed to reflect this primordial feature by being relatively circular, like the orbits in the Solar System. In order to account for the observed high eccentricities, several models have been suggested, and it seems that no single model can explain all cases. Chapter 3 presents some processes which can take part in creating these eccentricities.

1.4.4 Host-star metallicity

The chemical composition of a planet-hosting star is related to that of the primordial molecular cloud where the star formed. The *metallicity* is a measure of the relative amount of elements heavier than hydrogen and helium in the stellar atmosphere. Santos *et al.* (2004) have shown that the abundance of planets is strongly related to the metallicity of the host star. This effect is much stronger than the selection effect caused by the presence of more metal lines in the stellar spectrum (e.g., Murray and Chaboyer (2002)). This can plausibly be explained by the need to have enough solid material and ices in order to form planets in the protoplanetary disk, but a detailed explanation is still lacking.

1.4.5 Planetary systems and planets in binary systems

The Solar System is known to include eight major planets. The presence of more than one planet around a host star is easily explained by the protoplanetary disk paradigm (see Chapter 4). Thus, we expect extrasolar planets also to occur as multiple planet systems. The first extrasolar planetary *system* detected was a system of three planets orbiting the star υ Andromedae (Butler *et al.*, 1999). In such cases, the motion performed by the star is a combination of the various motions caused by all planets in the system (Figure 1.5).

A related issue is the existence of planets in systems of binary stars. A considerable proportion of stars are actually binary stars, where two stars are gravitationally bound and orbit their barycenter. Currently, about 20 planets are known to orbit components of binary stars (Udry *et al.*, 2004). The orbital characteristics of these planets seem to differ from those orbiting single stars, e.g., very massive planets can be found in very close proximity to the host star only in binary stellar systems (Zucker and Mazeh, 2002).

1.4.6 Planetary radii

Transiting planets offer the possibility to study the radii of extrasolar planets. The number of transiting planets is still too small to draw very significant statistical

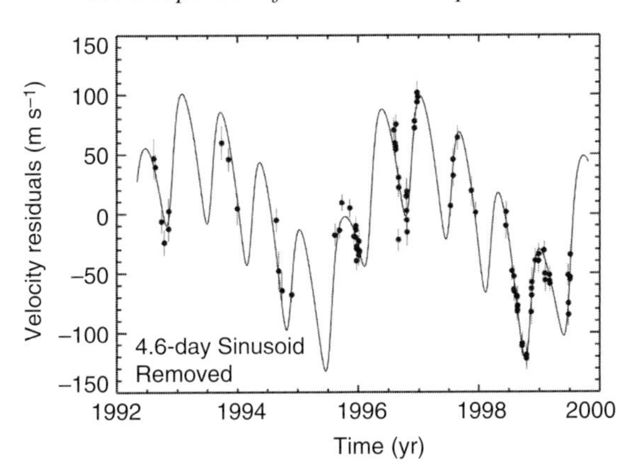

Fig. 1.5. The RV curve of υ Andromedae, after removing the motion caused by the shortest-period planet. Note the non-periodic nature of the motion, which is the sum of the motions caused by the two outer planets. From Butler *et al.* (1999).

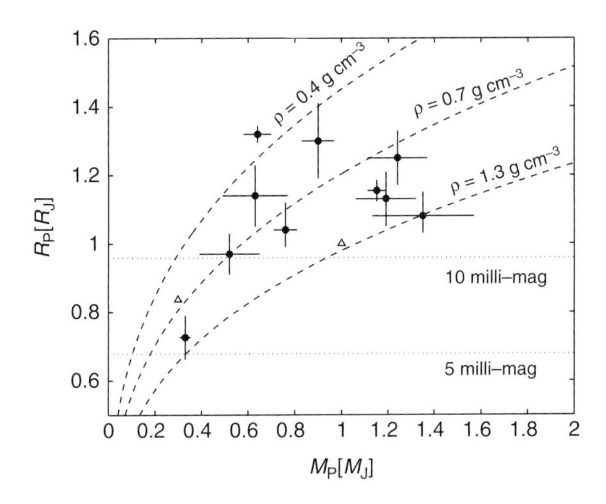

Fig. 1.6. Mass–radius relation for known transiting planets. The triangles represent Jupiter and Saturn. The dashed lines are constant density lines, for densities of 0.4, 0.7, and 1.3 g cm^{-3}. The dotted line represent levels of photometric precision required for detecting planets orbiting Sun-like stars.

conclusions, but it is worthwhile to examine the data. Figure 1.6 shows the mass–radius diagram of the ten transiting planets, together with Jupiter and Saturn for comparison, and representative isodensity lines. Early attempts to compare such diagrams with theoretical predictions seem promising (e.g., Guillot (2005), Laughlin *et al.* (2005)). It seems that most hot Jupiters have sizes close to that of Jupiter itself, and therefore their structure is not significantly perturbed by their proximity to the star.

1.5 Other methods of detection

So far, the RV and transit methods have contributed most of the observational knowledge on extrasolar planets around main-sequence stars. However, considerable efforts have been put into exploring other techniques, mainly in order to have a better coverage of the various configurations in which planets can be found. In the following subsections we give a very brief account of four such techniques.

1.5.1 Astrometry

In Section 1.2 we have shown that the star performs an orbital motion similar to that of the planet, but on a much smaller scale (Equation (1.1)). When we observe a star at a distance of d parsecs, we may be able to directly detect this orbital motion, scaled down, due to the distance, to a semi-major axis α:

$$\alpha = \frac{a_\star}{d} = \frac{a}{d}\left(\frac{M_\mathrm{p}}{M_\star}\right), \tag{1.6}$$

where α is measured in arcseconds. The *astrometric signature*, α, is therefore proportional to both the planet mass and the orbital radius, unlike the RV semi-amplitude K, which is inversely related to the period (and therefore to the orbital radius). Astrometric techniques aim to detect planets by measuring this motion.

The first claim of an extrasolar planet detected by astrometry was published in 1969 by van de Kamp (1969). Van de Kamp made astrometric observations of Barnard's Star – one of the closest stars to the Solar System (at a distance of only 1.8 parsecs). He observed the star between 1938 and 1969 and claimed that he had detected a planet around Barnard's Star, with a mass of $1.6\,M_\mathrm{J}$ and a period of 24 years. In the following years van de Kamp published a series of papers in which he refined his findings and claimed to have detected not one, but two planets. Eventually, van de Kamp's claims were refuted conclusively, e.g., by Benedict *et al.* (1999).

Astrometry does not suffer from the inclination ambiguity, and if an astrometric orbit is detected, the planet mass should be easily derived. The expected motion is extremely small, well below 1 milliarcsecond, and detecting it from the ground requires developing new techniques to overcome the various noise sources. The most promising technique is interferometry, which has already been proven to reach precisions of tens of microarcseconds in instruments in various stages of development, such as the Palomar Testbed Interferometer (Colavita and Shao, 1994), the Keck Interferometer (Colavita *et al.*, 1998), and ESO's VLTI (Mariotti *et al.*, 1998). Several useful astrometric observations of known extrasolar planets have been obtained using space-based observatories (Section 1.6).

1.5.2 Direct imaging

Various techniques have been designed to improve the chances of directly imaging extrasolar planets. These techniques include *coronagraphy*, which aims at suppressing the light from the star, *adaptive optics*, which tries to compensate for atmospheric turbulence, and simply using longer wavelengths where the contrast between the star and the planet is more favourable. A combination of coronagraphy and adaptive optics seems a very promising path towards direct detection, e.g., the NICI instrument on the Gemini South telescope, which is currently (2007) being commissioned (Toomey and Ftaclas, 2003).

Using NACO, an infrared adaptive optics instrument on the VLT, Chauvin *et al.* (2004) managed to image a giant planet orbiting the brown dwarf 2 M 1207 at an orbital distance of 55 AU. The planet is very faint, and its mass can only be inferred by comparing characteristics of its spectrum to models. Thus, Chauvin *et al.* estimate its mass at $5 \pm 1\,M_{\mathrm{J}}$. The large orbital distance implies a very long period, and therefore there is no dynamical measurement of the mass of 2 M 1207 b. This detection, being the first of its kind, like that of HD 209458, may open the gate to new kinds of observations which will be routinely performed on other planets that will be imaged directly.

1.5.3 Pulsar timing

Several planets were detected around pulsars a few years before the detection of 51 Peg b. In principle they were detected using RVs, but these RVs were not measured by usual spectroscopic means, but by the precise *timing* of the pulsar's pulses (Wolszczan and Frail, 1992; Wolszczan, 1994). Pulsars are neutron stars, i.e., they are no longer main-sequence stars but remnants of core-collapse supernova explosions. In view of the extremely violent process that forms pulsars, the existence of planets around them is very intriguing, but it is not the focus of our quest which is to seek planets around main-sequence stars.

1.5.4 Gravitational lensing

Due to a well-known phenomenon in General Relativity, when a faint relatively close star (the 'lens') passes in front of a very distant star (the 'source'), the light of the source undergoes a strong amplification. When there is a planet orbiting the 'lens' star, it may be detected by its effect on the amplification curve (Sackett, 1999). When only this magnification is observable, this phenomenon is commonly known as 'microlensing', to distinguish it from other lensing events which can be imaged in detail (e.g., lensing of galaxies by other galaxies). A few teams are routinely monitoring and collecting microlensing events, and already a few planet candidates have been detected this way (Bond *et al.*, 2004), but follow-up studies of those candidates are not feasible because the lens stars are much too faint (we know they exist only by the lensing events). This renders the main contribution of these detections a statistical one, regarding the frequency of planets in the Galaxy.

1.6 Future prospects for space missions

In the very near future two space missions designed to detect planetary transits are expected to be launched. Free from 'seeing' problems, caused by the atmospheric turbulences, these space telescopes are expected to monitor dense stellar fields in search of transiting planets. These missions are the French-led COROT satellite, which was launched on 27 December 2006 (Bordé *et al.*, 2003), and the American satellite Kepler (Borucki *et al.*, 2003) which is expected to be launched around 2008. A technological precursor is the Canadian MOST satellite, which was designed for asteroseismology studies, but requires the same kind of photometric precision (Walker *et al.*, 2003). Besides providing many detections of Jupiter-sized planets, these missions are also expected to detect Earth-sized planet candidates. The main challenge ahead will be the RV follow-up, which will be extremely difficult for the small planets.

In the early 1990s, the astrometric satellite Hipparcos was launched by ESA. Hipparcos, which operated for about three years, had an astrometric precision of the order of milliarcseconds (ESA, 1997). This was not enough to detect the motion caused by any known extrasolar planet. It was enough, though, to put upper limits on the planetary masses, thus proving the substellar nature of a few candidates (Pourbaix, 2001; Pourbaix and Arenou, 2001; Zucker and Mazeh, 2001a). Two astrometric space missions are now planned that will hopefully reach the astrometric precision required for detection: the European Gaia mission is scheduled for 2011 (Perryman *et al.*, 2001) and the American Space Interferometry Mission (SIM) for 2015 (Shao, 2004). Both missions are expected to reach astrometric precisions of the order of microarcseconds.

The prospects for the next decades are quite exciting – two missions are planned that will make use of innovative techniques to achieve the goal of imaging terrestrial planets! ESA is planning an infrared space interferometer – Darwin – that will comprise a flotilla of space telescopes (Fridlund, 2004). The American counterpart by NASA is the TPF – Terrestrial Planet Finder (Beichman *et al.*, 2004) – which is now planned to include two separate missions: an interferometer (TPF-I) and a coronagraph (TPF-C). Both projects are extremely ambitious, both in terms of the technology required and in terms of their goal – to directly image terrestrial planets and study them for signs of life.

Acknowledgements

I wish to thank the anonymous referee whose comments helped me improve this review. I am grateful to the Origins Institute at McMaster University, its director Ralph Pudritz, and its secretary Rosemary McNeice, for financial support and for organizing a productive and pleasant conference.

References

Aitken, R. G. (1938). Is the solar system unique? *Astron. Soc. Pac. Leaflet*, **112**, 98–106.
Alonso, R., Brown, T. M., Torres, G., *et al.* (2004). TrES-1: the transiting planet of a bright K0V star. *Astrophys. J.*, **613**, L153–L156.
Bakos, G., Noyes, R. W., Kovács, G., *et al.* (2004). Wide-field millimagnitude photometry with the HAT: a tool for extrasolar planet detection. *Publ. Astron. Soc. Pac.*, **116**, 266–277.
Baranne, A., Queloz, D., Mayor, M., *et al.* (1996). ELODIE: a spectrograph for accurate radial velocity measurements. *Astron. Astrophys. Sup.*, **119**, 373–390.
Beichman, C., Gómez, G., Lo, M., Masdemont, J., and Romans, L. (2004). Searching for life with the Terrestrial Planet Finder: Lagrange point options for a formation flying interferometer. *Adv. Space. Res.*, **34**, 637–644.
Benedict, G. F., McArthur, B., Chappell, D. W., *et al.* (1999). Interferometric astrometry of Proxima Centauri and Barnard's star using Hubble Space Telescope Fine Guidance Sensor 3: detection limits for substellar companions. *Astron. J.*, **118**, 1086–1100.
Bond, I. A., Udalski, A., Jaroszyński, M., *et al.* (2004). OGLE 2003-BLG-235/MOA 2003-BLG-53: a planetary microlensing event. *Astrophys. J.*, **606**, L155–L158.
Bordé, P., Rouan, D., and Léger, A. (2003). Exoplanet detection capability of the COROT space mission. *Astron. Astrophys.*, **405**, 1137–1144.
Borucki, W. J., Koch, D. G., Lissauer, J. J., *et al.* (2003). The Kepler mission: a wide-field photometer designed to determine the frequency of Earth-size planets around solar-like stars. *Soc. Photo-Opt. Instru.*, **4854**, 129–140.
Brown, T. M., Charbonneau, D., Gilliland, R. L., Noyes, R. W., and Burrows, A. (2001). Hubble Space Telescope time-series photometry of the transiting planet of HD 209458. *Astrophys. J.*, **552**, 699–709.
Burke, C. J., Gaudi, B. S., DePoy, D. L., Pogge, R. W., and Pinsonneault, M. H. (2004). Survey for transiting extrasolar planets in stellar systems I. Fundamental parameters of the open cluster NGC 1245. *Astron. J.*, **127**, 2382–2397.
Burrows, A., Marley, M., Hubbard, W. B., *et al.* (1997). A nongray theory of extrasolar giant planets and brown dwarfs. *Astrophys. J.*, **491**, 856–875.
Butler, R. P., Marcy, G. W., Fischer, D. A., *et al.* (1999). Evidence for multiple companions to upsilon Andromedae. *Astrophys. J.*, **526**, 916–927.
Butler, R. P., Marcy, G. W., Williams, E., McCarthy, C., and Vogt, S. S. (1996). Attaining Doppler precision of 3 m s^{-1}. *Publ. Astron. Soc. Pac.*, **108**, 500–509.
Campbell, B., and Walker, G. A. H. (1979). Precision radial velocities with an absorption cell. *Publ. Astron. Soc. Pac.*, **91**, 540–545.
Campbell, B., Walker, G. A. H., and Yang, S. (1988). A search for substellar companions to solar-type stars. *Astrophys. J.*, **331**, 902–921.
Chabrier, G., and Baraffe, I. (2000). Theory of low-mass stars and substellar objects. *Annu. Rev. Astron. Astr.*, **38**, 337–377.

Charbonneau, D., Brown, T. M., Latham, D. W., and Mayor, M. (2000). Detection of planetary transits across a sun-like star. *Astrophys. J.*, **529**, L45–L48.

Chauvin, G., Lagrange, A.-M., Dumas, C., *et al.* (2004). A giant planet candidate near a young brown dwarf. Direct VLT/NACO observations using IR wavefront sensing. *Astron. Astrophys.*, **425**, L29–L32.

Colavita, M. M., and Shao, M. (1994). Indirect planet detection with ground-based long-baseline interferometry. *Astrophys. Space. Sci*, **212**, 385–390.

Colavita, M. M., Boden, A. F., Crawford, S. L., *et al.* (1998). Keck interferometer. *Soc. Photo-Opt. Instru.*, **3350**, 776–784.

Cumming, A., Marcy, G. W., and Butler, R. P. (1999). The Lick planet search: detectability and mass thresholds. *Astrophys. J.*, **526**, 890–915.

Drake, A. J. (2003). On the selection of photometric planetary transits. *Astrophys. J.*, **589**, 1020–1026.

ESA (1997). *The Hipparcos and Tycho Catalogues*, ESA SP-1200. Noordwijk: ESA.

Fridlund, C. V. M. (2004). The Darwin mission. *Adv. Space. Res.*, **34**, 613–617.

Guillot, T. (2005). The interiors of giant planets: models and outstanding questions. *Annu. Rev. Earth. Pl. Sc.*, **33**, 493–530.

Halbwachs, J. L., Arenou, F., Mayor, M., Udry, S., and Queloz, D. (2000). Exploring the brown dwarf desert with Hipparcos. *Astron. Astrophys.*, **355**, 581–594.

Henry, G. W., Marcy, G. W., Butler, R. P., and Vogt, S. S. (2000). A transiting '51 Peg-like' planet. *Astrophys. J.*, **529**, L41–L44.

Horne, K. (2006). http://star-www.st-and.ac.uk/~kdh1/transits/table.html

Jorissen, A., Mayor, M., and Udry, S. (2001). The distribution of exoplanet masses. *Astron. Astrophys.*, **379**, 992–998.

Latham, D. W., Mazeh, T., Stefanik, R. P., Mayor, M., and Burki, G. (1989). The unseen companion of HD 114762: a probable brown dwarf. *Nature*, **339**, 38–40.

Laughlin, G., Aaron, W., Vanmunster, T., *et al.* (2005). A comparison of observationally determined radii with theoretical radius predictions for short-period transiting extrasolar planets. *Astrophys. J.*, **621**, 1072–1078.

Lin, D. N. C., Bodenheimer, P., and Richardson, D. C. (1996). Orbital migration of the planetary companion of 51 Pegasi to its present location. *Nature*, **380**, 606–607.

Marcy, G. W., and Butler, R. P. (1996). A planetary companion to 70 Virginis. *Astrophys. J.*, **464**, L147–L151.

Marcy, G. W., and Butler, R. P. (1998). Detection of extrasolar giant planets. *Annu. Rev. Astron. Astr.*, **36**, 57–98.

Marcy, G. W., and Butler, R. P. (2000). Planets orbiting other suns. *Publ. Astron. Soc. Pac.*, **112**, 137–140.

Marcy, G. W., Butler, R. P., Williams, E., *et al.* (1997). The planet around 51 Pegasi. *Astrophys. J.*, **481**, 926–935.

Mariotti, J.-M., Denise, C., Derie, F., *et al.* (1998). VLTI program: a status report. *Soc. Photo-Opt. Instru.*, **3350**, 800–806.

Mayor, M., and Queloz, D. (1995). A Jupiter-mass companion to a solar-type star. *Nature*, **378**, 355–359.

Mazeh, T., Latham, D. W., and Stefanik, R. P. (1996). Spectroscopic orbits for three binaries with low-mass companions and the distribution of secondary masses near the substellar limit. *Astrophys. J.*, **466**, 415–426.

Mazeh, T., Naef, D., Torres, G., *et al.* (2000). The spectroscopic orbit of the planetary companion transiting HD 209458. *Astrophys. J.*, **532**, L55–L58.

Murray, N., and Chaboyer, B. (2002). Are stars with planets polluted? *Astrophys. J.*, **566**, 442–451.

Naef, D., Latham, D. W., Mayor, M., *et al.* (2001). HD 80606 b, a planet on an extremely elongated orbit. *Astron. Astrophys.*, **375**, L27–L30.
Nelson, R. P., Papaloizou, J. C. B., Masset, F., and Kley, W. (2000). The migration and growth of protoplanets in protostellar discs. *Mon. Not. R. Astron. Soc.*, **318**, 18–36.
Pätzold, M., and Rauer, H. (2002). Where are the massive close-in extrasolar planets? *Astrophys. J.*, **568**, L117–L120.
Perryman, M. A. C., de Boer, K. S., Gilmore, G., *et al.* (2001). Gaia: composition, formation and evolution of the Galaxy. *Astron. Astrophys.*, **369**, 339–363.
Pont, F., Melo, C. H. F., Bouchy, F., *et al.* (2005). A planet-sized star around OGLE-TR-122. Accurate mass and radius near the hydrogen-burning limit. *Astron. Astrophys.*, **433**, L21–L24.
Pourbaix, D. (2001). The Hipparcos observations and the mass of sub-stellar objects. *Astron. Astrophys.*, **369**, L22–L25.
Pourbaix, D., and Arenou, F. (2001). Screening the Hipparcos-based astrometric orbits of sub-stellar objects. *Astron. Astrophys.*, **372**, 935–944.
Sackett, P. D. (1999). Searching for unseen planets via occultation and microlensing, in *Planets Outside the Solar System: Theory and Observations* (NATO-ASI), eds. J.-M. Mariotti and D. Alloin. Dordrecht: Kluwer, pp. 189–227.
Santos, N. C., Israelian, G., and Mayor, M. (2004). Spectroscopic [Fe/H] for 98 extra-solar planet-host stars. *Astron. Astrophys.*, **415**, 1153–1166.
Schneider, J. (2006). *The Extrasolar Planet Encyclopaedia*. http://exoplanet.eu/index.php
Seager, S., and Mallén-Ornellas, G. (2003). A unique solution of planet and star parameters from an extrasolar planet transit light curve. *Astrophys. J.*, **585**, 1038–1055.
Shao, M. (2004). Science overview and status of the SIM project. *Soc. Photo-Opt. Instru.*, **5491**, 328–333.
Struve, O. (1952). Proposal for a project of high-precision stellar radial velocity work. *Observatory*, **72**, 199–200.
Tingley, B. (2004). Using color photometry to separate transiting exoplanets from false positives. *Astron. Astrophys.*, **425**, 1125–1131.
Toomey, D. W., and Ftaclas, C. (2003). Near infrared coronagraphic imager for Gemini South. *Soc. Photo-Opt. Instru.*, **4841**, 889–900.
Trilling, D. E., Benz, W., Guillot, T., *et al.* (1998). Orbital evolution and migration of giant planets: modeling extrasolar planets. *Astrophys. J.*, **500**, 428–439.
Trilling, D. E., Lunine, J. I., and Benz, W. (2002). Orbital migration and the frequency of giant planet formation. *Astron. Astrophys.*, **394**, 241–251.
Udalski, A., Paczyński, B., Żebruń, K., *et al.* (2002). The optical gravitational lensing experiment. Search for planetary and low-luminosity object transits in the galactic disk. Results of 2001 campaign. *Acta Astron.*, **52**, 1–37.
Udalski, A., Szymański, M., Kubiak, M., *et al.* (2004). The optical gravitational lensing experiment. Planetary and low-luminosity object transits in the fields of galactic disk. Results of the 2003 OGLE observing campaigns. *Acta Astron.*, **54**, 313–345.
Udry, S., Eggenberger, A., Mayor, M., Mazeh, T., and Zucker, S. (2004). Planets in multiple-star systems: properties and detections, in *The Environment and Evolution of Double and Multiple Stars*, Proc. IAU Coll. 191, eds. C. Allen and C. Scarfe. *Rev. Mex. Astron. Astrof. Ser. Conf.*, **21**, pp. 207–214.
Udry, S., Mayor, M., Naef, D., *et al.* (2002). The CORALIE survey for southern extra-solar planets. VIII. The very low-mass companions of HD 149137, HD 162020, HD 168433 and HD 202206: brown dwarfs or 'superplanets'? *Astron. Astrophys.*, **390**, 267–279.

van de Kamp, P. (1969). Parallax, proper motion, acceleration, and orbital motion of Barnard's Star. *Astron. J.*, **74**, 238–240.

Walker, G., Matthews, J., Kuschnig, R., *et al.* (2003). The MOST asteroseismology mission: ultraprecise photometry from space. *Publ. Astron. Soc. Pac.*, **115**, 1023–1035.

Wolszczan, A. (1994). Confirmation of Earth-mass planets orbiting the millisecond pulsar PSR B 1257+12. *Science*, **264**, 538–542.

Wolszczan, A., and Frail, D. A. (1992). A planetary system around millisecond pulsar PSR 1257+12. *Nature*, **355**, 145–147.

Zucker, S., and Mazeh, T. (2001a). Analysis of the Hipparcos observations of the extrasolar planets and the brown dwarf candidates. *Astrophys. J.*, **562**, 549–557.

Zucker, S., and Mazeh, T. (2001b). Derivation of the mass distribution of extrasolar planets with MAXLIMA, a maximum likelihood algorithm. *Astrophys. J.*, **562**, 1038–1044.

Zucker, S., and Mazeh, T. (2002). On the mass-period correlation of the extrasolar planets. *Astrophys. J.*, **568**, L113–L116.

2

The atmospheres of extrasolar planets

L. Jeremy Richardson
NASA Goddard Space Flight Center
and
Sara Seager
Massachusetts Institute of Technology

2.1 Introduction

In this chapter we examine what can be learned about extrasolar planet atmospheres by concentrating on a class of planets that *transit* their parent stars. As discussed in the previous chapter, one way of detecting an extrasolar planet is by observing the drop in stellar intensity as the planet passes in front of the star. A transit represents a special case in which the geometry of the planetary system is such that the planet's orbit is nearly edge-on as seen from Earth. As we will explore, the transiting planets provide opportunities for detailed follow-up observations that allow physical characterization of extrasolar planets, probing their bulk compositions and atmospheres.

2.2 The primary eclipse

The vast majority of the currently known extrasolar planets have been detected using the radial velocity technique.[1] As detailed in the previous chapter, the radial velocity method searches for periodic motion of a star caused by the gravitational pull of an orbiting companion. Figure 1.1 shows a sketch of a typical periodic radial velocity signal and the basic geometry of the planetary system. This method is sensitive only to movement of the star towards and away from the observer, i.e., along the line of sight from the system to the observer on Earth. Thus, radial velocity observations provide only a determination of the *minimum* mass M of the planet, and the orbital inclination i of the system remains unknown, as in

$$M = M_{\rm p} \sin i, \tag{2.1}$$

where $M_{\rm p}$ is the *true* mass of the planet. (See Section 1.2 for further details.)

[1] An up-to-date reference and catalogue of all known extrasolar planets can be found at http://vo.obspm.fr/exoplanetes/encyclo/encycl.html

Planetary Systems and the Origins of Life, eds. Ralph E. Pudritz, Paul G. Higgs, and Jonathon R. Stone.
Published by Cambridge University Press.

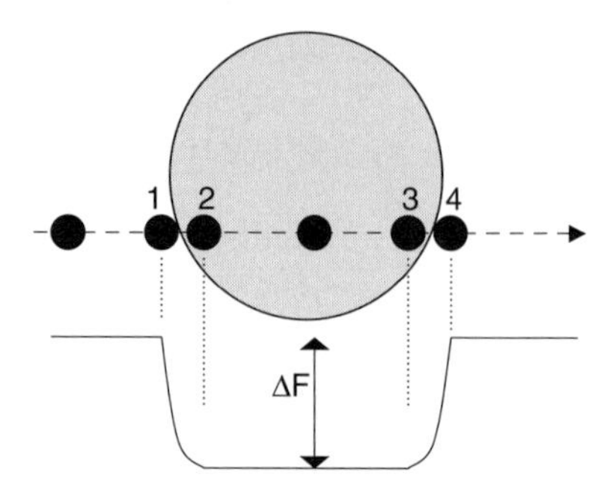

Fig. 2.1. Sketch showing a planet crossing the disk of its parent star. The transit light curve is shown below.

The *primary eclipse*, or transit, occurs when the planet's orbit happens to be nearly edge-on as seen from Earth. This means that the planet periodically crosses in front of the star as it orbits, and we detect this as a decrease in the light from the star that occurs once per planet revolution, as indicated schematically in Figure 2.1. The dimming is typically a few per cent or less for the currently known transiting planets. In this geometry, the orbital inclination is now known to be $\sim 90°$ (and can be determined precisely from the details of the transit light curve). We can therefore derive the true planetary mass, $M_{\rm p}$, from Equation (2.1).

A number of other physical parameters of the planet and star can be derived from the shape of the light curve (Seager and Mallén-Ornelas, 2003). The depth of the transit ΔF (i.e., the change in flux from outside transit to during transit, as shown in Figure 2.1) is directly proportional to the ratio of the area of the planetary disk to the area of the stellar disk. That is,

$$\Delta F \equiv \frac{F_{\text{out of transit}} - F_{\text{transit}}}{F_{\text{out of transit}}} = \frac{A_{\rm p}}{A_*} = \left(\frac{R_{\rm p}}{R_*}\right)^2, \tag{2.2}$$

where F represents the total flux, A is the area of the disk (planet or star), and R is the radius (planet or star). With a stellar mass–radius relation, it is possible to derive both the planetary and stellar radii simultaneously. With the planetary mass $M_{\rm p}$ and radius $R_{\rm p}$, one can immediately calculate the average density of the planet from

$$\rho = \frac{M_{\rm p}}{\frac{4}{3}\pi R_{\rm p}^3}. \tag{2.3}$$

The discovery of transiting planets allowed a direct measurement of the true mass, radius, and density of planets outside the Solar System for the first time. The planetary radius is key to determining the reflection and thermal emission of the planets from flux measurements. The density measurements derived from transit observations indicate that all but one of the transiting planets are hydrogen–helium gas giants, similar in bulk composition to Jupiter and Saturn in our own Solar System.

When the planet is in front of the star, the planet's atmosphere appears as an annulus surrounding the planetary disk, and some of the starlight passes through this annulus to the observer. The detection of starlight that has passed through the transiting planet's atmosphere in this manner is called *transmission spectroscopy*. By measuring how much starlight is transmitted as a function of wavelength, we can learn about the atomic and molecular species present in the planet's atmosphere, providing a much greater wealth of information than simply the average density and bulk composition. We introduce the broad study of spectroscopy in Section 2.5 and discuss observations of transiting planets using transmission spectroscopy in Section 2.7.

2.3 The secondary eclipse

A planet that crosses in front of its parent star will disappear behind the star later in its orbit. This disappearance is called the *secondary eclipse*. For a circular orbit, the secondary eclipse occurs exactly one-half of an orbital period after the primary eclipse. However, for a non-circular orbit, the secondary eclipse can occur earlier or later (depending on the eccentricity and the orientation of the orbit), and its duration can differ from that of the primary eclipse (Charbonneau, 2003). In addition to clues about the eccentricity of the planet's orbit from the secondary eclipse timing and duration, the secondary eclipse yields information about the nature of the planet's atmosphere.

For example, in visible light, the secondary eclipse probes the amount of starlight reflected by the planet's atmosphere (called the *albedo*). In the infrared, however, it measures the direct thermal emission (or intrinsic heat output) of the planet. In neither case does this imply imaging the planet; rather, the idea is to observe the total energy output of the system (star + planet) and attempt to detect a decrease as the planet is hidden from view.

Figure 2.2 illustrates this decrease in the total energy output of the system during secondary eclipse and shows that the thermal emission of the planet may be detectable at infrared wavelengths using this technique. The basic situation is that the incident starlight (which peaks in the visible for a Sun-like star) is absorbed and reprocessed by the planet's atmosphere, and some of that radiation is later emitted at infrared wavelengths. The figure shows the thermal emission of the planet HD 209458 b relative to its parent star. This calculation assumes that both the star and planet emit only blackbody radiation (Equation (2.6)), and it assumes that the planet emits uniformly in both hemispheres. In the visible region (solid

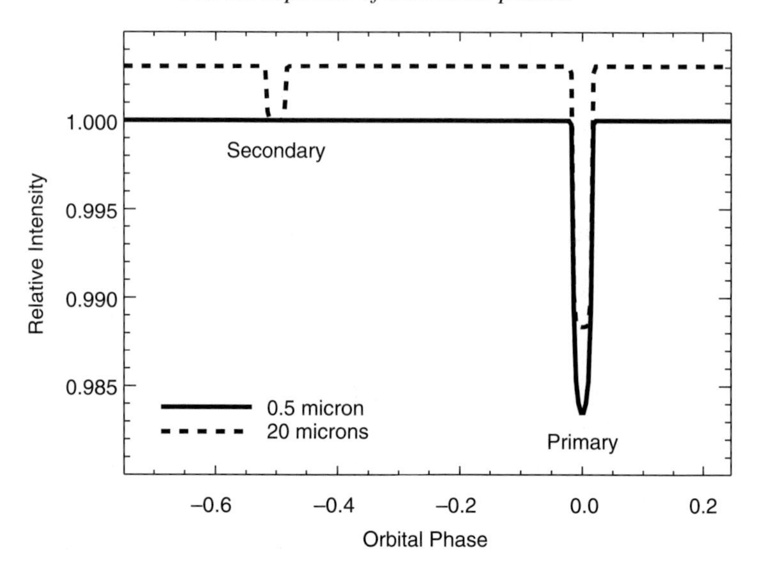

Fig. 2.2. Comparison of primary and secondary eclipses in the visible and infrared for the thermal emission of the planet HD 209458 b. These curves were calculated from a simple model that assumes the star and planet emit blackbody radiation only.

curve), the secondary eclipse is undetectable, both because the planet has virtually no emission at these wavelengths and because the reflected light from the planet is $\ll 0.01\%$ of the stellar output. However, as the figure shows, the situation is quite different at 20 μm. The total intensity *relative to the star* is higher outside of the eclipse, because the planet has a small but measurable intrinsic energy output at this wavelength. The secondary eclipse appears as a dip of $\sim 0.3\%$ in the total intensity as the planet is hidden by the star.

The eclipse depths at visible and infrared wavelengths can be estimated with the following flux ratios. For reflected light,

$$\frac{F_{\mathrm{p}}}{F_*} = A_{\mathrm{g}} \left(\frac{R_{\mathrm{p}}}{a} \right)^2, \tag{2.4}$$

where A_{g} is the geometric albedo (the fraction of incident radiation scattered back into space when the planet is in full phase), R_{p} is the planetary radius, and a is the orbital semi-major axis. For thermal emission,

$$\frac{F_{\mathrm{p}}}{F_*} = \frac{T_{\mathrm{p}}}{T_*} \left(\frac{R_{\mathrm{p}}}{R_*} \right)^2, \tag{2.5}$$

where T_p and T_* are the planet and star effective temperatures (see Equation (2.19) for an estimate of T_p). Here we have used the approximation for the Wien tail of the blackbody flux, whereby the flux ratio translates into a temperature ratio.

2.4 Characteristics of known transiting planets

A total of *ten* transiting extrasolar planets have been discovered as of May 2006. Their physical characteristics are given in Table 2.1, and they are plotted in Figure 2.3. The upper panel (period vs mass) illustrates the two groups of transiting planets. The 'hot Jupiters' (to the upper left of the plot) have masses smaller than that of Jupiter and orbital periods greater than ~2.5 days. This name is something of a misnomer, since the so-called hot Jupiters are quite different from our own Jupiter – because of the fact that they orbit at such small orbital distances, they are much hotter and therefore have different chemical species present in their atmospheres. The other group, often called the 'very hot Jupiters', is characterized by planets that orbit much closer to their parent stars (with orbital periods less than 2.5 days) and are more massive than Jupiter. These two dynamically distinct groups of planets may have different evolutionary histories, possibly resulting from different migration mechanisms (Gaudi *et al.*, 2005), and thus could potentially have

Table 2.1. *Physical properties of transiting extrasolar planets*

Planet	Period (days)	Radius (R_J)	Mass (M_J)	T_* (K)	T_{eq} [1] (K)
OGLE-TR-56b[2,3]	1.212	1.25 ± 0.08	1.24 ± 0.13	6119	1929
OGLE-TR-113b[2,4]	1.432	1.09 ± 0.10	1.08 ± 0.28	4804	1234
OGLE-TR-132b[5,6]	1.690	1.13 ± 0.08	1.19 ± 0.13	6411	1933
HD 189733 b[7,8]	2.219	1.154 ± 0.032	1.15 ± 0.04	5050	1096
HD 149026 b[9,10]	2.876	$0.726^{+0.062}_{-0.066}$	0.33 ± 0.023	6147	1593
TrES-1[2,11,12]	3.030	1.08 ± 0.05	0.729 ± 0.036	5226	1059
OGLE-TR-10b[2,13,14]	3.101	1.14 ± 0.09	0.63 ± 0.14	6075	1402
HD 209458 b[2,15,16]	3.525	1.320 ± 0.025	0.657 ± 0.006	6117	1363
XO-1b[17]	3.942	1.30 ± 0.11	0.90 ± 0.07	5750	1148
OGLE-TR-111b[2,18]	4.016	0.97 ± 0.06	0.52 ± 0.13	5044	935

[1]Calculated from Equation (2.19) with $f = 1$ and $A_B = 0.3$; [2](Santos *et al.*, 2006); [3](Torres *et al.*, 2004); [4](Konacki *et al.*, 2004); [5](Bouchy *et al.*, 2004); [6](Moutou *et al.*, 2004); [7](Bouchy *et al.*, 2005); [8](Bakos *et al.*, 2006); [9](Sato *et al.*, 2005); [10](Charbonneau *et al.*, 2005); [11](Laughlin *et al.*, 2005); [12](Alonso *et al.*, 2004); [13](Konacki *et al.*, 2005); [14](Holman *et al.*, 2005); [15](Knutson *et al.*, 2006); [16](Winn *et al.*, 2005); [17](McCullough *et al.*, 2006); [18](Pont *et al.*, 2004).

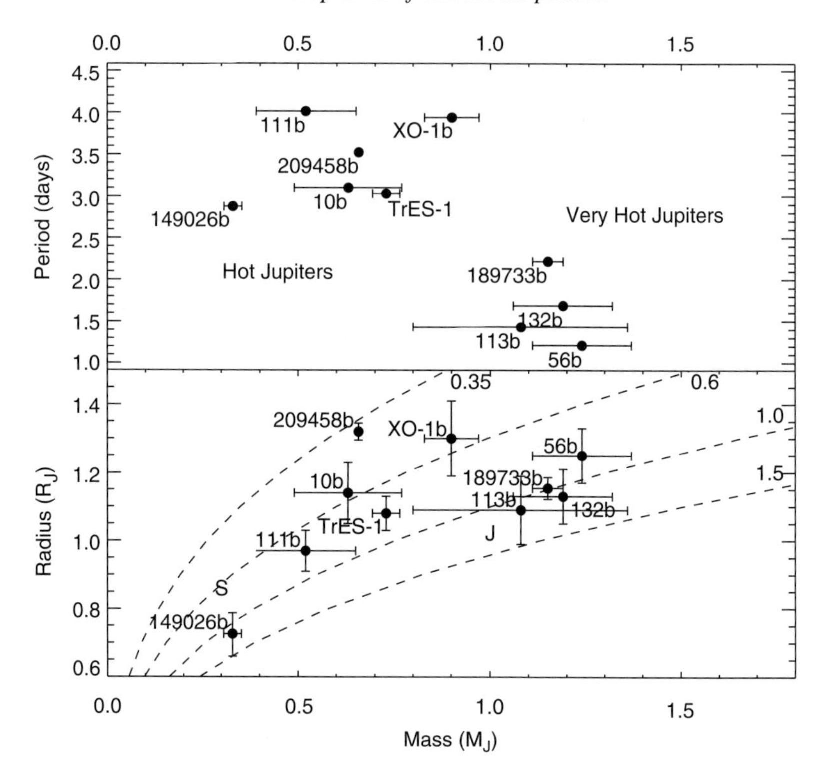

Fig. 2.3. The ten currently known transiting planets, as a function of planetary mass. The upper panel shows the period vs mass, and the lower panel shows radius vs mass. Dashed curves indicate curves of constant density in grams per cubic centimetre. For comparison, Jupiter and Saturn are shown, marked by 'J' and 'S', respectively.

very different atmospheric properties. We now have one bright planet from each group – HD 209458 b and HD 198733 b – allowing us to compare observations of the two planets and gain insights into their atmospheric structure and evolutionary history.

The lower panel of Figure 2.3 shows the radius of each planet vs mass. The dashed curves indicating constant density provide context for understanding the bulk composition of the planets. For example, most of the transiting planets are similar in average density to Jupiter ($\rho = 1.33$ g cm^{-2}) and Saturn ($\rho = 0.69$ g cm^{-2}). However, these 'close-in' extrasolar planets are likely to be quite different from our own Jupiter, due to the fact that they are much closer to their stars. At orbital distances of $a < 0.1$ AU (by comparison, Mercury is at $a \sim 0.38$ AU), these planets are bombarded by radiation from their parent stars and are therefore expected to be hot ($T > 1000$ K). Jupiter, at 5 AU from the Sun, has a blackbody temperature of only 110 K. Because of the large temperature difference, we expect the atmospheric composition of the hot Jupiters to be significantly different from that of Jupiter. For example, at low temperatures ($T < 1000$ K), chemical equilibrium calculations show that carbon is mostly present in the form of CH_4, while at higher temperatures it appears as CO (Burrows and Sharp, 1999).

Finally, from Figure 2.3, we note that not all of the known transiting planets have densities similar to Jupiter and Saturn. The most extreme example is HD 209458 b, with an average density of $\sim$0.35 g cm^{-2}. This planet was the first one found to exhibit a transit (Charbonneau *et al.*, 2000; Henry *et al.*, 2000), although it was originally detected using the radial velocity method (Mazeh *et al.*, 2000). Since then, it has been observed extensively from the ground and from space, as we shall discuss further in Section 2.7. From the beginning, it was unclear why the planet appears to have a larger radius than is predicted by theory (Burrows *et al.*, 2000, 2003), and this remains one of the unanswered questions in the field today (Winn and Holman, 2005).

2.5 Spectroscopy

We now turn from a general discussion of the transiting planets to the specific topic of spectroscopy and the radiative transfer equation. By studying the spectroscopy of extrasolar planets, we can gain key insights into the atmospheric composition, temperature, and structure of these planets. We begin this section by introducing the Planck blackbody law, which describes the thermal emission of an object in the absence of scattering or absorbing particles, and move to the radiative transfer equation, which does account for the effects of scattering and absorption. The radiative transfer equation governs the interaction of energy (in the form of emitted or absorbed radiation) with matter (in this case the particles that make up the planetary atmosphere). In Section 2.6 we give an overview of how models of planetary atmospheres are computed. Spectra derived from such models help us to interpret observational results and facilitate the design of future planet atmosphere detection instruments.

At the most basic level we can approximate the star and planet as blackbodies. In that case, where we ignore the details of the atmosphere, the emission is given by Planck's blackbody law:

$$B_\lambda(T) = \frac{2hc^2}{\lambda^5(e^{hc/\lambda kT} - 1)}, \tag{2.6}$$

where λ is wavelength, c is the speed of light in vacuum, k is Boltzmann's constant, T is the temperature of the blackbody, and h is Planck's constant. For a stellar or planetary atmosphere which contains a variety of different species that absorb, emit, and scatter radiation, however, the blackbody law is not sufficient to describe the resulting spectrum, and it becomes necessary to understand how matter interacts with the radiation.

Following convention, we begin by considering a pencil of radiation travelling through a medium. The energy in the beam is given by

$$\mathrm{d}E_\nu = I_\nu \cos\theta \, \mathrm{d}A \, \mathrm{d}\Omega \, \mathrm{d}\nu \, \mathrm{d}t, \tag{2.7}$$

where ν is wavenumber $(1/\lambda)$, I_ν is the monochromatic (spectral) *radiance* (sometimes called intensity), θ is the angle from the normal to the surface, $\mathrm{d}A$ is a differential area element intercepting the beam, Ω is the solid angle in steradians, and t is time.

Next, we explicitly describe how the radiation beam changes as it interacts with matter (of density ρ), travelling through a distance $\mathrm{d}s$:

$$\mathrm{d}I_\nu = -k_\nu \rho I_\nu \mathrm{d}s + j_\nu \rho \mathrm{d}s. \tag{2.8}$$

The first term on the right hand side represents the amount of radiation removed from the beam (extinction cross-section k_ν) and the second term represents the amount of radiation added to the beam (emission cross-section j_ν). Defining the source function S_ν as the ratio of the emission cross-section to the extinction cross-section, we have

$$\frac{\mathrm{d}I_\nu}{\rho k_\nu \mathrm{d}s} = -I_\nu + S_\nu. \tag{2.9}$$

This is the *radiative transfer equation* and it governs the fundamental physics at work in the atmosphere. The simplistic form of the radiative transfer equation hides its true complexity. The main problem lies in the non-linearity of the equation. The solution of I_ν depends on j_ν, but if there is scattering in the atmosphere j_ν also depends on I_ν. A second problem lies in the definition of the source function $S_\nu = j_\nu / k_\nu$. The opacities that make up k_ν and j_ν can be composed of millions of lines for molecular species, and in the case of cloud opacities can involve a number of free parameters.

Finally, we can write Equation (2.9) in a more conventional form by making a few more definitions. If we consider a plane-parallel atmosphere, we are interested only in radiation flowing in the vertical direction z. We can define

$$\mu = \cos\theta, \tag{2.10}$$

where θ is again the angle measured from the vertical, or the zenith angle. The distance $\mathrm{d}s$ can be projected along the vertical axis, as in

$$\mathrm{d}s = \cos\theta \, \mathrm{d}z = \mu \, \mathrm{d}z. \tag{2.11}$$

We can now define the *optical depth* τ as

$$\tau_\nu(z) = -\int_z k_\nu(z)\,\rho(z)\,\mathrm{d}z. \tag{2.12}$$

The minus sign appears because the optical depth is by convention measured from the top of the atmosphere increasing downward. Using the definition of the optical depth, we can rewrite Equation (2.9) as

$$\mu \frac{\mathrm{d}I_\nu}{\mathrm{d}\tau} = I_\nu - S_\nu. \tag{2.13}$$

The detailed solution of this equation is beyond the scope of this work, and we refer the interested reader to more comprehensive works that describe the solution and application of the radiative transfer equation (e.g., Mihalas (1970), Liou (2002), Salby (1996)).

Fortunately, under specific assumptions, the solution to Equation (2.13) becomes simple. As discussed in Section 2.2, during transit the planet passes in front of the star, and some starlight passes through the annulus of the planetary atmosphere before reaching the observer. At visible wavelengths (where the thermal emission is negligible) the starlight is attenuated by the absorbing gases in the planet atmosphere. In this case, we take the emission, and thus the source function S_ν, to be zero, and Equation (2.13) reduces to

$$\mu \frac{\mathrm{d}I_\nu}{\mathrm{d}\tau} = I_\nu, \tag{2.14}$$

which can easily be integrated to obtain

$$I_\nu = I_\nu(0)\,\mathrm{e}^{-\tau_\nu/\mu}. \tag{2.15}$$

This equation is known as Beer's law or Lambert's law (Liou, 2002). It describes the dissipation of radiation as it travels through a medium. Because atoms and molecules absorb at specific wavelengths, the amount of starlight that is transmitted through the planetary atmosphere changes with wavelength.

Another physical situation with a simple solution to the radiative transfer equation is the case of thermal emission and no scattering. This situation would hold at infrared wavelengths if clouds (i.e., scattering particles) were not present. In this case of thermal emission, the source function is simply the blackbody function:

$$S_\nu = B_\nu. \tag{2.16}$$

The radiative transfer equation (2.13) then reduces to a linear form:

$$\mu \frac{\mathrm{d}I_\nu}{\mathrm{d}\tau} = I_\nu - B_\nu. \tag{2.17}$$

The solution is

$$I_\nu(z) = \int_0^\pi \frac{1}{\mu} \int_0^\infty B_\nu(\tau) e^{-\tau_\nu(z)/\mu} d\tau d\mu. \tag{2.18}$$

With a given vertical temperature and pressure profile, the opacities and hence B_ν can be computed, and the right hand side of the above equation is straightforward to integrate.

2.6 Model atmospheres

A full model atmosphere computation is needed to understand the details of the planetary spectrum. Usually the models assume that the planetary atmosphere is one-dimensional and plane-parallel (no curvature). The models produce the temperature and pressure as a function of altitude and the radiation field (i.e., the emergent flux from the atmosphere) as a function of altitude and wavelength (see Seager *et al.* (2005) and references therein). To derive these three quantities, three equations are solved: the radiative transfer equation (Equation (2.13)), the equation of hydrostatic equilibrium, and the radiative and convective equilibrium. The boundary conditions are the incident stellar radiation at the top of the atmosphere and the interior energy (assumed) at the bottom of the atmosphere. With this type of calculation, only the planetary surface gravity and the incident stellar radiation are known with certainty. Although the physics governing the model is relatively simple, a number of physical processes must be included in order for the calculation to proceed, including (Seager *et al.*, 2005; Marley *et al.*, 2006):

- atmospheric chemistry (including elemental abundances, non-equilibrium chemistry, and photochemistry);
- cloud properties;
- atmospheric circulation;
- internal heat flow; and
- gaseous opacities.

Most published model results have used solar elemental abundances (i.e., having the same relative concentrations as the Sun.) Of course, this assumption is limited in that different stars will have relative abundances different from the Sun. Even the relative abundances of the elements in the Sun remain somewhat uncertain. More importantly, the Solar System giant planets are enriched in carbon, and Jupiter and Saturn are also enriched in nitrogen relative to the solar composition (see Marley *et al.* (2006) and references therein). With the assumed elemental abundances, chemical equilibrium calculations determine the abundances of the different atomic, molecular, and liquid or solid species as a function of temperature and pressure. For example, given the elemental abundances of carbon and oxygen relative to hydrogen, the relative concentrations of methane (CH_4) and carbon monoxide (CO) can be computed. CH_4 and CO are particularly interesting molecules for the hot Jupiters (Seager *et al.*, 2000) because it is unclear which one is the dominant form of carbon due to the uncertain temperatures and metallicities of the hot Jupiters. At higher temperatures and higher C to O ratios, we expect CO to be the dominant form of carbon, while at lower temperatures CH_4 is the dominant form of carbon.

With the computed abundances of chemical species, the opacities can be determined. The *opacity* represents the amount of radiation that a given species can absorb as a function of wavelength. The opacities of the expected chemical species in the model atmosphere play a pivotal role in determining the structure of the resulting spectrum. In particular, water, methane, ammonia (NH_3), sodium, and potassium all have significant spectral signatures for hot gas giant planets and are expected to be present in the atmospheres of these planets. Opacities are particularly sensitive to choices of metallicity, which species (atomic and molecular) are included, and whether equilibrium or non-equilibrium chemistry is considered. Absorption due to collisions between molecules (called collision-induced absorption) also has a measurable effect, and modellers typically have to account for interactions between H_2–H_2 and H_2–He.

Cloud structure plays a critical role in controlling the resulting atmospheric spectra. Unfortunately, clouds are extremely difficult to model and represent one of the greatest uncertainties in the atmospheric models. The structure, height, and composition of the clouds depend on the local conditions in the atmosphere as well as the transport (horizontal and vertical) of the condensates present in the atmosphere. In 'ad hoc' cloud models, the type of condensates, the degree of condensation, and the particle size distribution are all free parameters in defining the cloud structure. One-dimensional cloud models use cloud microphysics to compute these parameters (Ackerman and Marley, 2001; Cooper *et al.*, 2003). All extrasolar planet atmosphere models currently in the literature further assume that the clouds are uniformly distributed over the entire planet.

Since the hot Jupiters are likely to be tidally locked (meaning the same hemisphere of the planet always faces the star), atmospheric circulation is key for redistributing absorbed stellar energy and determining the temperature gradients across the planet atmosphere. Atmospheric circulation models (e.g., Showman and Guillot (2002), Cho *et al.* (2003), Cooper and Showman (2005)) have not yet been coupled with radiative transfer models. In this absence, the atmospheric circulation has been parameterized by a parameter f: a value of $f = 1$ implies that the incident stellar radiation is emitted into 4π steradians (meaning the heat is evenly redistributed throughout the planet's atmosphere), while $f = 2$ implies that the incident stellar radiation is emitted into only 2π steradians (i.e., only the day side absorbs and emits the radiation, and there is no transport to the night side). This parameter is a way of quantifying the atmospheric dynamics, and it is used in the models to interpret the observed spectra (see Section 2.7). In model atmospheres this factor f is used in reducing the incident stellar radiation. It is also used in estimating the equilibrium temperature T_{eq}, defined as

$$T_{eq} = T_* \sqrt{\frac{R_*}{2a}} [f(1 - A_B)]^{1/4}, \tag{2.19}$$

where T_* is the stellar temperature, R_* is the stellar radius, a is the orbital semi-major axis, and A_B is the (unknown) Bond albedo, which is the fraction of incident stellar radiation scattered back into space in all directions by the planet. This relation was used to derive the values listed in Table 2.1, assuming $f = 1$ and $A_B = 0.3$.

We now turn to a discussion of the specific spectroscopic and photometric observations of extrasolar planets that have been conducted.

2.7 Observations

In this section, we summarize the important spectroscopic and photometric observations of transiting planets that have been conducted, during both primary and secondary eclipse. Most of these observations have been performed on the planet HD 209458 b, since it was detected first. We conclude by describing how the model calculations have helped to interpret these results.

As discussed in Section 2.2, the planetary spectrum can be probed during transit using a method called transmission spectroscopy. Although the planetary spectrum is $\sim$10 000 times fainter than that of the star, the differential nature of the measurement makes it possible to achieve this precision. Several detections and useful upper limits have been obtained on HD 209458 b:

- sodium doublet detected (Charbonneau *et al.*, 2002);
- hydrogen Lyman-α detected (Vidal-Madjar *et al.*, 2003);
- carbon monoxide upper limit (Deming *et al.*, 2005a).

The amount of sodium detected was approximately a factor of 3 smaller than expected from simple models of the atmosphere, suggesting the presence of a high cloud that masks the true sodium abundance. The detection of the transit in H Lyman-α was very significant – a 15% drop in stellar flux during transit, 10 times greater than the transit depth at visible wavelengths. This implies an extended atmosphere of 3 or 4 Jupiter radii, and suggests that the planet is losing mass over its lifetime. The CO non-detection further reinforces the notion of a high cloud in the planet's atmosphere.

The complementary technique during secondary eclipse is called *occultation spectroscopy*. Briefly, this involves taking spectra of the system when the planet is out of eclipse (when both the star and planet are visible) and comparing these with spectra recorded when the planet is hidden during secondary eclipse. By carefully differencing these spectra, one can, in principle, derive the spectrum of the planet itself. Although this technique has not yet been successfully conducted on extrasolar planets, early attempts have yielded some useful information:

- upper limit on emission near 2.2 μm (Richardson *et al.*, 2003b);
- upper limit on methane abundance (Richardson *et al.*, 2003a).

Both of these limits were derived from ground-based observations, which are often limited by variations in the terrestrial atmosphere, making detection of spectral features difficult.

We now turn to photometric observations of the secondary eclipse that have occurred most recently. Although measurable, the effect due to the secondary eclipse is small, e.g., ~0.3% for HD 209458 b at 20 μm (see Figure 2.2), and decreasing for smaller wavelengths. NASA's Spitzer Space Telescope[2] was responsible for the first detection of a secondary eclipse of a transiting planet. Spitzer, with an 85-cm aperture, has three instruments on board that together perform photometry and spectroscopy at infrared wavelengths. In March 2005, two independent research groups announced detections of the secondary eclipse of two different planets using two Spitzer instruments. Observations of HD 209458 b with the Multiband Imaging Photometer for Spitzer (MIPS) detected the secondary eclipse at 24 μm (Deming *et al.*, 2005b), while TrES-1 was observed in two wavelengths (4.5 and 8 μm) with the Infrared Array Camera (IRAC) (Charbonneau *et al.*, 2005). These observations represent the first *direct* detection of an extrasolar planet. More recently, the secondary eclipse of HD 189733 b was observed at 16 μm using the Infrared Spectrograph (IRS), although the observation was performed photometrically, not spectroscopically, using a detector that is normally used only to align the star on the slit (Deming *et al.*, 2006).

The secondary eclipse detections provide a measurement of the *brightness temperature* of the planets, at the respective wavelengths. The brightness temperature is the blackbody temperature of an object at a particular wavelength; given the irradiance, the blackbody function (see Equation (2.6)) can be inverted to solve for temperature. For HD 209458 b the brightness temperature at 24 μm is 1130 ± 150 K (Deming *et al.*, 2005b), and for TrES-1 it is 1010 ± 60 K at 4.5 μm and 1230 ± 110 K at 8 μm (Charbonneau *et al.*, 2005). HD 189733 b has a brightness temperature of 1117 ± 42 K at 16 μm (Deming *et al.*, 2006). Although models

[2] http://ssc.spitzer.caltech.edu/

have predicted the effective temperature of the atmospheres of extrasolar planets, these are the first observational measurements of the temperature of an extrasolar planetary atmosphere.

With the Spitzer photometry of several transiting planets, as well as ground-based spectroscopic observations, we can now compare the observational results to theoretical calculations and begin to construct a comprehensive picture of the atmospheres of the transiting planets. In the wake of the three initial photometric detections of thermal emission from two extrasolar planets (Deming *et al.*, 2005b; Charbonneau *et al.*, 2005), *four* theory papers (Seager *et al.*, 2005; Barman *et al.*, 2005; Fortney *et al.*, 2005; Burrows *et al.*, 2005) appeared within a few months to explain the results! Some of these even have conflicting conclusions. One conclusion on which all of the explanations agree is that the planets are hot, as predicted. (We note that this was not a given; a planet with a high Bond albedo, for example, would reflect much of the incident stellar irradiation and therefore could be much cooler, as seen in Equation (2.19).)

The second point on which all modellers agree is that the TrES-1 data points at 4.5 and 8.0 μm are not consistent with the assumption of solar abundances, because the 8.0 μm flux is too high. Beyond these two conclusions, the interpretations diverge.

Seager *et al.* (2005) conclude that a range of models remain consistent with the data. They include the 2.2 μm upper limit reported by Richardson *et al.* (2003b) (which has been largely ignored by modellers), as well as an upper limit on the albedo from the Canadian MOST satellite (Rowe *et al.*, 2005), and are able to eliminate the models for HD 209458 b that are on the hot and cold ends of the plausible temperature range for the planet. Their work suggests that an intermediate value for f (see Equation (2.19)) is most likely, indicating that the atmospheric circulation is somewhere between the two extremes (efficient redistribution vs none at all). The interpretation by Seager *et al.* (2005) and the observational results for HD 209458 b are shown in Figure 2.4.

Fortney *et al.* (2005) show that standard models using solar abundances are consistent with HD 209458 b but only marginally consistent (within 2σ at 8.0 μm) for TrES-1. For both planets, their best fit models assume that the incident stellar radiation is redistributed efficiently throughout the atmosphere (i.e., $f = 1$). On the other hand, Burrows *et al.* (2005) conclude that the $f = 2$ case is more likely, indicating that the day side is significantly brighter in the infrared than the night side. They also infer the presence of CO and possibly H_2O. The resolution of this discrepancy awaits further Spitzer observations.

Finally, with the recent Spitzer detection of HD 189733 b during secondary eclipse at 16 μm (Deming *et al.*, 2006) and detections by IRAC and MIPS under analysis (D. Charbonneau, private communication), we have more data available for comparison with theoretical spectra. In addition, observations are being analysed

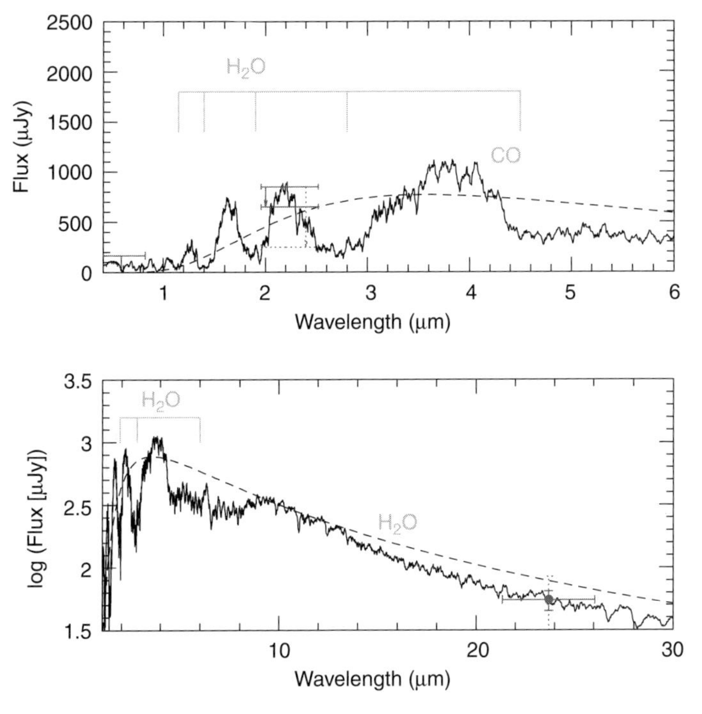

Fig. 2.4. Theoretical spectrum of HD 209458 b with data points and upper limits. The solid curve is a cloud-free model with solar abundances and $f = 2$, characterized by deep water vapour absorption features. From left to right the data are: the MOST upper limit (Rowe *et al.*, 2005), a constraint on the H_2O band depth (Richardson *et al.*, 2003a; Seager *et al.*, 2005), and the Spitzer/MIPS thermal emission point at 24 μm (Deming *et al.*, 2005b). The solid lines show 1 σ error bars or upper limits and the dashed lines show 3 σ values. Note the linear flux scale on the upper panel and the log flux scale on the lower panel.

or planned to detect a mid-infrared emission spectrum of HD 209485 b and HD 189733 b, respectively, both using Spitzer/IRS. These observations would be the first observed emission spectra of an extrasolar planet and will advance our understanding beyond the few photometric data points we have now.

2.8 Future missions

The spectroscopic and photometric observations of hot Jupiters have provided a wealth of information on their physical characteristics and led to insights about their atmospheric structure. What about extrasolar planets similar to the Earth? Although detection of such rocky planets remains just beyond the limits of current techniques, a few short-period planets with masses only 5–15 times that of the Earth (sometimes called 'hot super-massive Earths') have been discovered

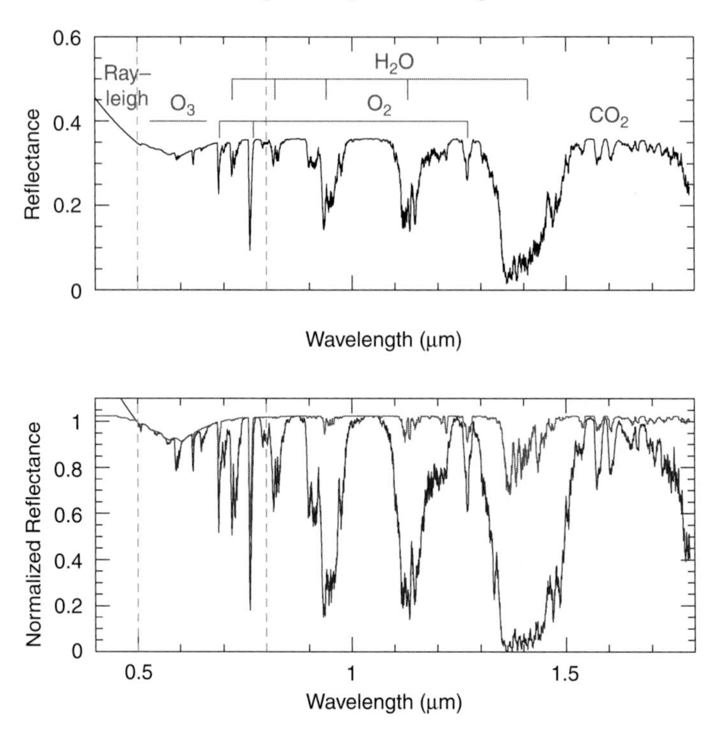

Fig. 2.5. Theoretical spectra of Earth. Upper panel: theoeretical model that matches the Woolf *et al.* (2002) Earthshine data. The dashed vertical lines show the nominal wavelength range of TPF-C. Lower panel: normalized models of Earth showing effects of clouds. The top curve is for uniform high cloud coverage showing weaker water vapour features. The bottom curve shows the case with no clouds, resulting in deep absorption features.

(Santos *et al.*, 2004; Beaulieu *et al.*, 2006), pushing the detection limit to ever smaller planets. We close this chapter with a brief discussion of how we might search for Earth-like planets around other stars and what future missions are being planned to tackle this fundamental question.

The goal of directly imaging an Earth-like planet is to search for *biosignatures*, which are spectral features that can be used as diagnostics to search for the presence of life as we know it. The Earth has several such biosignatures that are indicative of habitability or life. Figure 2.5 shows two of these species: O_2 and its photolytic product O_3, two of the most reliable biosignature gas indicators of life. O_2 is highly reactive and therefore will remain in significant quantities in the atmosphere only if it is continually produced. There are no abiotic continuous sources of large quantities of O_2 and only rare false positives that in most cases could likely be ruled out by other planetary characteristics. N_2O is a second gas produced by life – albeit in small quantities – during microbial oxidation-reduction reactions. N_2O has a very weak spectroscopic signature.

In addition to atmospheric biosignatures, the Earth has one very strong and very intriguing biosignature on its surface: vegetation. The reflection spectrum of photosynthetic vegetation has a dramatic sudden rise in albedo around 750 nm by almost an order of magnitude! (This effect is not included in the model plotted in Figure 2.5.) Vegetation has evolved this strong reflection feature, known as the 'red edge', as a cooling mechanism to prevent overheating which would cause chlorophyll to degrade. On Earth, this feature is likely reduced by a few per cent due to clouds. A surface biosignature might be distinguished from an atmospheric signature by observing time variations; i.e., as the continents, for example, rotate in and out of view, the spectral signal will change correspondingly. Other spectral features, although not biosignatures because they do not reveal direct information about life or habitability, can nonetheless provide significant information about the planet. These include CO_2 (which is indicative of a terrestrial atmosphere and has a very strong mid-infrared spectral feature) and CH_4 (which has both biotic and abiotic origins). A range of spectral features is needed to characterize Earth-like planet atmospheres.

The James Webb Space Telescope (JWST) (e.g., Gardner *et al.* (2006)), tentatively scheduled for launch after 2013, will pick up where Spitzer leaves off, in terms of extrasolar planet characterization by primary and secondary eclipse studies. JWST is an infrared telescope with an aperture 6.5 m in diameter, representing a factor of $\sim$60 greater collecting area over Spitzer's 0.85 m diameter aperture. JWST will not only be able to detect thermal emission spectra from hot Jupiters, but also may be able to see emission from hot, super-massive Earths. It may also be possible to perform transmission spectroscopy on such planets with JWST.

NASA's Terrestrial Planet Finder (TPF) missions and ESA's Darwin mission seek to find and characterize Earth-like planets orbiting nearby stars. TPF is split into two separate missions, a visible coronagraph (TPF-C) and an infrared nulling interferometer (TPF-I). Although scheduling and budgets for TPF are tentative, these missions would provide direct imaging of planets and thus low-resolution spectra of a wide variety of planet sizes and semi-major axes. One major goal of these missions would be to search the observed spectra for the biosignature features described above, in the hope of finding evidence for life on another world.

2.9 Summary

The transiting extrasolar planets have provided new opportunities to characterize the atmospheres and bulk compositions of worlds beyond the Solar System. The geometry of these systems, in which the planet periodically crosses in front of its parent star (primary eclipse) and disappears behind the star (secondary eclipse), has allowed measurements of the true mass, radius, density, and (in a few cases) the brightness temperature of these planets for the first time. This chapter has also presented a brief overview of spectroscopy, summarized how model atmospheres are computed, and described the notable observations of transiting planets. Finally, this chapter has addressed the detection and characterization of Earth-like planets around other stars and summarized a few missions being planned to accomplish this.

References

Ackerman, A. S., and Marley, M. S. (2001). Precipitating condensation clouds in substellar atmospheres. *Astrophys. J.*, **556**, 872–884.

Alonso, R., Brown, T. M., Torres, G., *et al.* (2004). TrES-1: the transiting planet of a bright K0 V star. *Astrophys. J. Lett.*, **613**, L153–L156.

Bakos, G. A., Pal, A., Latham, D. W., Noyes, R. W., and Stefanik, R. P. (2006). A stellar companion in the HD 189733 system with a known transiting extrasolar planet. *Arxiv astrophysics e-prints*, Feb.

Barman, T. S., Hauschildt, P. H., and Allard, F. (2005). Phase-dependent properties of extrasolar planet atmospheres. *Astrophys. J.*, **632**, 1132–1139.

Beaulieu, J.-P., Bennett, D. P., Fouqué, P., *et al.* (2006). Discovery of a cool planet of 5.5 Earth masses through gravitational microlensing. *Nature*, **439**, 437–440.

Bouchy, F., Pont, F., Santos, N. C., *et al.* (2004). Two new 'very hot Jupiters' among the OGLE transiting candidates. *Astron. Astrophys.*, **421**, L13–L16.

Bouchy, F., Udry, S., Mayor, M., *et al.* (2005). ELODIE metallicity-biased search for transiting hot Jupiters. II. A very hot Jupiter transiting the bright K star HD 189733. *Astron. Astrophys.*, **444**, L15–L19.

Burrows, A., and Sharp, C. M. (1999). Chemical equilibrium abundances in brown dwarf and extrasolar giant planet atmospheres. *Astrophys. J.*, **512**, 843–863.

Burrows, A., Guillot, T., Hubbard, W. B., *et al.* (2000). On the radii of close-in giant planets. *Astrophys. J. Lett.*, **534**, L97–LL100.

Burrows, A., Hubeny, I., and Sudarsky, D. (2005). A theoretical interpretation of the measurements of the secondary eclipses of TrES-1 and HD 209458b. *Astrophys. J. Lett.*, **625**, L135–L138.

Burrows, A., Sudarsky, D., and Hubbard, W. B. (2003). A theory for the radius of the transiting giant planet HD 209458b. *Astrophys. J.*, **594**, 545–551.

Charbonneau, D. (2003). HD 209458 and the power of the dark side. *Scientific Frontiers in Research on Extrasolar Planets*. ASP Conf. Ser. 294, pp. 449–456. San Francisco: ASP.

Charbonneau, D., Allen, L. E., Megeath, S. T., *et al.* (2005). Detection of thermal emission from an extrasolar planet. *Astrophys. J.*, **626**, 523–529.

Charbonneau, D., Brown, T. M., Latham, D. W., and Mayor, M. (2000). Detection of planetary transits across a sun-like star. *Astrophys. J. Lett.*, **529**, L45–L48.

Charbonneau, D., Brown, T. M., Noyes, R. W., and Gilliland, R. L. (2002). Detection of an extrasolar planet atmosphere. *Astrophys. J.*, **568**, 377–384.

Cho, J. Y.-K., Menou, K., Hansen, B. M. S., and Seager, S. (2003). The changing face of the extrasolar giant planet HD 209458b. *Astrophys. J. Lett.*, **587**, L117–L120.

Cooper, C. S., and Showman, A. P. (2005). Dynamic meteorology at the photosphere of HD 209458b. *Astrophys. J. Lett.*, **629**, L45–L48.

Cooper, C. S., Sudarsky, D., Milsom, J. A., Lunine, J. I., and Burrows, A. (2003). Modeling the formation of clouds in brown dwarf atmospheres. *Astrophys. J.*, **586**, 1320–1337.

Deming, D., Brown, T. M., Charbonneau, D., Harrington, J., and Richardson, L. J. (2005a). A new search for carbon monoxide absorption in the transmission spectrum of the extrasolar planet HD 209458 b. *Astrophys. J.*, **622**, 1149–1159.

Deming, D., Harrington, J., Seager, S., and Richardson, L. J. (2006). Strong infrared emission from the extrasolar planet HD189733 b. *Arxiv astrophysics e-prints*, Feb.

Deming, D., Seager, S., Richardson, L. J., and Harrington, J. (2005b). Infrared radiation from an extrasolar planet. *Nature*, **434**, 740–743.

Fortney, J. J., Marley, M. S., Lodders, K., Saumon, D., and Freedman, R. (2005). Comparative planetary atmospheres: models of TrES-1 and HD 209458 b. *Astrophys. J. Lett.*, **627**, L69–L72.

Gardner, J. P., Mather, J. C., Clampin, M., *et al.* (2006). The James Webb Space Telescope. *Arxiv astrophysics e-prints*, June.

Gaudi, B. S., Seager, S., and Mallen-Ornelas, G. (2005). On the period distribution of close-in extrasolar giant planets. *Astrophys. J.*, **623**, 472–481.

Henry, G. W., Marcy, G. W., Butler, R. P., and Vogt, S. S. (2000). A transiting '51 Peg-like' planet. *Astrophys. J. Lett.*, **529**, L41–L44.

Holman, M. J., Winn, J. N., Stanek, K. Z., *et al.* (2005). High-precision transit photometry of OGLE-TR-10. *Arxiv astrophysics e-prints*, June.

Knutson, H., Charbonneau, D., Noyes, R. W., Brown, T. M., and Gilliland, R. L. (2006). Using stellar limb-darkening to refine the properties of HD 209458b. *Arxiv astrophysics e-prints*, Mar.

Konacki, M., Torres, G., Sasselov, D. D., *et al.* (2004). The transiting extrasolar giant planet around the star OGLE-TR-113. *Astrophys. J. Lett.*, **609**, L37–L40.

Konacki, M., Torres, G., Sasselov, D. D., and Jha, S. (2005). A transiting extrasolar giant planet around the star OGLE-TR-10. *Astrophys. J.*, **624**, 372–377.

Laughlin, G., Wolf, A., Vanmunster, T., *et al.* (2005). A comparison of observationally determined radii with theoretical radius predictions for short-period transiting extrasolar planets. *Astrophys. J.*, **621**, 1072–1078.

Liou, K. N. (2002). *An Introduction to Atmospheric Radiation*. Second edn. San Diego: Academic Press.

Marley, M. S., Fortney, J., Seager, S., and Barman, T. (2006). Atmospheres of extrasolar giant planets. *Arxiv astrophysics e-prints*, Feb.

Mazeh, T., Naef, D., Torres, G., *et al.* (2000). The spectroscopic orbit of the planetary companion transiting HD 209458. *Astrophys. J. Lett.*, **532**, L55–L58.

McCullough, P. R., Stys, J. E., Valenti, J. A., *et al.* (2006). A transiting planet of a sun-like star. *Arxiv astrophysics e-prints*, May.

Mihalas, D. (1970). *Stellar Atmospheres*. San Francisco: W. H. Freeman and Company.

Moutou, C., Pont, F., Bouchy, F., and Mayor, M. (2004). Accurate radius and mass of the transiting exoplanet OGLE-TR-132b. *Astron. Astrophys.*, **424**, L31–L34.

Pont, F., Bouchy, F., Queloz, D., *et al.* (2004). The 'missing link': A 4-day period transiting exoplanet around OGLE-TR-111. *Astron. Astrophys.*, **426**, L15–L18.

Richardson, L. J., Deming, D., and Seager, S. (2003b). Infrared observations during the secondary eclipse of HD 209458 b. II. Strong limits on the infrared spectrum near 2.2 microns. *Astrophys. J.*, **597**, 581.

Richardson, L. J., Deming, D., Wiedemann, G., *et al.* (2003a). Infrared observations during the secondary eclipse of HD 209458b. I. 3.6 micron occultation spectroscopy using the very large telescope. *Astrophys. J.*, **584**, 1053–1062.

Rowe, J. F., Matthews, J. M., Seager, S., *et al.* (2005). MOST spacebased photometry of the transiting exoplanet system HD 209458: I. Albedo measurements of an extrasolar planet. *American Astronomical Society meeting abstracts*, **207**, 1339.
Salby, M. L. (1996). *Fundamentals of Atmospheric Physics*. Academic Press.
Santos, N. C., Bouchy, F., Mayor, M., *et al.* (2004). The HARPS survey for southern extra-solar planets. II. A 14 Earth-masses exoplanet around μ Arae. *Astron. Astrophys.*, **426**, L19–L23.
Santos, N. C., Pont, F., Melo, C., *et al.* (2006). High resolution spectroscopy of stars with transiting planets. The cases of OGLE-TR-10, 56, 111, 113, and TrES-1. *Astron. Astrophys.*, **450**, 825–831.
Sato, B., Fischer, D. A., Henry, G. W., *et al.* (2005). The N2K consortium. II. A transiting hot Saturn around HD 149026 with a large dense core. *Astrophys. J.*, **633**, 465–473.
Seager, S., and Mallén-Ornelas, G. (2003). A unique solution of planet and star parameters from an extrasolar planet transit light curve. *Astrophys. J.*, **585**, 1038–1055.
Seager, S., Richardson, L. J., Hansen, B. M. S., *et al.* (2005). On the dayside thermal emission of hot Jupiters. *Astrophys. J.*, **632**, 1122–1131.
Seager, S., Whitney, B. A., and Sasselov, D. D. (2000). Photometric light curves and polarization of close-in extrasolar giant planets. *Astrophys. J.*, **540**, 504–520.
Showman, A. P., and Guillot, T. (2002). Atmospheric circulation and tides of '51 Pegasus b-like' planets. *Astron. Astrophys.*, **385**, 166–180.
Torres, G., Konacki, M., Sasselov, D. D., and Jha, S. (2004). New data and improved parameters for the extrasolar transiting planet OGLE-TR-56b. *Astrophys. J.*, **609**, 1071–1075.
Vidal-Madjar, A., des Etangs, A. L., Désert, J.-M., *et al.* (2003). An extended upper atmosphere around the extrasolar planet HD209458b. *Nature*, **422**, 143–146.
Winn, J. N., and Holman, M. J. (2005). Obliquity tides on hot Jupiters. *Astrophys. J. Lett.*, **628**, L159–L162.
Winn, J. N., Noyes, R. W., Holman, M. J., *et al.* (2005). Measurement of spin–orbit alignment in an extrasolar planetary system. *Astrophys. J.*, **631**, 1215–1226.
Woolf, N. J., Smith, P. S., Traub, W. A., and Jucks, K. W. (2002). The spectrum of earthshine: a pale blue dot observed from the ground. *Astrophys. J.*, **574**, 430–433.

3

Terrestrial planet formation

Edward W. Thommes

Canadian Institute for Theoretical Astrophysics

3.1 Introduction

Kant (1755) and Laplace (1796) laid the foundations of our theory of planet formation, arguing that the Solar System formed in a flattened disk orbiting the Sun. Today we know such disks to be a part of the star formation process (see Chapter 4), and we have come a long way toward understanding the details of how a protostellar disk turns itself into planets. Most importantly, for ten years now we have been finding planets orbiting other stars; the current count of detected exoplanets exceeds 180, so that the Solar System now constitutes only a small minority of all known planets. The bad news is that Mercury, Venus, Earth, and Mars remain the only examples of terrestrial planets we know, since such bodies are still well below the mass threshold for detection around ordinary stars. Thus, apart from a pair of bodies orbiting the pulsar PSR1257 + 12 (Wolszczan and Frail, 1992), all known exoplanets are giants: gaseous bodies like Jupiter or Saturn with a couple of potentially Neptune-like examples recently added to the menagerie. However, impressive as the 'king of the planets' and its kind are, it is a terrestrial planet – a modest rocky body, three-thousandths Jupiter's mass – which actually harbours the life in our Solar System. And as we contemplate the possibility of life elsewhere in the Universe, it is inevitably upon terrestrial planets that we focus our attention. It is thus especially important to attempt to understand the process of terrestrial planet formation, so that even in the absence of direct observations, we can assess the likelihood of other planets like our own existing around other stars. At the same time, such work allows us to plan ahead for future terrestrial planet-hunting missions like KEPLER, COROT, and (if it is resurrected) the Terrestrial Planet Finder (TPF).

This chapter gives an overview of the theory of terrestrial planet formation as it currently stands. We begin with the formation of the basic building blocks, planetesimals, in Section 3.2. In Section 3.3, we look at how protoplanets grow from

Planetary Systems and the Origins of Life, eds. Ralph E. Pudritz, Paul G. Higgs, and Jonathon R. Stone. Published by Cambridge University Press.

the agglomeration of planetesimals. This brings us to the final stage of terrestrial planet formation in Section 3.4, wherein the protoplanets collide to produce bodies like Earth and Venus. We examine the 'standard mode' of how this stage plays out, together with some more recent variations, as well as the constraints cosmochemistry places on the the timeline of the Earth's formation. In Section 3.5, we summarize the current state of the impact model for the origin of the Earth–Moon system. In Section 3.6, we consider terrestrial planet formation and evolution from an astrobiology perspective, including the concept of the habitable zone and the puzzle posed by the Late Heavy Bombardment. A summary is given in Section 3.7.

3.2 The formation of planetesimals

In the first phase of terrestrial planet formation, solids condense out of the protostellar gas disk as dust grains, settle to the disk midplane, and eventually form kilometre-sized planetesimals. The actual mechanism of planetesimal formation remains unclear. Goldreich and Ward (1973) put forward a model in which the dust layer becomes sufficiently thin that it undergoes gravitational instability, contracting (in the terrestrial region) into kilometre-sized clumps and thus producing planetesimals directly. However, Weidenschilling (1995) pointed out that turbulent stirring of the dust layer would likely prevent gravitational instability. The dust particles are on Keplerian orbits while the gas disk, being partly supported by its own internal pressure, revolves at slightly less than the Keplerian rate. The resulting vertical shear in velocity at the dust–gas interface induces turbulence in the dust layer, preventing it from becoming thin enough for the onset of gravitational instability.

Because of this problem, the most widely held current view is that dust particles build planetesimals by pairwise accretion. Laboratory experiments suggest that growth of dust to centimetre-sized aggregates proceeds quite easily, probably in $\ll 10^6$ y (Paraskov *et al.* (2005) and references therein). At metre size and larger, the process is less well understood. For one thing, it is unclear how readily larger bodies stick together. Turbulent eddies in the gas disk may play a role in bringing about local particle concentrations (e.g., Barranco and Marcus (2005)). However, orbital decay due to aerodynamic gas drag from the nebula seems to present a problem. Due to the sub-Keplerian rotation of the gas, a solid body orbiting with the disk experiences a headwind, loses energy, and experiences continual orbital shrinking. Orbital decay rates are peaked for metre-sized bodies; for typical nebula models, such planetesimals spiral from 1 AU (= astronomical unit, the mean radius of the Earth's orbit around the Sun, which is 1.5×10^{13} cm) into the central star on a timescale of order 100 years (Weidenschilling, 1977). Thus, it appears growth through this size regime must be extremely rapid in order to avoid the loss of all solids. Another possibility is that fluctuations in nebular gas density, turbulent or otherwise, may locally stall the migration.

3.3 The growth of protoplanets

Once the solids in the protoplanetary disk are mostly locked up in kilometre-sized bodies, their interactions become dominated by gravity, and in a sense the dynamics becomes simpler. We will begin with a qualitative description of what happens next. The planetesimals are on nearly circular coplanar Keplerian orbits within the disk, although mutual perturbations induce slight deviations from this. In terms of orbital elements, the planetesimals' interactions induce in each other non-zero orbital eccentricities (thus making neighbouring orbits cross) and inclinations (thus vertically 'puffing up' the disk). Since the self-stirring of the disk is gravitational, the upper limit to the relative velocities induced in the planetesimals is the surface escape velocity from the locally largest bodies. Therefore initially, when two planetesimals collide, they usually do so at an impact velocity low enough that accretion dominates over shattering. Larger-mass bodies have larger collision cross-sections, so accretion eventually becomes a runaway process: a relatively small subset of larger bodies, or protoplanets, grows very rapidly in the disk, consuming the smaller neighbours. However, runaway growth eventually runs out of steam when it becomes too successful; once the largest protoplanets are massive enough to dominate the gravitational stirring of the nearby disk, the feedback for accretion switches from positive to negative, and larger protoplanets grow more slowly than smaller ones. Thus neighbouring protoplanets converge toward similar sizes. For the little planetesimals it is now too late; with the large bodies having taken over the stirring, they now smash into each other at impact velocities too high for much accretion to occur, and they are reduced to serving as building material for the protoplanets. With the population of similar-sized protoplanets now firmly in charge, this regime is called *oligarchic growth*. The protoplanets keep sweeping up planetesimals, and occasionally two neighbours merge, until there is nothing but protoplanets left.

We can gain some additional insight from a more quantitative approach. To begin with, a universally important quantity is the *Hill radius* (or Hill sphere), which gives a measure of gravitational reach of a body orbiting a central mass. For a companion of mass M orbiting a primary of mass M_* at a distance r, it is defined as

$$r_H = \left(\frac{M}{3M_*}\right)^{1/3} r. \tag{3.1}$$

This is the radius within which a third body (with mass $\ll M$) is able to orbit around M. Thus a moon must be within its parent planet's Hill sphere. Of course, this is a necessary but not sufficient condition; a would-be satellite approaching a planet too fast will simply pass through the Hill sphere with some deflection, rather than be captured.

If the relative velocities between neighbouring planetesimals are dispersion-dominated – i.e., the main contribution to their relative velocities is their mutual perturbations, rather than the difference in Keplerian speed between neighbouring orbits – their growth rate can be estimated using a simple 'particle in a box' approach (Lissauer (1993) and references therein):

$$\frac{\mathrm{d}M}{\mathrm{d}t} = \rho_{\mathrm{plsmls}} \sigma v_{\mathrm{rel}}, \tag{3.2}$$

where ρ_{plsmls} is the density with which planetesimals are distributed in the planetesimal disk, σ is the collisional cross-section of a planetesimal, and v_{rel} is the average relative velocity between nearby planetesimals. The collisional cross-section of a planetesimal is given by

$$\sigma = \pi R^2 \left(1 + \frac{v_{\mathrm{esc}}^2}{v_{\mathrm{rel}}^2}\right) \equiv \pi R^2 f_{\mathrm{g}}, \tag{3.3}$$

where R is the planetesimal's physical radius, $v_{\mathrm{esc}} = \sqrt{2GM/R}$ is the escape velocity from its surface, and f_{g} is the enhancement of the cross-section by gravitational focusing. With a few substitutions and simplifications, we can gain more insight from Equation (3.2). To begin with, we will assume that gravitational focusing is effective, so that $f_{\mathrm{g}} \gg 1$. Also, $\rho_{\mathrm{plsmls}} = \Sigma_{\mathrm{plsmls}}/2H_{\mathrm{plsmls}}$, where Σ_{plsmls} is the surface density of the planetesimal disk, and H_{plsmls} is the disk's scale height. If the planetesimals have a characteristic random velocity v, then the vertical component is $\approx v/\sqrt{3}$, and so $H \approx (rv)/(\sqrt{3}v_{\mathrm{Kep}})$. Finally, $v_{\mathrm{Kep}} \propto r^{-1/2}$, $R \propto M^{1/3}$ and $v_{\mathrm{esc}} \propto M^{1/3}$, and so the growth rate has the form

$$\frac{\mathrm{d}M}{\mathrm{d}t} \propto \frac{\Sigma_{\mathrm{plsmls}} M^{4/3}}{v^2 r^{3/2}}. \tag{3.4}$$

As shown by Wetherill and Stewart (1989), planetesimal accretion is subject to positive feedback, which results in runaway growth: the largest bodies grow the fastest, rapidly detaching themselves from the size distribution of the planetesimals. The reason for this can be seen from Equation (3.4): $\mathrm{d}M/\mathrm{d}t \propto M^{4/3}$, and so the growth timescale is

$$t_{\mathrm{grow}} \equiv \frac{M}{\mathrm{d}M/\mathrm{d}t} \propto M^{-1/3}, \tag{3.5}$$

thus larger bodies grow faster than smaller ones.

Ida and Makino (1993) showed that runaway growth only operates temporarily; once the largest protoplanets are massive enough to dominate the gravitational stirring of the nearby planetesimals, the mode of accretion changes. From this point on, near a protoplanet of mass M, the planetesimal random velocity is approximately proportional to the surface escape speed from the protoplanet, and so $v \propto M^{1/3}$. Therefore, we now have

$$\frac{\mathrm{d}M}{\mathrm{d}t} \propto \frac{\Sigma_{\mathrm{plsmls}} M^{2/3}}{r^{3/2}}, \tag{3.6}$$

which makes

$$t_{\mathrm{grow}} \propto M^{1/3}. \tag{3.7}$$

Thus growth becomes orderly; larger bodies grow more slowly, and we have the onset of oligarchic growth (Kokubo and Ida, 1998). In the terrestrial region, the transition from runaway to oligarchic growth already happens when the largest protoplanets are still many orders of magnitude below an Earth mass (Thommes *et al.*, 2003). As a result, protoplanets spend almost all of their time growing oligarchically.

As long as nebular gas is present, the random velocity v of the planetesimals is set by two competing effects: gravitational stirring acts to increase v, while aerodynamic gas drag acts to reduce it. One can estimate the equilibrium random velocity by equating the two rates (Kokubo and Ida, 1998). In this way, one can eliminate v and write the protoplanet growth rate (recall that there is essentially no more planetesimal–planetesimal growth by this time) in terms of planetesimal and gas disk properties. Also, for simplicity we approximate the planetesimal population as having a single characteristic mass m. The result is (for details see Thommes *et al.* (2003))

$$\frac{\mathrm{d}M}{\mathrm{d}t} \approx A \Sigma_{\mathrm{plsmls}} M^{2/3}, \tag{3.8}$$

where

$$A = 3.9 \frac{b^{5/2} C_{\mathrm{D}}^{2/5} G^{1/2} M_*^{1/6} \rho_{\mathrm{gas}}^{2/5}}{\rho_{\mathrm{plsml}}^{4/15} \rho_M^{1/3} r^{1/10} m^{2/15}}. \tag{3.9}$$

C_{D} is a dimensionless drag coefficient ~ 1 for kilometre-sized or larger planetesimals, M_* is the mass of the central star, ρ_{gas} is the density of the gas disk, and ρ_M and ρ_m are the material densities of the protoplanet and the planetesimals, respectively. The parameter b is the spacing between adjacent protoplanets in units of their Hill radii (3.1). An equilibrium between mutual gravitational scattering on the one hand and recircularization by dynamical friction on the other keeps $b \sim 10$ (Kokubo and Ida, 1998).

Protoplanet growth by sweep-up of planetesimals ceases when all planetesimals are gone. This leads to the notion of the *isolation mass*, which is the mass at which a protoplanet has consumed all planetesimals within an annulus of width $br_{\rm H}$ centred on its orbital radius. The isolation mass is given by

$$M_{\rm iso} = \frac{\left(2\pi b \Sigma_{\rm plsmls} r^2\right)^{3/2}}{\sqrt{3M_*}}. \tag{3.10}$$

Note that when $\Sigma_{\rm plsmls} \propto r^{-2}$, $M_{\rm iso}$ is constant with radius from the star; the increasing gravitational reach at larger distance from the star exactly balances the falloff in the density of planetesimals. For any shallower surface density profile, $M_{\rm iso}$ increases with r.

In order to obtain values for protoplanet accretion times and final masses, a useful starting point is the 'minimum-mass Solar Nebula' (MMSN) model (Hayashi, 1981), which is obtained by 'smearing out' the refractory elements contained in the Solar System planets into a power-law planetesimal disk, then adding gas to obtain solar abundance:

$$\Sigma_{\rm gas}^{\rm min} = 1700 \left(\frac{r}{1\ {\rm AU}}\right)^{-3/2} \ {\rm g\,cm^{-2}}, \tag{3.11}$$

$$\rho_{\rm gas}^{\rm min} \approx \Sigma_{\rm gas}/2H, \quad H = 0.0472 \left(\frac{r}{1\ {\rm AU}}\right)^{5/4} \ {\rm AU}, \tag{3.12}$$

and

$$\Sigma_{\rm solid}^{\rm min} = 7.1 F_{\rm SN} \left(\frac{r}{1\ {\rm AU}}\right)^{-3/2} \ {\rm g\,cm^{-2}}, \tag{3.13}$$

where

$$F_{\rm SN} = \begin{cases} 1, & r < r_{\rm SN} \\ 4.2, & r > r_{\rm SN} \end{cases} \tag{3.14}$$

is the 'snow line' solids enhancement factor: beyond $r_{\rm SN}$ (= 2.7 AU in the Hayashi model), water freezes out, thus adding to the surface density of solids.

Since the exponent of the surface density power law is greater than -2, we know from Equation (3.10) that the isolation mass will increase with distance from the star (though from Equation (3.8), the time to finish growing also increases with r). Using Equation (3.10) with $b = 10$, the MMSN yields an isolation mass of 0.07 $M_\oplus$ at 1 AU. If we assume a material density of 2 $\rm g\,cm^{-3}$ for planetesimals and protoplanets, and a characteristic planetesimal size of ~1 km (with a corresponding

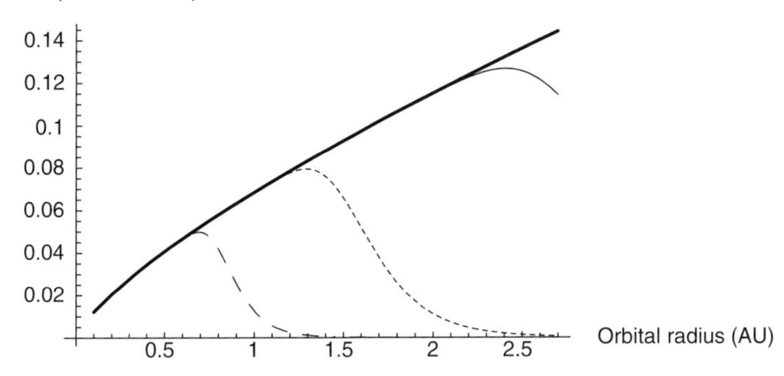

Fig. 3.1. Protoplanet accretion during the 'oligarchic growth' phase in an MMSN. The average protoplanet mass as a function of distance from the star is plotted at 10^4 (dashed), 10^5 (dotted), and 10^6 (solid) years, showing how growth proceeds as a 'wave' from inside out through the disk. The upper limit, or isolation mass, is also plotted (thick). By a million years, growth has finished, i.e., essentially all planetesimals have been agglomerated into protoplanets, well out into the region of the present-day asteroid belt.

mass of $10^{-12}\ M_{\oplus}$), the growth timescale for an isolation-mass body at 1 AU is $t_{\rm grow} \equiv M/(\mathrm{d}M/\mathrm{d}t) \approx 6 \times 10^4$ y. Thus, the sweep-up of planetesimals into protoplanets proceeds very rapidly in the terrestrial region, concluding long before the nebular gas dissipates, which takes a few million years (Haisch *et al.*, 2001). Figure 3.1 plots the solution to Equation (3.8) at different times, showing the progression of protoplanet growth through the terrestrial region during the first million years of the disk's lifetime.

3.4 The growth of planets

With the protoplanets (or 'oligarchs') now starved of planetesimals, the final stage of terrestrial planet formation requires them to agglomerate with one another. For this to happen, their orbits have to cross. This is referred to as the giant-impact (or chaotic) phase. We will begin with a description of how this unfolds in the 'standard model', and then touch on some more recent efforts to tie up the loose ends.

Traditionally, N-body simulations have been the tool of choice for characterizing the giant-impact phase. The problem is particularly well suited to this approach, because once all the planetesimals have been locked up in perhaps tens to hundreds of protoplanets, N is no longer an intractably large number. The first direct N-body calculations (Cox and Lewis, 1980; Lecar and Aarseth, 1986) revealed a number of key features of the giant-impact phase. They showed that an ensemble of lunar-to

Martian-sized bodies, allowed to agglomerate with one another in collisions, naturally formed Solar-System-like terrestrial planets. They also demonstrated that this process is highly stochastic. Because only a few bodies remain at the end, in any single simulation the notion of a characteristic planet mass as a function of distance from the star becomes meaningless, and the approach we use in Section 3.3 can no longer be applied.

Taken together, the work of Chambers and Wetherill (1998), Agnor *et al.* (1999), and Chambers (2001) can be said to delineate the 'standard model' of how the final stage of terrestrial planet formation plays out: simulations begin with a population of 10^2–10^3 lunar- to-Martian-sized protoplanets, the endproducts of the previous oligarchic growth phase. Different realizations of a particular initial condition typically vary only by the (randomly-generated) initial orbital phases of the protoplanets; the stochastic nature of these systems ensures that this is enough to make the details of the evolution completely different from one simulation to the next. Nevertheless, averaging over enough simulations, a number of common features emerge. On a timescale of $\sim$(1–3) $\times$ 10^8 years, these bodies collide with each other and accrete to form planetary systems which, on average, resemble the Solar System's terrestrial planets in mass, number, and orbital radius. Mercury and Mars, given their small sizes, seem to fit into the picture as essentially primordial oligarchs that managed to come through the whole process without undergoing any major accretional collisions. The spin of each endproduct is dominated by the last few major impacts on it, and to lowest order is randomly oriented. This contrasts with the present-day Solar System, in which all terrestrial planet spins are approximately perpendicular to their orbital plane. However, the obliquities of the terrestrial planets have evolved significantly by processes such as tidal interactions and spin–orbit coupling, thus their spins are unlikely to be primordial. A more puzzling issue is that of the final orbital eccentricities and inclinations. Their time-averaged values are almost universally larger ($e, i \gtrsim 0.1$, for i in radians) than those of Earth and Venus today ($e, i \approx 0.03$) (Chambers and Wetherill, 1998; Nagasawa *et al.*, 2006).

Because numerical models of late-stage terrestrial planet formation are otherwise so successful, the problem of reproducing the low e and i of Earth and Venus has been the subject of much effort; it really is a case of 'so close and yet so far!' Considering that we are demanding that the terrestrial region evolve directly from wildly crossing orbits and giant impacts, to a very quiescent state of nearly circular, coplanar orbits, this failure is not really surprising. In the most general terms, to solve this problem an additional, dissipative physical process must be invoked, which will damp away the planets' non-circular motions. A number of different mechanisms have been put forth and, as yet, nothing approaching a consensus has been reached among theorists. Below, we look at three proposed models.

Perhaps the most obvious candidate is the planets' own leftover building material. If a significant mass in planetesimals is left in the system, this can exert *dynamical friction* on the planets (e.g., Wetherill and Stewart (1989)): a population of gravitationally interacting bodies with different masses evolves toward a state where the energy in random velocities (relative to circular Keplerian orbits; see Section 3.3) is the same for all bodies, regardless of mass. This makes smaller bodies gain eccentricity and inclination at the expense of the larger ones, so that the latter have their e and i damped. The problem is that the timescale for sweeping up the planetesimals is, as we saw in Section 3.3, only $\sim 10^5$ years. Thus, a significant reservoir of planetesimals is unlikely to persist long enough to affect the dynamics after 10^8 years or more. Alternatively, a large number of small bodies, right down to dust size, may be generated collisionally during the planet growth process (Goldreich *et al.*, 2004).

As long as the nebular gas persists, it also acts as a source of damping for protoplanet eccentricities and inclinations. Though bodies larger than kilometre size are not strongly affected by aerodynamic gas drag, their gravitational interaction with the gas disk can have a significant effect on them. Each planet launches density waves into the interior and exterior part of the disk, exchanging energy and angular momentum with it. The net result is that planets migrate inward while having their eccentricities damped (Goldreich and Tremaine, 1980; Artymowicz, 1993; Ward, 1997; Papaloizou and Larwood, 2000). The approximate orbital and eccentricity decay timescales for a planet of mass M orbiting a star of mass M_* with initial semi-major axis a_{p} and eccentricity e_{p} are (Artymowicz, 1993; Papaloizou and Larwood, 2000)

$$t_a \equiv -\frac{a_{\mathrm{p}}}{\dot{a}_{\mathrm{p}}} \sim \left(\frac{M}{M_*}\right)^{-1} \left(\frac{2\pi a_{\mathrm{p}}^2 \Sigma_{\mathrm{gas}}}{M_*}\right)^{-1} \left(\frac{H}{a_{\mathrm{p}}}\right)^2 T_{\mathrm{orb}} \tag{3.15}$$

and

$$t_e \equiv \frac{e_{\mathrm{p}}}{\dot{e}_{\mathrm{p}}} \sim \left(\frac{H}{a_{\mathrm{p}}}\right)^2 t_a, \tag{3.16}$$

where T_{orb} is the orbital period at a distance a_{p} from the star. Note that the quantities in brackets are all unitless; the first is the ratio of planet to star mass, the second is (roughly) the ratio of disk mass interior to a_{p} to star mass, and the third is the ratio of disk scale height to radius at radius a_{p} (i.e., the disk's vertical aspect ratio). For an Earth-mass body in MMSN gas disk (with Σ_{gas} and H given by Equations (3.11) and (3.12)) around the Sun, these ratios are 3×10^{-6}, 1.2×10^{-3}, and 0.05 respectively. The orbital period at 1 AU is 1 y, so the body suffers orbital decay on a timescale of $\sim 7 \times 10^5$ y, and has its eccentricity damped on a timescale of $\sim 2 \times 10^4$ y. Because of this difference in timescales, even when the gas disk has become so tenuous that only very little planet migration is occurring, the planets can still be subject to significant eccentricity damping.

Kominami and Ida (2002, 2004) performed simulations of this scenario. They found, first of all, that the giant-impact stage tends to be delayed until the gas is strongly depleted, such that $t_e \gtrsim 10^6$ y, approximately the crossing time for adjacent oligarchs. Therefore in this scenario, the terrestrial region is essentially frozen in the end state of oligarchic growth until the gas disk is tenuous enough that dynamical evolution can resume. Kominami and Ida demonstrated that low final eccentricities of Earth and Venus analogues can be generated in this way, but fine tuning of the disk properties, specifically the gas dispersal time, is required. Even then, on average six–seven bodies are left in the end, significantly more than the present-day number of Solar System terrestrial planets. The fundamental problem is that making the oligarchs' orbits cross and damping their eccentricities are two conflicting objectives. Also, they found that including the gravitational perturbations on the terrestrial region due to Jupiter and Saturn tended to make the model work more poorly (in the sense that eccentricities and inclinations became higher), leading them to speculate that the terrestrial planets might have formed at a time when Jupiter and Saturn had not yet reached their final masses.

Nagasawa *et al.* (2005) (see also Nagasawa *et al.* (2006)) developed a model for late-stage terrestrial planet formation via a large-scale dynamical shakeup. The model likewise hinges on the dissipating gas disk; however, it also includes the effect of the gas disk on the precession of all the embedded orbits, with the result that things play out in a dramatically different way. This is due to the effect of a *secular resonance* of Jupiter, a detailed discussion of secular resonances is given in, e.g., Murray and Dermott (1999). Here, we will restrict ourselves to a simplified, qualitative description: A secular resonance between two bodies occurs (approximately) when the precession frequencies of their orbits are equal. In this case, we are looking at the precession rate of terrestrial oligarchs relative to that of Jupiter. The former precess due to the gravitational influence of the gas disk, Jupiter, Saturn and, to a lesser extent, each other. Jupiter, meanwhile, is made to precess by the influence of Saturn and the gas disk, as well as a small contribution from the protoplanets themselves. When a given oligarch's precession matches that of Jupiter and a resonance occurs, a rapid exchange of angular momentum takes place between the two bodies. Being much more massive, Jupiter's orbit undergoes little change. The orbit of the small protoplanet, however, rapidly gains eccentricity. For a given gas disk density distribution, there is one particular orbital radius interior (and exterior, though this plays no role in the model) to Jupiter at which a body will be in secular resonance. As the gas disk dissipates (through a combination of accreting onto the star, photoevaporation, and stripping by stellar winds), its gravitational potential changes and so too do the precession rates of the embedded (proto)planets and thus the location of the interior secular resonance. The resonance begins near Jupiter in a massive gas disk, finally ending up – once the gas is all gone – at its present-day location at 0.6 AU, just inside the current orbit of Venus (Lecar and Franklin, 1997; Nagasawa and Ida, 2000). Along the way it thus sweeps the asteroid belt and much of the terrestrial region and one after another the oligarchs have their eccentricity sharply raised, thus crossing their neighbours' orbits. In this way, the giant-impact phase is kick-started when there is still a significant amount of gas left in the disk. Nagasawa *et al.* (2005) performed numerical simulations of this process and found that accretion among oligarchs can finish rapidly in this way, leaving enough time for the last remnants of the gas disk to damp the eccentricities and inclinations of the end products. Another effect in evidence is a net inward migration of material during the resonance sweeping process, since bodies forced to high eccentricities in the presence of a damping force lose orbital energy as their eccentricity is damped again, thus spiralling inward (Lecar and Franklin, 1997). This offers a way to both clear the asteroid belt region (which is highly depleted relative to even the MMSN planetesimal disk) and deliver water-rich building material for the Earth (see Section 3.6).

Parent–daughter pairs of radioactive isotopes can be exploited as chronometers to trace the timelines of processes in the early Solar System, potentially helping us to decide among the different formation models. Two such pairs are Hf–W and U–Pb. For both of these, the parent element is a lithophile that is retained in silicate reservoirs during planetary accretion, whereas the daughter element is segregated into the core. Since each of the large late-stage impacts would 'reset the clock' by remixing core and mantle, measurements of relative abundances of these elements in the Earth can be used to estimate how long ago the giant-impact phase concluded. It is thought that the last impact was the Moon-forming one (see Section 3.5). Measurements of Hf–W have been used to derive a growth time of the Earth ranging from about 15 to 50 My (Yin *et al.*, 2002; Kleine *et al.*, 2002; Wood and Halliday, 2005), intriguingly similar to the short timescales obtained by Nagasawa *et al.* (2005). However, the U–Pb chronometer implies a later formation time, $\sim$65–85 My (Halliday, 2004). It may be that U–Pb traces the last stages of core segregation, while Hf–W traces the time for *most* of the core to finish forming (Sasaki and Abe, 2005). Given the difficult nature of these measurements, it should not be surprising that the different groups' results have still not been completely reconciled with each other. Future developments in this area may yet allow us to determine with some certainty whether the formation of the terrestrial planets was rapid or drawn out.

3.5 The origin of the Earth–Moon system

The leading model for the formation of the Moon is that a Mars-mass impactor struck the proto-Earth, with the Moon accreting out of the resulting debris disk (Hartmann and Davis, 1975; Cameron and Ward, 1976). This scenario naturally results in a Moon depleted in volatiles and iron-poor (since the impact would have preferentially splashed out mantle material from the proto-Earth). The giant-impact model was explored further via a succession of smoothed-particle hydrodynamics (SPH) simulations, greatly increasing over time in resolution as computing power advanced (Benz *et al.*, 1986; Cameron and Benz, 1991; Canup and Asphaug, 2001; Canup, 2004). As mentioned in Section 3.4, this impact is thought to have been the last major accretion event in the Earth's history.

The latest simulations reproduce the Earth–Moon system best with an impactor just over 0.1 $M_\oplus$ striking proto-Earth at about a 45° angle. This ejects about a lunar mass of material exterior to Earth's Roche radius (the distance inside which a strengthless, self-gravitating body would be disrupted by Earth's tidal forces) at about 3 Earth radii ($R_\oplus$). The vast majority of this material comes from the impactor itself. Initially, 10–30% of it is in the form of silicate vapour. The orbital period at the Roche radius is only about 7 hours, thus subsequent evolution is very fast. Ida *et al.* (1997) simulated the accretion of the Moon in the protolunar disk, showing that it takes between a month and a year to get to a single remaining lunar-mass body. The Moon's current orbital radius is $\approx 60\ R_\oplus$, thus its orbit has expanded significantly from the time it formed. This is effected by the Moon's tidal interaction with the Earth, which transfers angular momentum from the Earth's spin to the Moon's orbit. Early on, interaction with the last remnants of the protolunar disk likely also made a large contribution to the outward migration of the Moon (Ward and Canup, 2000).

It has been suggested that the Moon may have actually aided the development of life on Earth (Benn (2001) and references therein): the Moon stabilizes the Earth against chaotic obliquity variations, and thus against large variations in climate. Also, without the Moon-forming impact, Earth may have retained its thick primordial atmosphere and ended up more similar to Venus. Finally, a more closely-orbiting early Moon would have raised substantial tides on the Earth's surface, which may have aided the formation of life in tidal pools.

3.6 Terrestrial planets and life

The term 'habitable zone' (HZ) goes back to Huang (1959). The criteria adopted for what constitutes a potentially life-supporting planet have varied widely. The most durable has proven to be simply the requirement that liquid water can exist on the planet surface (Rasool and de Bergh, 1970; Kasting *et al.*, 1988). In fact, 'Follow the water!' has become one of the guiding principles of astrobiology. Kasting *et al.* (1993) used a climate model to calculate this zone for stars later than F0, which have lifetimes exceeding 2 Gy, thus giving life ample time to evolve. The inner edge of the HZ is set by water loss via photolysis and hydrogen escape; the outer edge is set by the formation of CO_2 clouds, which cool a planet by raising its albedo.

For G stars like the Sun, Kasting *et al.* (1993) conservatively estimated that the HZ stretches from 0.95 to 1.37 AU. F stars, being hotter, have a more distant HZ, while later type stars have a closer-in HZ (their Fig. 16). They point out that the logarithmic width of HZs is approximately constant across stellar types. This is significant because the spacing between adjacent planets ought to be set at least in part by their Hill radii; since $r_H \propto r$ (Equation (3.1)) this implies (for a given planetary mass) logarithmic spacing. Thus, the number of planets which fit into an HZ can be expected to be very roughly constant across stellar types. However, the closer-in HZ of later type stars does lead to a potential obstacle to life: for M type stars, planets in the HZ will become tidally locked into synchronous rotation with the central star over a time less than the star's age. The resultant eternally-dark planetary hemisphere would likely be frozen and uninhabitable.

Water is the basis of life on Earth, but tracing the origin of the Earth's water is also central to understanding terrestrial planet formation. Most models of the protosolar nebula have too high a temperature at 1 AU for water to condense out of the gas (Nagasawa *et al.* (2006) and references therein). Hence, models of terrestrial planet formation generally invoke a way to deliver water from reservoirs at larger heliocentric distances. This idea is supported by the deuterium to hydrogen (D/H) ratio of the Earth's oceans, which is many times higher than expected in the protosolar nebula at 1 AU. Still uncertain is what fraction of such material came from the asteroid belt (e.g., Morbidelli *et al.* (2000)) vs the trans-Jupiter region (e.g., Levison *et al.* (2001), Lunine *et al.* (2003), Raymond *et al.* (2004), Gomes *et al.* (2005)), i.e., 'wet asteroids' vs comets. Observational evidence appears to favour the asteroid belt, since the D/H ratio of carbonaceous chondrites is closer to that of the Earth's oceans than is that of comets (Balsiger *et al.*, 1995; Meier *et al.*, 1998; Bockelee-Morvan *et al.*, 1998; Dauphas *et al.*, 2000; Drake and Righter, 2002). One way to systematically deliver large amounts of icy material from beyond the snow line to the terrestrial region is via the late-stage formation model of Nagasawa *et al.* (2005) outlined in Section 3.4.

The cratering record on the Moon seems to suggest a spike in the impact rate about 700 My after the formation of the Solar System (e.g., Hartmann *et al.* (2000)). This cataclysmic event in our early history turns out to be a puzzle from the point of view of both celestial mechanics and the origin of life. Insofar as the latter is concerned, there exists some (controversial) carbon isotopic evidence for the existence of life 3.7–3.85 Gy ago (Wells *et al.* (2003) and references therein), right around the time of the Late Heavy Bombardment (LHB). This is surprising since the LHB is likely to have sterilized the entire Earth (Sleep *et al.*, 1989). One explanation put forward by Maher and Stevenson (1988) is that around this time, life might simply have arisen repeatedly and on fairly short timescales, only to be continuously frustrated by catastrophic impacts. Thus as soon as the impacts ceased, life arose one last time and took hold. Another possibility is that life found a way to weather the impacts, either deep underground or in impact ejecta (Sleep *et al.*, 1998). Such ejecta may even have seeded life elsewhere in the Solar System, such as Mars – a sort of cosmic enactment of not having all the eggs in one basket.

From a dynamics point of view, the timing of the LHB has also been the cause of much study and speculation. As discussed in Section 3.3, the sweep-up of planetesimals in the terrestrial region proceeds on a timescale well below a million years. Where, then, do all the impactors for the LHB come from? Several different scenarios have been proposed. Zappala *et al.* (1998) suggested the breakup of an asteroid; however, to provide enough material for the LHB, the asteroid's size must have been an order of magnitude greater than Ceres (which, with a diameter of 950 km, is the largest body in the asteroid belt today). Chambers and Lissauer (2002) showed that an extra planet beyond Mars, perhaps around 2 AU, could have been the culprit. Such a planet could have become unstable on a 700 My timescale, become eccentric, and crossed as-yet undepleted planetesimal reservoirs in the inner asteroid belt. These planetesimals would then have been perturbed onto orbits crossing the terrestrial region. The chief difficulty lies in preserving a significant planetesimal population between 2 AU and Jupiter for this length of time. Finally, Gomes *et al.* (2005) devised a model in which the LHB is triggered directly by Jupiter and Saturn. The giant planets almost certainly underwent a period of gradual, divergent migration as they exchanged angular momentum via the scattering of planetesimals (Fernández and Ip, 1996; Hahn and Malhotra, 1999). Gomes *et al.* (2005) posited an initially compact configuration for the giant planets, with Uranus and Neptune between 11 and 15 AU, and Jupiter and Saturn just inside their 1:2 mean-motion resonance. As the planetesimals are scattered back and forth among the giant planets they migrate apart, and at the moment when Jupiter and Saturn cross their 1:2 resonance divergently, both receive a kick in eccentricity, as previously shown by Chiang *et al.* (2002). This excitation, then, serves as the trigger for the LHB, delivering a large flux of planetesimals from the outer disk onto orbits that cross the terrestrial region. To a lesser extent, material remaining in the asteroid belt may also participate.

The trick is to make this event happen so late; Gomes *et al.* (2005) showed that the timing of the migration can be controlled by the distance between the outermost giant planet and the inner edge of the planetesimal disk. Since many planetesimals will have already been accreted or scattered during the giant planets' formation, a distance of more than 1 AU between the giant planets and the inner disk edge, as in their best-fit model, is not implausible. However, although this model constitutes a clever way to obtain an essentially arbitrarily long evolution followed by a cataclysm, the initial locations of the giant planets need to be rather finely tuned, without any obvious cosmogonical justification.

In Section 3.4, we looked at one way in which the formation process of terrestrial planets in the Solar System may have been driven by the dynamical influence of Jupiter and Saturn. We now consider other, more general, ways in which giant planets may affect terrestrial planets, with an emphasis on the issue of habitability.

While Section 3.4 describes a scenario in which the giant planets may help terrestrial planet formation, it is just as possible for giant planets to hinder the formation of terrestrial planets. In a mature planetary system, the existence of a Jovian planet in or near the HZ will make it unlikely terrestrial planets will have survived in the same region. Even a stable HZ is no guarantee of a friendly environment for terrestrial planets. During the lifetime of the gas disk in the first few million years, substantial migration of planets, especially giant planets, is likely to occur due to gravitational planet–disk interactions (Goldreich and Tremaine, 1980; Lin and Papaloizou, 1986; Ward, 1997). Numerous gas giants planets may in fact be lost during this time by migrating inside a disk gap ('Type II' migration) all the way into the central star (Trilling *et al.*, 2002). If gas giants form by core accretion (Pollack *et al.*, 1996), growing giant planet cores, $\sim 10\ M_\oplus$ in mass, likely migrate even faster than gas giants, so that even a system in which no gas giant ever penetrates the terrestrial region may suffer an early period of being repeatedly transited by Neptune-sized bodies (Thommes and Murray, 2006). Dynamical studies do suggest, though, that even under such adverse conditions, some fraction of the (proto)planets in the terrestrial region can survive (Mandell and Sigurdsson, 2003; Fogg and Nelson, 2005). Finally, it is worth mentioning that even in systems where a gas giant ends up right in the HZ, biology may have another recourse: if the giant has one or more large moons, these may provide a suitable setting for life to arise. Laughlin *et al.* (2004) argued that the formation of gas giants ought to be unlikely around low-mass M dwarfs. They assumed that the disk mass scales with the stellar mass; this would make for a longer growth timescale and a lower final mass for planets grown about such stars ($M \leq 0.4\ M_\odot$). If gas giants grow by core accretion, this makes it less likely that cores of large enough mass to accrete the requisite massive gas envelope will grow within the time that the gas disk persists, $\lesssim$10 My (GJ 876, $M_* = 0.3\ M_\odot$, orbited by two planets with $M \sin i = 0.6\ M_{\rm Jup}$ and $1.9\ M_{\rm Jup}$, is a notable counterexample). However, growing a terrestrial planet is a much less demanding proposition, and effectively the only time limit is the age of the system. Thus terrestrial planets ought to still be common around M dwarfs. In fact, with fewer gas giants to threaten their stability, it is conceivable that terrestrial planets are actually *more* common on average around lower-mass stars. Since M dwarfs are by far the most common stars, making up about 70% by number of the stars in the Galaxy, the total number of HZ-dwelling terrestrial planets is potentially very high.

Although a system with no giant planets at all may be a safer haven for terrestrial planet formation and survival, a lack of gas giants may negatively impact the habitability of the planets which do form, even those that end up in their respective HZ. In the Solar System, Jupiter intercepts a large fraction of the comet flux which would otherwise cross the terrestrial region. It is thus often argued that without such a dynamical barrier, life would never have arisen and survived on our planet, and that a gas giant exterior to the HZ is an additional necessary condition for habitability. However, the issue is perhaps not quite as clear-cut as it appears at first glance; after all, Jupiter's perturbations also play a large role in producing short-period comets from the Kuiper belt and long-period comets from the Oort cloud in the first place. In fact, outward scattering of planetesimals by Jupiter plays the main role in actually building the Oort cloud (Duncan *et al.*, 1987). Quantifying the difference in comet flux into the HZ of a mature planetary system with and without exterior gas giants is thus an important area for future work, and one which needs to be undertaken before we can properly assess the potential for low-mass stars to harbour life.

3.7 Summary

From the theorist's perspective of how terrestrial planets form, perhaps the most important aspect is that it seems to be relatively *easy*, much more so than making giant planets. Numerical simulations have no problem producing such bodies in the inner 1–2 AU of a system. Assuming a disk manages to produce planetesimals (notwithstanding the problems outlined in Section 3.2), their sweep-up proceeds rapidly in the terrestrial region, taking typically much less than 1 My. The simplest scenario, in which the resultant protoplanets then interact and collide in a gas-free environment, results in the formation of Earth-mass planets on a timescale of $\sim 10^8$ y for a 'minimum-mass' distribution of solid material, i.e., one which begins with essentially just the mass of the present-day terrestrial planets. This is because planet formation in the inner few astronomical units is very efficient; with no systematic migration, almost all of the mass one puts in is built into planets. Thus terrestrial planet formation is fundamentally a very robust process, so much so that it may actually be difficult to *prevent* it from happening in any given system.

This being said, there are still a number of questions about the details of how terrestrial planets grew in the Solar System and, by extension, how they form around other stars. The low eccentricities of Earth and Venus seem to require processes beyond the standard model. It is certainly possible that these low eccentricities will turn out to be a peculiarity of the Solar System; eccentricities < 0.1 are not in general necessary for long-term stability. Neither are they required for habitability. However, this 'small detail' may in fact be pointing to a fundamentally different course of events in the last stage of formation, such as the secular-resonance 'shakeup' model outlined in Section 3.4. Another puzzle is the nature of the LHB. What initiated it? Did life survive this event, or did it have to wait until the end of the LHB to arise? How likely are analogues to the LHB in other planetary systems? The probable ubiquity of terrestrial planet formation means that any given star ought to have a good chance of harbouring one in its HZ. This prospect may be ruined if a giant planet exists close enough to the HZ to destabilize bodies within it. At the same time, though, the presence of (exterior) giant planets may be required in order to protect the HZ against comet bombardment. The HZ is defined by the presence of liquid surface water, but the source of that water is itself still not fully understood.

The discovery of the first extrasolar terrestrial planet will certainly be a momentous event. Until then, no amount of modelling can change the fact that our sample size of terrestrial planet systems is one. As with the gas giants, surprises may await which overturn our currently-held picture of how such planets form. And, of course, the search for exo-terrestrial planets is made especially exciting by the possibility that another life-bearing world is waiting to be discovered.

Acknowledgements

It is a pleasure to thank Doug Lin, Makiko Nagasawa, Scott Kenyon, John Chambers and Stein Jacobsen for stimulating and informative interactions, which were a valuable source of inspiration for this chapter.

References

Agnor, C. B., Canup, R. M., and Levison, H. F. (1999). On the character and consequences of large impacts in the late stage of terrestrial planet formation. *Icarus*, **142**, 219–237.

Artymowicz, P. (1993). Disk-satellite interaction via density waves and the eccentricity evolution of bodies embedded in disks. *Astrophys. J.*, **419**, 166.

Balsiger, H., Altwegg, K., and Geiss, J. (1995). D/H and O-18/O-16 ratio in the hydronium ion and in neutral water from in situ ion measurements in comet Halley. *J. Geophys. Res.*, **100**, 5827–5834.

Barranco, J. A., and Marcus, P. S. (2005). Three-dimensional vortices in stratified protoplanetary disks. *Astrophys. J.*, **623**, 1157–1170.

Benn, C. R. (2001). The moon and the origin of life. *Earth Moon Planets*, **85**, 61–66.

Benz, W., Slattery, W. L., and Cameron, A. G. W. (1986). The origin of the moon and the single-impact hypothesis. I. *Icarus*, **66**, 515–535.

Bockelee-Morvan, D., Gautier, D., Lis, D. C., *et al.* (1998). Deuterated water in comet C/1996 B2 (Hyakutake) and its implications for the origin of comets. *Icarus*, **133**, 147–162.

Cameron, A. G. W., and Benz, W. (1991). The origin of the Moon and the single impact hypothesis. IV. *Icarus*, **92**, 204–216.

Cameron, A. G. W., and Ward, W. R. (1976). The origin of the Moon. *Lunar and Planetary Institute Conf. Abstr.*, **7**, 120.

Canup, R. M. (2004). Simulations of a late lunar-forming impact. *Icarus*, **168**, 433–456.

Canup, R. M., and Asphaug, E. (2001). Origin of the Moon in a giant impact near the end of the Earth's formation. *Nature*, **412**, 708–712.

Chambers, J. E. (2001). Making more terrestrial planets. *Icarus*, **152**, 205–224.

Chambers, J. E., and Lissauer, J. J. (2002). A new dynamical model for the lunar Late Heavy Bombardment. *33rd Annual Lunar and Planetary Sci. Conf.*, abstract no. 1093.

Chambers, J. E., and Wetherill, G. W. (1998). Making the terrestrial planets: N-body integrations of planetary embryos in three dimensions. *Icarus*, **136**, 304–327.

Chiang, E. I., Fischer, D., and Thommes, E. (2002). Excitation of orbital eccentricities of extrasolar planets by repeated resonance crossings. *Astrophys. J.*, **564**, L105–L109.

Cox, L. P., and Lewis, J. S. (1980). Numerical simulation of the final stages of terrestrial planet formation. *Icarus*, **44**, 706–721.

Dauphas, N., Robert, F., and Marty, B. (2000). The late asteroidal and cometary bombardment of earth as recorded in water deuterium to protium ratio. *Icarus*, **148**, 508–512.

Drake, M. J., and Righter, K. (2002). Determining the composition of the Earth. *Nature*, **416**, 39–44.

Duncan, M., Quinn, T., and Tremaine, S. (1987). The formation and extent of the solar system comet cloud. *Astron. J.*, **94**, 1330.

Fernández, J. A., and Ip, W.-H. (1996). Orbital expansion and resonant trapping during the late accretion stages of the outer planets. *Planet. Space Sci.*, **44**, 431–439.

Fogg, M. J., and Nelson, R. P. (2005). Oligarchic and giant impact growth of terrestrial planets in the presence of gas giant planet migration. *Astron. Astrophys.*, **441**, 791–806.

Goldreich, P., and Tremaine, S. (1980). Disk-satellite interactions. *Astrophys. J.*, **241**, 425–441.

Goldreich, P., and Ward, W. R. (1973). The formation of planetesimals. *Astrophys. J.*, **183**, 1051–1062.

Goldreich, P., Lithwick, Y., and Sari, R. (2004). Planet formation by coagulation: a focus on Uranus and Neptune. *Annu. Rev. Astron. Astr.*, **42**, 549–601.

Gomes, R., Levison, H. F., Tsiganis, K., and Morbidelli, A. (2005). Origin of the cataclysmic Late Heavy Bombardment period of the terrestrial planets. *Nature*, **435**, 466–469.

Hahn, J. M., and Malhotra, R. (1999). Orbital evolution of planets embedded in a planetesimal disk. *Astron. J.*, **117**, 3041–3053.

Haisch, K. E., Lada, E. A., and Lada, C. J. (2001). Disk frequencies and lifetimes in young clusters. *Astrophys. J.*, **553**, L153–L156.

Halliday, A. (2004). Geochemistry: The clock's second hand. *Nature*, **431**, 253–254.

Hartmann, W. K., and Davis, D. R. (1975). Satellite-sized planetesimals and lunar origin. *Icarus*, **24**, 504–514.

Hartmann, W. K., Ryder, G., Dones, L., and Grinspoon, D. (2000). The time-dependent intense bombardment of the primordial Earth/Moon system, in *Origin of the Earth and Moon*, eds. R. M. Canup and K. Righter. Tucson: University of Arizona Press, pp. 493–512.

Hayashi, C. (1981). Structure of the solar nebula, growth and decay of magnetic fields and effects of magnetic and turbulent viscosities on the nebula. *Prog. Theoret. Phys.*, **70**, 35–53.

Huang, S.-S. (1959). The problem of life in the Universe and the mode of star formation. *PASP*, **71**, 421.

Ida, S., Canup, R. M., and Stewart, G. R. (1997). Lunar accretion from an impact-generated disk. *Nature*, **389**, 353–357.

Ida, S., and Makino, J. (1993). Scattering of planetesimals by a protoplanet – slowing down of runaway growth. *Icarus*, **106**, 210.

Kant, I. (1755). *Allgemeine Naturgeschichte und Theorie des Himmels*. Königsberg, Prussia.

Kasting, J. F., Toon, O. B., and Pollack, J. B. (1988). How climate evolved on the terrestrial planets. *Sci. Am.*, **258**, 90–97.

Kasting, J. F., Whitmire, D. P., and Reynolds, R. T. (1993). Habitable zones around main sequence stars. *Icarus*, **101**, 108–128.

Kleine, T., Münker, C., Mezger, K., and Palme, H. (2002). Rapid accretion and early core formation on asteroids and the terrestrial planets from Hf–W chronometry. *Nature*, **418**, 952–955.

Kokubo, E., and Ida, S. (1998). Oligarchic growth of protoplanets. *Icarus*, **131**, 171–178.

Kominami, J. and Ida, S. (2002). The effect of tidal interaction with a gas disk on formation of terrestrial planets. *Icarus*, **157**, 43–56.

Kominami, J. and Ida, S. (2004). Formation of terrestrial planets in a dissipating gas disk with Jupiter and Saturn. *Icarus*, **167**, 231–243.

Laplace, P. S. (1796). *Exposition du Systè'me du Monde*. Paris.

Laughlin, G., Bodenheimer, P., and Adams, F. C. (2004). The core accretion model predicts few Jovian-mass planets orbiting red dwarfs. *Astrophys. J.*, **612**, L73–L76.

Lecar, M., and Aarseth, S. J. (1986). A numerical simulation of the formation of the terrestrial planets. *Astrophys. J.*, **305**, 564–579.

Lecar, M., and Franklin, F. (1997). The solar nebula, secular resonances, gas drag, and the asteroid belt. *Icarus*, **129**, 134–146.

Levison, H. F., Dones, L., Chapman, C. R. *et al.* (2001). Could the lunar 'Late Heavy Bombardment' have been triggered by the formation of Uranus and Neptune? *Icarus*, **151**, 286–306.

Lin, D. N. C., and Papaloizou, J. (1986). On the tidal interaction between protoplanets and the protoplanetary disk. III Orbital migration of protoplanets. *Astrophys. J.*, **309**, 846–857.

Lissauer, J. J. (1993). Planet formation. *Annu. Rev. Astron. Astr.*, **31**, 129–174.

Lunine, J. I., Chambers, J., Morbidelli, A., and Leshin, L. A. (2003). The origin of water on Mars. *Icarus*, **165**, 1–8.

Maher, K. A., and Stevenson, D. J. (1988). Impact frustration of the origin of life. *Nature*, **331**, 612–614.

Mandell, A. M., and Sigurdsson, S. (2003). Survival of terrestrial planets in the presence of giant planet migration. *Astrophys. J.*, **599**, L1114.

Meier, R., Owen, T. C., Jewitt, D. C., *et al.* (1998). Deuterium in comet C/1995 O1 (Hale-Bopp): detection of DCN. *Science*, **279**, 1707.

Morbidelli, A., Chambers, J., Lunine, J. I., *et al.* (2000). Source regions and time scales for the delivery of water to Earth. *Meteorit. Planet. Sci.*, **35**, 1309–1320.

Murray, C. D., and Dermott, S. F. (1999). *Solar System Dynamics*. Cambridge: Cambridge University Press.

Nagasawa, M., and Ida, S. (2000). Sweeping secular resonances in the Kuiper Belt caused by depletion of the solar nebula. *Astron. J.*, **120**, 3311–3322.

Nagasawa, M., Lin, D. N. C., and Thommes, E. (2005). Dynamical shake-up of planetary systems. I. Embryo trapping and induced collisions by the sweeping secular resonance and embryo-disk tidal interaction. *Astrophys. J.*, **635**, 578–598.

Nagasawa, M., Thommes, E., Kenyon, S., Bromley, B., and Lin, D. N. C. (2006). The diverse origins of terrestrial-planet systems, in *Protostars and Planets V*, eds. B. Reipurth, D. Jewitt, and K. Keil. Tuscon, AZ: University of Arizona Press, in press.

Papaloizou, J. C. B., and Larwood, J. D. (2000). On the orbital evolution and growth of protoplanets embedded in a gaseous disc. *MNRAS*, **315**, 823–833.

Paraskov, G. B., Wurm, G., and Krauss, O. (2005). Planetesimal growth in high velocity impacts, in *Protostars and Planets V*, proceedings of the conference held October 24–28, (2005) in Hilton Waikoloa Village, Hawai'i. LPI Contribution No. 1286., Tucson: University of Arizona Press, p. 8318.

Pollack, J. B., Hubickyj, O., Bodenheimer, P., *et al.* (1996). Formation of the giant planets by concurrent accretion of solids and gas. *Icarus*, **124**, 62–85.

Rasool, S. I., and de Bergh, C. (1970). The runaway greenhouse and accumulation of CO_2 in the Venus atmosphere. *Nature*, **226**, 1037–1039.

Raymond, S. N., Quinn, T., and Lunine, J. I. (2004). Making other earths: dynamical simulations of terrestrial planet formation and water delivery. *Icarus*, **168**, 1–17.

Sasaki, T., and Abe, Y. (2005). Imperfect equilibration of Hf–W system by grant impacts, in *Protostars and Planets V*, proceedings of the conference held October 24–28, (2005) in Hilton Waikoloa Village, Hawai'i. LPI Contribution No. 1286. Tucson: University of Arizona Press, p. 8221.

Sleep, N. H., Zahnle, K. J., Kasting, J. F., and Morowitz, H. J. (1989). Annihilation of ecosystems by lare asteroid impacts on the early Earth. *Nature*, **342**, 139–142.

Thommes, E. W., and Murray, N. (2006). Giant planet accretion and migration: surviving the type I regime. *Astrophys. J.*, **644**, 1214–1222.

Thommes, E. W., Duncan, M. J., and Levison, H. F. (2003). Oligarchic growth of giant planets. *Icarus*, **161**, 431–455.

Trilling, D. E., Lunine, J. I., and Benz, W. (2002). Orbital migration and the frequency of giant planet formation. *Astron. Astrophys.*, **394**, 241–251.

Ward, W. R. (1997). Protoplanet migration by nebula tides. *Icarus*, **126**, 261–281.

Ward, W. R., and Canup, R. M. (2000). Satellite recoil from a circumplanetary disk. *Lunar and Planetary Conf. Abstr.*, 2050.

Weidenschilling, S. J. (1977). Aerodynamics of solid bodies in the solar nebula. *MNRAS*, **180**, 57–70.

Weidenschilling, S. J. (1995). Can gravitation instability form planetesimals? *Icarus*, **116**, 433–435.

Wells, L. E., Armstrong, J. C., and Gonzalez, G. (2003). Reseeding of Early Earth by impacts of returning ejecta during the late heavy bombardment. *Icarus*, **162**, 38–46.

Wetherill, G. W., and Stewart, G. R. (1989). Accumulation of a swarm of small planetesimals. *Icarus*, **77**, 330–357.

Wolszczan, A., and Frail, D. A. (1992). A planetary system around the millisecond pulsar PSR1257 + 12. *Nature*, **355**, 145–147.

Wood, B. J., and Halliday, A. N. (2005). Cooling of the Earth and core formation after the giant impact. *Nature*, **437**, 1345–1348.

Yin, Q., Jacobsen, S. B., Yamashita, K., *et al.* (2002). A short timescale for terrestrial planet formation from Hf–W chronometry of meteorites. *Nature*, **418**, 949–952.

Zappala, V., Cellino, A., Gladman, B. J., Manley, S., and Migliorini, F. (1998). NOTE: asteroid showers on Earth after family breakup events. *Icarus*, **134**, 176–179.

4

From protoplanetary disks to prebiotic amino acids and the origin of the genetic code

Paul G. Higgs and Ralph E. Pudritz
McMaster University

4.1 Introduction

The robust formation of planets as well as abundant sources of water and organic molecules are likely to be important prerequisites for the wide-spread appearance of life in the cosmos. The nebular hypothesis of Kant and Laplace was the first to propose that the formation of planets occurs in gaseous disks around stars. The construction of new infrared and submillimetre observatories over the last decade and a half has resulted in the discovery of protoplanetary discs around most, if not all, forming stars regardless of their mass (e.g., reviews by Meyer *et al.* (2006), Dutrey *et al.* (2006)). The recent discoveries of extrasolar planets in over a hundred planetary systems provides good evidence that Jovian planets, at least, may be relatively abundant around solar-like stars (see Chapter 1). These results beg the question of whether protoplanetary disks are also natural settings for the manufacture of the molecular prerequisites for life. Life requires water and organic molecules such as amino acids, sugars, nucleobases, and lipids as building blocks out of which biological macromolecules and cellular structures are made, and many of these can be manufactured in protoplanetary disks.

In the first part of this chapter we review the properties of protoplanetary disks and how planets are believed to form within them. We then consider the evidence that these disks may be a major source of the water and biomolecules available for the earliest life, as on the Earth. We focus on amino acids because they are the key components of proteins and more is known about prebiotic synthesis of amino acids than most other biomolecules. In the second part of this review, we compare the different environments that have been proposed for amino acid synthesis – including the atmosphere, hydrothermal vents in the deep ocean, and protoplanetary disks – and show that there is considerable consensus on which amino acids can be formed prebiotically, even if there is still disagreement on the location.

Planetary Systems and the Origins of Life, eds. Ralph E. Pudritz, Paul G. Higgs, and Jonathon R. Stone.
Published by Cambridge University Press.

We will argue that the amino acids that were least thermodynamically costly to form were the ones that were most abundant before life arose. Early organisms could make use of existing amino acids in the first proteins. Later organisms evolved biochemical pathways to synthesize additional amino acids that were not common prebiotically, thus increasing the diversity and functional specificity of the proteins they could make.

The synthesis of specific proteins is only possible after the origin of the genetic code (i.e., the mapping between codons in RNA and amino acids in proteins). We will discuss the origin of the genetic code in the context of our understanding of prebiotic amino acid frequencies. We will also consider the evidence that the genetic code is optimized to reduce translational error and discuss how it came to be this way.

It is likely that early proteins were composed of a smaller set of amino acids than the 20 used currently. We will discuss several experimental studies of proteins composed of deliberately reduced amino acid sets that indicate that properly folded and functional structures could have formed in such proteins. Bioinformatics techniques can also be used to make predictions about early proteins. For example, using molecular phylogenetics it is possible to estimate the most likely ancestral sequence for a given set of related proteins. Several studies indicate that the abundances of amino acids in the ancestral proteins differed from average abundances in modern organisms. It is possible that this is a relic of a time when only a subset of amino acids was used.

4.2 Protoplanetary disks and the formation of planet systems

Stars and planets form within dense, cold clouds of gas and dust, known as molecular clouds. Surveys of many molecular clouds show that most stars form as members of entire star clusters. These stellar nurseries arise from dense regions within molecular clouds known as clumps which extend over about 1 pc in physical scale (or about 3 light years, or 205 000 astronomical units – AU – where 1 AU = 1.5×10^{13} cm is the distance between the Earth and the Sun) and typical initial temperatures around 20 K. The gas in clouds and clumps is stirred vigorously by supersonic turbulence which also generates its filamentary structure. A single star, or a binary stellar system, forms within a small, dense subregion of a clump (known as a core) that, for stars like the Sun, extends over a scale of 0.04 pc and has a (particle) number density ranging from 10^4 to 10^8 cm^{-3}.

Numerical simulations of supersonic turbulent gas under the observed conditions show how stars and their protoplanetary disks form. Supersonic turbulence first sweeps up the gas into systems of shocked sheets and filaments. Dense core-like regions are produced as such flows develop. Also, the shock waves that produce the cores are oblique and hence impart spin to the cores (see MacLow and Klessen (2004) for a review). The eventual gravitational collapse of such slowly spinning cores under their own weight preserves most of their angular momentum, resulting in the formation of protoplanetary disks (e.g., Tilley and Pudritz (2004)).

Surveys of star-forming regions have established that disks are universal around solar mass stars. Stars range over more than three decades in mass, and for all of them, one has evidence for disks. The least massive of these is the disk around an object that has only 15 M_J (Jovian masses)! (Note, the mass of Jupiter is one thousandth the mass of the sun; $M_J = 0.001\ M_{\odot}$.) Disks are commonly observed around solar-like stars as they form. At the most massive end, rotating disks have been found around B stars that are up to 10 $M_{\odot}$ (e.g., Schreyer *et al.* (2005)).

Most studies generally do not resolve a disk around a star but infer its presence from the excess of infrared emission that is seen in the spectrum of the star. Disks around forming solar-type stars typically extend out to more than 100 AU, with the temperature varying as a function of the radius r. In our Solar System, the orbital radius of the Earth around the Sun is 1 AU, while that of the most massive outer planet, Neptune, is 30 AU. The total emission from this collection of rings of gas and dust, each of which emits radiation like a blackbody, adds up to the observed excess infrared emission (known as the spectral energy distribution or SED).

For solar-type stars, the disk temperature that can be inferred from such observations scales with disk radius as $T \propto r^{-0.6}$, with a temperature at 100 AU of about 30 K. Similarly, one can deduce that the column density of the disk (which is the volume density integrated over the disk thickness at a given radius) varies as $\Sigma \propto r^{-1.5}$ and has a value at 100 AU of 0.8 g cm^{-2}. Disks seen at these later stages of their evolution are far less massive than are their central stars, having typically $10^{-2} M_{\odot}$ of gas and dust (e.g., Dutrey *et al.* (2006)). Hence, their dynamics are governed by the gravitational field of the central star, and the material is expected to be in nearly Keplerian orbit about the star (i.e., $v_{\mathrm{Kep}}(r) = \sqrt{(}GM_*/r))$.

Occasionally, one can also spatially resolve the disk around a young star. The rotation curves of the disks can be measured under such conditions, as has been done in molecular observations (Simon *et al.*, 2000), and these have been found to be nearly Keplerian. At a slightly later stage (a few hundred thousand years), after the surrounding cloud is blown away by the young massive star, stars and their disks still in formation can be seen in optical images. This is seen in the Hubble Space Telescope (HST) image in Figure 4.1, where we see several images of young stars that are forming in the Orion Nebula Cluster. A massive star has created a strongly illuminated background of glowing nebular gas. The HST image shows a disk around each young star which is seen as a silhouette against this bright background (McCaughrean and Stauffer, 1994). Disks are heated primarily by the young star that they surround as well as by a massive nearby star in the case of

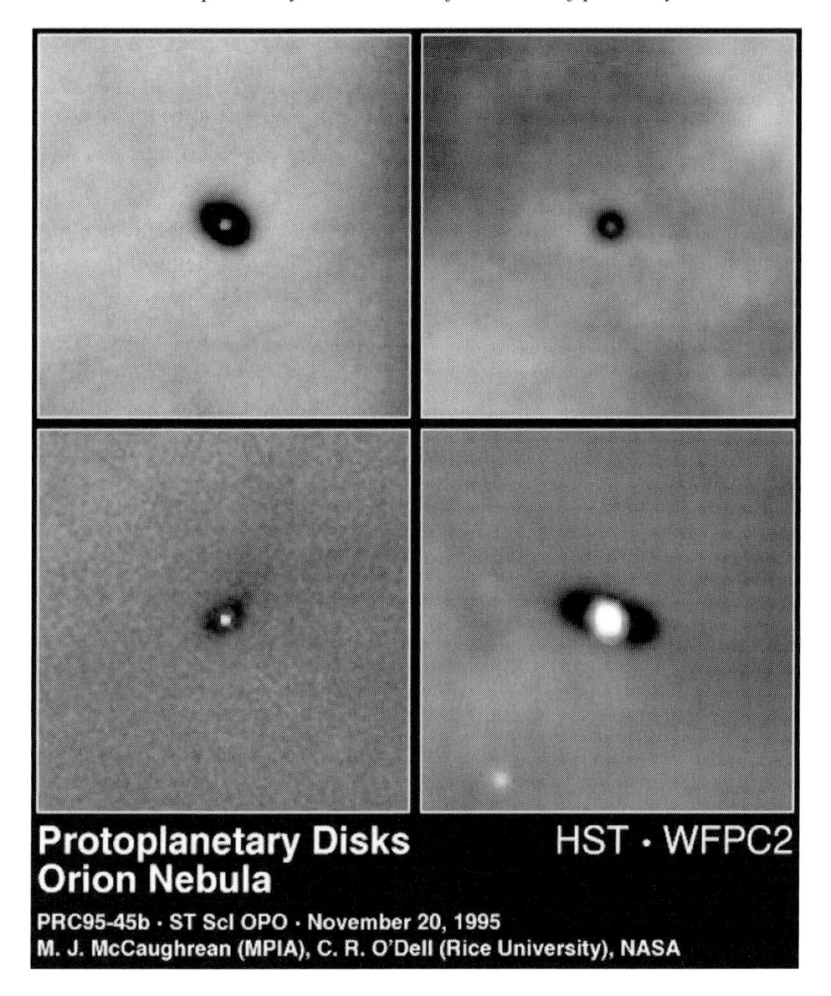

Fig. 4.1. Hubble images of four protoplanetary disks in the Orion Nebula. These are seen as silhouettes against the bright background of the nebula.

clustered star formation. They are also illuminated by ultraviolet radiation as well as X-rays from the central star, and this drives the production of organic molecules (as we discuss in the next section).

Studies of the frequency of disks in Orion also allow us to determine their lifetimes. One can calibrate the lifetimes of protoplanetary disks because one can measure the ages of the young stars that they encompass. The result is that 50% of disks are gone by 3 million years after formation, and 90% are gone by 5 million years. This lifetime for disks translates into a hard upper limit for the time for the formation of massive Jovian planets. While Jupiter has an 8–10 Earth mass core, its bulk consists of gas that was accreted from the protoplanetary disk in which it formed.

The building of planets and molecules depends, in part, on the balance between carbon and oxygen in the molecular gas within protoplanetary disks (e.g., Gaidos and Selsis (2006)). These two elements are the most abundant in the interstellar medium after hydrogen and helium and are predominantly bound in the form of CO in molecular clouds. During gravitational collapse, however, the presence of high pressure can shift this balance in favour of the formation of water and methane (CH_4) from CO and molecular hydrogen. Thus, if there is an excess of O over C, then the extra O is taken up in the formation of water (through combination with abundant molecular hydrogen). In the converse case, an excess of C over O is used up in the formation of graphite and organic molecules. The balance between C and O depends on the condensation and sedimentation of grains.

While the bulk of the mass of disks is dominated by gas, dust grains play several important roles. Dust grains in more diffuse gas consist of a mixture of silicates and carbon, with sizes ranging typically from 100 Å to 0.2–0.3 μm. Once dust starts to collect in denser environments such as protostellar disks, it grows by agglomeration via collisions. This growth continues up to the scale of kilometre-sized objects (called planetesimals) and, eventually, planets. Dust grains, while still small, are the dominant form of opacity in protostellar disks and effectively absorb the ultraviolet radiation that falls on disks from their central and surrounding stars (e.g., Natta *et al.* (2006)).

Models of terrestrial planet formation agree that the process starts as dust settles onto the midplane of protoplanetary disks (see Chapter 3). In the earliest phases, dust grains condense and grow by collisions with other dust grains, gradually settling to the disk midplane as they become more massive. After the formation of planetesimals, an 'oligarchic' growth phase occurs wherein gravitational interactions lead to focusing of orbits and collisions of planetesimals resulting in large numbers of Moon- to Mars-sized objects. At a radial distance of 1 AU from the central star, this process occurs within a million years (e.g., Kokubo and Ida (2002)). The gaseous disk is still present at this phase. The late collisions of these oligarchs over 100 My produces a sparse system of massive planets.

If the Earth formed at around 1 AU in such a disk, it must have formed in a very dry zone without much water. This is because the snow-line for water – that radius in the protostellar disk which separates a hot inner region of the disk, where water cannot condense, from an outer region, where the temperature is low enough for water to freeze out and condense in grains and planetesimals – occurs at 2.5 AU. The Earth's ocean probably arose as a consequence of the delivery of water by bodies that formed beyond 2.5 AU and that later collided with the Earth. It is now thought that water ice in comets is unlikely to be the source, as the deuterium to hydrogen ratio (D/H) of comets is twice that found in Earth's water. Perhaps only 10% of the Earth's ocean came from a cometary source. It has been suggested (Morbidelli *et al.*, 2000) that water arrived from the asteroid belt through the impact of water-laden planetary embryos, such as those discussed above.

Massive planets, such as Jupiter, play a major role in this process. Numerical simulations (Raymond *et al.*, 2004) show that the presence of a Jovian mass planet stirs up planetesimals and perturbs them into ever more eccentric orbits. Such orbits carry the planetary embryos into the hot, dry inner regions of the disk where they can undergo collisions with the Earth. This is a stochastic process, and the simulations show that the majority of Earth-like planets that formed in the 'habitable zone' (0.8–1.5 AU) have water contents that are within an order of magnitude of that of the Earth. The water content has a broad range, however, extending from some dry worlds to a few watery ones that have more than 100 times more water than there is on Earth.

In fact, massive gas-giant (Jovian) planets play a dominant role in the formation and evolution of entire planetary systems. How they form is still is unclear, but there are two major contemporary models that are under intense investigation. The more popular of these is the core accretion picture wherein a rocky core that grows to about 5–10 Earth masses can then rapidly accrete gas from the disk. This may take a few million years, however – uncomfortably close to the observational limit for disk lifetimes (Pollack *et al.*, 1996). The alternative picture is that Jovian planets form as a consequence of rapid gravitational instability in a fairly massive disk (e.g., Mayer *et al.* (2002)). Whatever the mechanism of its formation, the final mass of a giant planet is dependent strongly on the structure of the disk (its pressure scale height) as well as the magnitude of its turbulent viscosity. Jovian planets cease to grow when their mass (known as the gap-opening mass) is large enough to exert a torque on their surrounding disks in the face of turbulent viscosity, thereby opening a significant gap in it. The disappearance of the corresponding ring of emission from the gas at this radius translates into a notch-like feature in the SED, which can be searched for in the spectra of forming stars. The advent of the highly sensitive Atacama Large Millimeter Array (ALMA) in the latter part of this decade will allow such gaps to be imaged directly, as will the emission from the forming Jupiter (e.g., Wolf and D'Angelo (2005)).

Planets exert significant tidal forces on their natal disks and this results in the inward migration of the planet (Lin and Papaloizou, 1993). This process is quite general and explains why the newly discovered extrasolar planets observed around solar-type stars have orbits well within 5 AU and sometimes at the equivalent radius of the orbit of Mercury (see Chapter 1). Jupiter mass planets are thought to have formed at much larger disk radii, where gas is more abundant, and migrated to their present positions. Models show that planets can migrate very quickly – within a million years or so – and are liable to be swallowed by the star. This presents a problem as to why there are any planetary systems in the first place. Also, the migration of massive planets will strongly perturb the smaller terrestrial planets and this is of profound significance for the origin of life in such systems.

Several mechanisms have been proposed for saving planetary systems from such rapid destruction by their central stars. One of possible universal applicability is the role of so-called dead zones that occur in disks (Matsumura *et al.*, 2007). Dead zones are regions that are too poorly ionized to support the kind of hydromagnetic turbulence in disks that makes them 'viscous'. Such zones may extend up to 15 AU or so and are marked by low viscosity. A massive planet that migrates into such a region will readily open a large gap in the disk. Its radial inward migration is then locked to the slow rate at which gas in the dead zone drifts inwards.

4.3 Protoplanetary disks and the formation of biomolecules

Well over 100 molecules, many of them organic, have been identified in interstellar gas. The clumps and cores in molecular clouds out of which stars form have densities ranging from 10^4 cm^{-3} to 10^8 cm^{-3} and low temperatures. In this state, cold gas chemistry can account for the formation of simple molecules (CO, N_2, O_2, C_2H_2, C_2H_4, and HCN). The surfaces of dust grains play host to the formation of more complex organic molecules, which include nitriles, aldehydes, alcohols, acids, ethers, ketones, amines, amides, and long-chain hydrocarbons.

For decades, our knowledge of the chemistry in the solar nebula was based on studies of planetary atmospheres, meteorites, and comets. Current observations allow us to study directly the chemistry of disks around other stars. This work reveals that active chemistry occurs near the surface regions of protoplanetary disks. This chemistry is in disequilibrium and is similar observationally to that found in dense regions directly exposed to ultraviolet and X-ray irradiation (Bergin *et al.*, 2006).

Chemistry and reaction rates in protoplanetary disks depend on the local gas density, temperature, and radiation field. Disks are not flat but are instead observed to be flared. Their surfaces, therefore, are exposed to radiation from the central star. The vertical structure of disks arises from the fact that the pressure gradient of the gas at any radius and scale height in the disk supports the weight of the material above it. The force that must be balanced is the vertical component of the gravitational field of the central star (e.g., Dubruelle *et al.* (2006)).

Self-consistent calculations show that the heating of such flaring disks is due primarily to the ultraviolet radiation from the central star that is absorbed by dust grains in such surface layers (Chiang and Goldreich, 1997; D'Alessio *et al.*, 1998). The dust in the disk envelope then reradiates this energy in the form of infrared and submillimetre photons – about half of which escape vertically away from the disk,

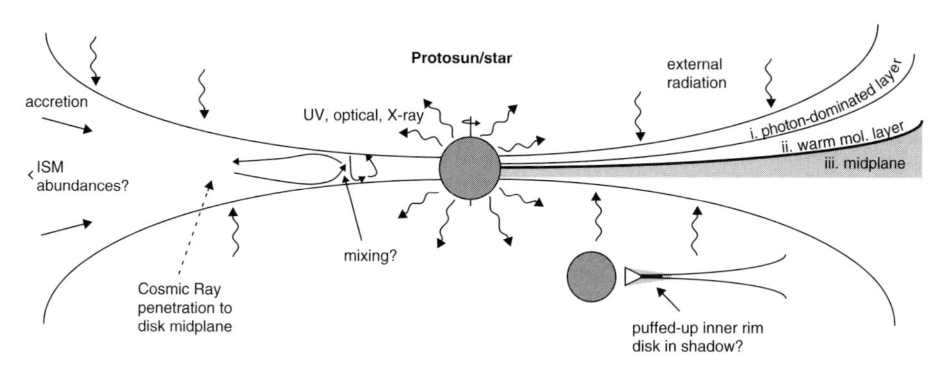

Fig. 4.2. Illustration of the structure of protostellar disks in response to heating and ionization from their central stars and cosmic rays. Note the general appearance of a molecular layer. Adapted from Bergin *et al.* (2006), with permission.

while the remainder radiate downwards towards the disk midplane. These infrared photons are then absorbed by the deeper lying dust and ultimately radiate out of the disk. The gas is heated by collisions with the warm dust and cools by radiating this energy in the form of molecular vibration–rotation emission lines that can be observed. This overall process produces a disk that is hottest near the surface layers and cooler towards the midplane. By balancing all of these heating and cooling rates, one can determine the local disk temperature distribution and hence the disk's SED. The prediction that the temperature should scale as $T \propto r^{-0.5}$ is close to that observed.

The vertical structure of disks at disk radii beyond 100 AU consists of three layers; the radiation-dominated surface layer or 'photon-dominated region' (PDR), which consists mainly of atomic and ionized species, a molecular layer at greater depth beyond which most ultraviolet radiation has been absorbed by the grains, and the cold midplane layer noted above. This midplane region turns out to be so cold (around 20 K) that most heavy species, such as CO, freeze out on the dust grains. Figure 4.2 shows a schematic of the disk. It is the molecular layer that is of primary interest for the synthesis of complex organic molecules.

Observations of molecules so far are still limited by the sensitivity of telescopes as the emission is so weak. By far, the most abundant molecule is molecular hydrogen, H_2, but it is generally hard to detect. The most abundant species that have been observed to date in disks include HCO^+, CN, CS, HCN, H_2CO, and DCO^+ (e.g., Dutrey *et al.* (1997), van Zadelhoff *et al.* (2001)). Silicates such as Mg_2SiO_4 (forsterite) have been detected in the hot surface layers of the inner regions of the disk ($r \leq$ 1–10 AU). This is important because the shape and strength of silicate features in disk spectra allow one to track the evolution and growth of dust grains in disks with time. Ices of H_2O, CO_2, and CO are observed in the outer regions of the disks, where temperatures drop to less than 100 K. The most abundant organic molecules observed are the polycyclic aromatic hydrocarbons, or PAHs. Their abundance per H atom is of the order of 10^{-7}. Up to 50% of carbon may be locked up in such carbonaceous solids. They are important for disk chemistry because they absorb ultraviolet radiation and act as potential sites for H_2 formation.

Dust grains that are coated with simple icy mantles warm up as they are mixed and transported to dense, active protostellar regions. This can occur in regions such as the so-called hot cores that are hosts to massive star formation or the innermost regions of protoplanetary disks. These regions are particularly rich in organic molecules. Ultraviolet irradiation, or perhaps X-ray bombardment, then breaks bonds, and allows reactions that can produce organic molecules. However, evaporation of these simple dust grain surfaces can drive gas phase production of organics.

It is now well established that organic molecules of extra terrestrial origin are found in carbonaceous chondrite meteorites. The best-studied of these is the Murchison meteorite, in which 8 of the 20 biological amino acids have been found (Engel and Nagy, 1982). Meteorites also contain many other organic molecules that are relevant in astrobiology (Sephton, 2002), including lipids that have been shown to be capable of membrane formation (see Chapter 5). The organic molecules in meteorites may have originally arisen on dust grains, which were then incorporated into comets and meteorites. The possibility that dust grains with icy mantles are a site for amino acid synthesis has been investigated experimentally. Bernstein *et al.* (2002) and Munoz-Caro *et al.* (2002) have shown that amino acids such as glycine, alanine, and serine may result from the exposure of icy mantles consisting of HCN, ammonia (NH_3), and formaldehyde (H_2CO) to ultraviolet radiation.

Alternatively, it has been proposed that amino acids such as glycine and alanine, and the base adenine can form during the gravitational collapse of a core. Chakrabarti and Chakrabarti (2000) added reactions that make amino acids (e.g., by the Strecker process – wherein an aldehyde starting material combines with HCN in ammonia to form the corresponding aminonitrile, which upon hydrolysis forms the amino acid) to the existing UMIST database of chemical reactions in molecular clouds. This created a reaction network involving 421 molecular species. As an example, they found that the glycine that was formed had a mass fraction of order 10^{-12}. It peaks at a radius of approximately 100 AU and a time of a million years after the initiation of the collapse. The astronomical detection of interstellar amino acids, such as glycine, would strengthen the link between the chemistry of disks and observed amino acids in meteorites. A spectroscopic survey for amino acids in clouds and disks will be possible with ALMA.

Organic molecules contained in comets and meteorites are delivered to Earth when impact events occur, and it is known that impacts were frequent in the early history of the Solar System. Impacts release huge amounts of energy which might cause thermal degradation of macromolecules. However, Pierazzo and Chyba (1999) predicted that molecules such as amino acids can survive such impacts and that the rate of delivery of biomolecules from space could exceed the rate of synthesis on Earth.

4.4 Measurements and experiments on amino acid synthesis

The previous section presents the case for synthesis of biomolecules in protoplanetary disks, but several mechanisms of prebiotic synthesis on Earth have also been debated for many years. In this section, we show that these results combine to give a coherent picture of which amino acids were most abundant on the early Earth. We focus on amino acids because of their fundamental importance in biology and because they have been detected in a wide range of non-biological contexts relevant to prebiotic synthesis.

In Table 4.1, each column contains observed amino acid concentrations, normalized relative to glycine unless otherwise stated. In the following discussion, column numbers refer to this table. Column M1 is the H_2O extract from the Murchison meteorite (Engel and Nagy, 1982). Column M2 is the Murray meteorite (Cronin and Moore, 1971), and column M3 is the interior, hydrolysed sample of the Yamato meteorite (Shimoyama *et al.*, 1979). Several amino acids that are not used biologically are found to be as abundant in meteorites as the biological ones (Kvenvolden *et al.*, 1971). We will not consider non-biological amino acids in this chapter. However, the question of why certain amino acids came to be used in proteins and others did not is an important one, and in many cases reasons have been proposed as to why the non-biological ones were avoided (Weber and Miller, 1981). Column I1 shows measurements from an experiment simulating the chemistry on icy grains (Munoz-Caro *et al.*, 2002).

Miller and Urey showed that amino acids could be synthesized by exposure of a mixture of reducing gases to ultraviolet radiation or electrical discharge. These experiments were intended to show that synthesis was possible in the atmosphere of the early Earth, which was presumed to be reducing. Column A1 shows the results with an atmosphere of CH_4, NH_3, H_2O, and H_2 (Miller and Orgel, 1974), and column A2 shows the results with an atmosphere of CH_4, N_2, H_2O, and traces of NH_3 (Miller and Orgel, 1974). Although yields are lower in non-reducing atmospheres, column A3 shows an experiment using proton irradiation of an atmosphere of CO, N_2, and H_2O (Miyakawa *et al.*, 2002), which is not a strongly reducing mixture.

Another competing theory is that life originated in high-temperature, high-pressure environments in the deep sea (Amend and Shock, 1998). Chemical

Table 4.1. *Frequencies of amino acids observed in non-biological contexts. In the column headings, M denotes meteorites, I denotes icy grains, A denotes atmospheric synthesis, H denotes hydrothermal synthesis, and S denotes other chemical syntheses (details in text).* R_{obs} *is the mean rank derived from these observations. The final six amino acids are not observed: N asparagine, Q glutamine, C cysteine, Y tyrosine, M methionine, W tryptophan*

	M1	M2	M3	I1	A1	A2	A3	H1	H2	S1	S2	S3	R_{obs}
G glycine	1.00	1.0	1.00	1.000	1.000	1.000	1.000	18	12	1.000	1.000	40	1.1
A alanaine	0.34	0.4	0.38	0.293	0.540	1.795	0.155	15	8	0.473	0.097	20	2.8
D aspartic acid	0.19	0.5	0.035	0.022	0.006	0.077	0.059	10	10		0.581	30	4.3
E glutamic acid	0.40	0.5	0.11		0.010	0.018		6	11			20	6.8
V valine	0.19	0.3	0.10	0.012		0.044		1		0.006		2	8.5
S serine			0.003	0.072		0.011	0.018	8	11		0.154		8.6
I isoleucine	0.13		0.06			0.011		8	9		0.002	4	9.1
L leucine	0.04		0.035			0.026		3		0.001	0.002	7	9.4
P proline	0.29	0.1		0.001		0.003		9				2	10.0
T threonine			0.003			0.002		2			0.002	1	11.7
K lysine								7				14	12.6
F phenylalanine								4				1	13.2
R arginine												15	13.3
H histidine												15	13.3
NQCYMW													14.2

syntheses of amino acids in hydrothermal conditions have been reported. Column H1 is from Marshall (1994), and H2 is from Hennet *et al.* (1992). Each of these columns gives the number of times that the amino acid was observed in greater than trace amounts in a large number of experiments. The final three columns are miscellaneous chemical synthesis experiments: S1 is shock synthesis from a gaseous mixture (Bar-Nun *et al.*, 1970); S2 is synthesis from ammonium cyanide (Exp. 2 of Lowe *et al.*, 1963); S3 is synthesis from CO, H_2, and NH_3 at high temperature in the presence of catalysts (Yoshino *et al.*, 1971). This column shows the number of positive identifications of the amino acid from many experiments.

It is possible that the organic molecules that were present when life began were synthesized by several different mechanisms. A central issue is to estimate what fraction originated on Earth and what fraction was brought to Earth after synthesis in protoplanetary disks (Whittet, 1997). The yields on Earth are dependent on the mixture of gases in the atmosphere. If the atmosphere is non-reducing, yields on Earth are low, and delivery of molecules from space is more relevant (Chyba and Sagan, 1992; Kasting, 1993; Pierazzo and Chyba, 1999). Although there is now a reasonable consensus that the atmosphere was not strongly reducing, the importance of synthesis in the atmosphere is maintained by some (Lazcano and Miller, 1996; Miyakawa *et al.*, 2002), and this type of synthesis still works to some degree in intermediate atmospheres. Despite this uncertainty, the results in Table 4.1 are surprisingly consistent as regards *which* amino acids can be synthesized, even if they are in disagreement about *where and how* they were formed.

In his studies of the origin of the genetic code, Trifonov (2000, 2004) collected 60 criteria by which the amino acids can be ranked. A consensus order was obtained by averaging the rankings. This procedure incorporates diverse criteria that are not all consistent with one another, and some of these must be misleading. Nevertheless, the consensus order that emerges appears to be a useful one. We shall call the average ranks derived from genetic code criteria R_{code}. In this section we derive our own ranking, R_{obs}, based only on observed frequencies of amino acids in non-biological syntheses, i.e., we use non-speculative criteria.

We ranked the amino acids according to each criterion in Table 4.1 separately. For example, for criterion M2, G is most frequent and is given rank 1. The next two, D and E, are equally frequent, and are both given rank 2.5. The next three are A, V, and P, with ranks 4, 5, and 6. All the remaining amino acids are not observed, and are given an equal bottom rank 13.5 (the average of the numbers between 7 and 20). R_{obs} is the mean of the ranks obtained from the 12 columns.

Several points are worth noting from Table 4.1. G is the most abundant amino acid in all but column A2, where it is second. A, D, and E are also abundant in most experiments. The next six (V, S, I, L, P, T) are all found in the Miller–Urey experiment (A2), at least one of the meteorites, and several of the other cases. Therefore, there is good evidence for prebiotic synthesis of all the amino acids down as far as T on the list. The following four (K, F, R, H) are not found in the Miller experiments or in meteorites and occur only in one or two of the other chemical synthesis experiments. The remaining amino acids (N, Q, C, Y, M, W) are not formed in any of the experiments. In our view, the data are not sufficient to make a definite distinction between K, F, R, H and N, Q, C, Y, M, W.

4.5 A role for thermodynamics

Thermodynamic arguments can help us to understand what the mixtures of amino acids available to early organisms may have been like, and why there is considerable consensus among the experiments considered in the previous section. Table 4.2 lists the amino acids and their properties according to the ranking R_{obs} derived above. We will refer to the top ten amino acids as 'early' and the bottom ten as 'late'. Early amino acids were easily synthesized by non-biological means and were, therefore, available for use by early organisms. Late amino acids were not available in appreciable quantities until organisms evolved a means to synthesize

Table 4.2. *Thermodynamic and evolutionary properties of amino acids*

		R_{obs}	R_{code}	ΔG_{surf}	ΔG_{hydro}	ATP	MW	dp/dt	Δp
G	Gly	1.1	3.5	80.49	14.89	11.7	75	−0.0063	−0.5
A	Ala	2.8	4.0	113.66	−12.12	11.7	89	−0.0239	−3.3
D	Asp	4.3	6.0	146.74	32.78	12.7	133	−0.0039	−0.4
E	Glu	6.8	8.1	172.13	−1.43	15.3	147	−0.0137	0.7
V	Val	8.5	6.3	178.00	−70.12	23.3	117	0.0098	−1.6
S	Ser	8.6	7.6	173.73	69.47	11.7	105	0.0167	−0.6
I	Ile	9.1	11.4	213.93	−96.40	32.3	131	0.0089	−0.9
L	Leu	9.4	9.9	205.03	−105.53	27.3	131	−0.0017	3.2
P	Pro	10.0	7.3	192.83	−38.75	20.3	115	−0.0139	0.0
T	Thr	11.7	9.4	216.50	53.51	18.7	119	0.0091	−1.4
K	Lys	12.6	13.3	258.56	−28.33	30.3	146	−0.0065	1.6
F	Phe	13.2	14.4	303.64	−114.54	52.0	165	0.0042	1.5
R	Arg	13.3	11.0	409.46	197.52	27.3	174	0.0038	−0.2
H	His	13.3	13.0	350.52	154.48	38.3	155	0.0073	−1.3
N	Asn	14.2	11.3	201.56	83.53	14.7	132	0.0073	−0.6
Q	Gln	14.2	11.4	223.36	44.03	16.3	146	0.0020	1.3
C	Cys	14.2	13.8	224.67	60.24	24.7	121	0.0067	0.5
Y	Tyr	14.2	15.2	334.20	−59.53	50.0	181	−0.0005	1.6
M	Met	14.2	15.4	113.22	−174.71	34.3	149	0.0088	−0.3
W	Trp	14.2	16.5	431.17	−38.99	74.3	204	0.0002	0.6

them. The early–late distinction is relevant in the discussion of the origin of the genetic code in the following section. Our ranking, R_{obs}, is close to the ranking, R_{code}, taken from Table V of Trifonov (2004). The top three are the same, and nine amino acids are in the top ten of both orders. We prefer R_{obs} on the grounds that it is derived directly from experimental observables; however, the conclusions drawn from both rankings are very similar.

The ranking can be interpreted on thermodynamic grounds. Amend and Shock (1998) have calculated the free energy of formation of the amino acids from CO_2, NH_4^+, and H_2 in two sets of conditions. ΔG_{surf} in Table 4.2 corresponds to surface seawater conditions (18 °C, 1 atm), and ΔG_{hydro} corresponds to deep-sea hydrothermal conditions (100 °C, 250 atm). Figure 4.3 shows that R_{obs} is closely related to ΔG_{surf}. For the ten early amino acids, there is a strong correlation between the two ($r = 0.96$).

The late amino acids have significantly higher ΔG_{surf} than the early ones. The means and standard deviations of the groups are 169.3 ± 42.0 and 285.0 ± 94.1 kJ/mol. Table 4.2 also lists the molecular weight (MW) of each amino acid and the ATP cost, which is the number of ATP molecules that must be expended to synthesize the amino acids using the biochemical pathways in *E. coli* bacteria

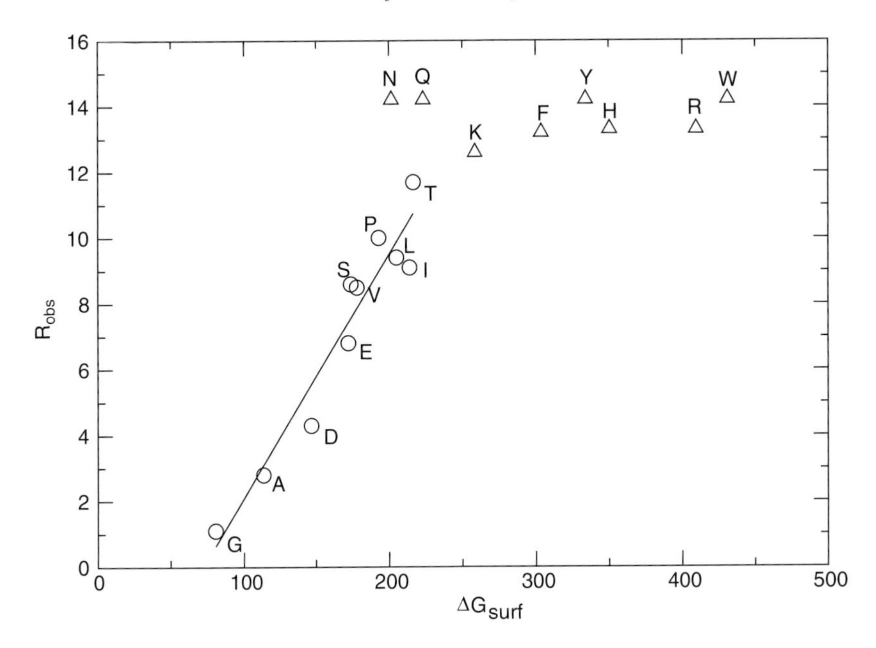

Fig. 4.3. Relationship between the rank of the amino acid and the free energy of formation: circles – early amino acids; triangles – late amino acids. The line is the linear regression for the early amino acids.

(Akashi and Gojobori, 2002). The means and standard deviations of MW for early and late groups are 116.2 ± 20.6 and 157.3 ± 23.2 Da, and the figures for ATP cost are 18.5 ± 6.9 and 36.2 ± 17.3. Thus it is clear that the late group are larger and more thermodynamically costly.

We have omitted the two sulphur-containing amino acids, Met and Cys, from Figure 4.3. Several of the experiments in Table 4.1 do not include sulphur in the reaction mixture, so these amino acids are bound to be absent from the products. Thus, the value of R_{obs} for Met and Cys is unclear. Nevertheless, Met and Cys are absent from the meteorites, and are late according to R_{code}, so it seems reasonable to classify them in the late group. Additionally, there are some uncertainties associated with Met and Cys in the data used in the calculation of ΔG_{surf} by Amend and Shock (1998): the concentration of Cys in seawater was not reported; Met and Val were not distinguishable, which makes the concentration of Met uncertain; and the concentration of H_2S was below detection (this affects ΔG_{surf} for Met and Cys but not the other amino acids). The value of $\Delta G_{surf} = 113.22$ kJ mol^{-1} for Met is puzzling. This is lower than for any of the other late amino acids, and is particularly surprising as it is lower than the figure for Cys (Met is a larger molecule than Cys). For all these reasons it seemed preferable to omit Met and Cys from this figure.

All the formation reactions are endergonic ($\Delta G_{surf} > 0$). The amino acids with the smallest ΔG_{surf} should be formed most easily, as they require the least free energy input. Figure 4.3 demonstrates this. The values of ΔG_{surf} are predictive of what we see in a wide range of meteorite and prebiotic synthesis experiments. If the mixture of compounds were in equilibrium, then we would expect the concentrations to depend exponentially on ΔG via the Boltzmann factor $\exp(-\Delta G/kT)$. The ranking procedure linearizes this relationship and makes the correlation easier to see. The exponential dependence explains why the amino acids with high ΔG are not seen in experiment: their concentration would be too low to detect. Observed concentrations also depend on the rates of formation and not just on equilibrium thermodynamics. The middle-ranking amino acids show considerable fluctuation between the columns of Table 4.1, being present in some cases and not others. This may reflect differences in rates of synthesis between different experimental conditions. The ranking procedure averages out these fluctuations.

The picture becomes less clear when we consider the free energy of formation under hydrothermal conditions. The central message of Amend and Shock (1998) is that many of the formation reactions are exergonic ($\Delta G_{hydro} < 0$) under hydrothermal conditions, and that even the endergonic ones are less positive than they are at the surface. They use this to explain why hydrothermal vents might be a good place for current life, and to support the idea that the first organisms might have been deep-sea chemosynthesizers. However, there is no correlation between rank and ΔG_{hydro}, and no significant difference in ΔG_{hydro} between early and late amino acids. The two experiments designed to simulate hydrothermal systems (H1 and H2) give results that agree fairly well with the combined ranking from the other data. The ΔG_{hydro} values do not seem to agree with the H1 and H2 experiments any better than they agree with the overall ranking. Although this does not rule out the possibility of a hydrothermal origin of life, what does appear clear is that the ten early amino acids identified by the ranking procedure can be predicted on the basis of ΔG_{surf} and not ΔG_{hydro}.

4.6 The RNA world and the origin of the genetic code

One of our main motivations for discussing prebiotic amino acid formation above is that we are interested in the origin of the genetic code and protein synthesis. The previous section gives us a good idea which were the early amino acids available at the time the genetic code evolved. However, many other types of biomolecules would have been around as well as amino acids, and life must already have reached a rather advanced stage before the genetic code and protein synthesis could have evolved. In this chapter, we will not attempt to discuss the origin of life itself, but we will present our reasons for believing that an RNA world stage existed in the early history of life, that the genetic code arose in the context of an RNA world, and that the mixture of prebiotic amino acids available influenced the structure of this code.

Current life is based on DNA and proteins: proteins carry out the majority of catalytic roles in the cell, and DNA stores the genetic information. Proteins are made using the information specified in the DNA sequence (genome) of the cell. Proteins do not have their own hereditary mechanism. Double-stranded DNA does have a hereditary mechanism because each strand can act as a template for synthesis of the complementary strand. In current organisms, DNA synthesis is catalysed by protein enzymes. Thus, DNA and proteins are mutually dependent, and it is difficult to see how one could have existed without the other. The RNA world hypothesis was proposed as a way out of this chicken-and-egg dilemma. The RNA world is a period thought to have existed in the early stages of life in which RNA molecules carried out both catalytic and genetic roles. This is supported by the fact that there are many viruses in which the genome is RNA (although no longer any cellular organisms), and by the increasing repertoire of catalytic RNAs that are being synthesized by in vitro selection techniques (Joyce, 2002).

It is thought that when organisms evolved the ability to synthesize specific proteins, these proteins took over most of the catalytic roles formerly performed by RNA. Also DNA took over the role of information storage from RNA at some stage. Nevertheless, RNA was retained as an intermediate between DNA and proteins: information in the DNA gene sequence is first transcribed into an RNA sequence and then translated into a protein sequence. There are also a certain number of types of RNAs with key biochemical roles that occur widely in all domains of life and that are thought to be relics of the RNA world (Jeffares *et al.*, 1998). Joyce (2002) likened this situation to a primitive civilization that existed before recorded history but which left its mark on the modern civilization that followed.

There are still important gaps in our understanding of the RNA world, and it is not known if complex self-replicating RNAs could have originated *de novo*. It may be that there was no single self-replicating RNA, and that life began with a network of reactions of many molecular components (Shapiro, 2006). These components could have been RNAs or other types of biomolecules. Nevertheless, we are persuaded that life on Earth passed through an RNA world stage, even if something preceded this. Translation is the process by which the sequence information in an RNA molecule is used to make a specific protein. Translation depends on RNA in several ways. Firstly, the messenger RNA contains the sequence information; secondly, ribosomal RNA is the active component of the ribosome – the complex of macromolecules that catalyses formation of proteins (Moore and Steitz, 2002); and thirdly, the transfer RNAs are the 'adaptors' that make the link between nucleic acids

Table 4.3. *The canonical genetic code. Each of the 64 codons is assigned to one of the 20 amino acids (or to a Stop signal). The first-position base is shown on the left, the second-position base at the top, and the third-position base at the right*

	U	C	A	G	
U	UUU Phe (F)	UCU Ser (S)	UAU Tyr (Y)	UGU Cys (C)	U
	UUC Phe (F)	UCC Ser (S)	UAC Tyr (Y)	UGC Cys (C)	C
	UUA Leu (L)	UCA Ser (S)	UAA Stop	UGA Stop	A
	UUG Leu (L)	UCG Ser (S)	UAG Stop	UGG Trp (W)	G
C	CUU Leu (L)	CCU Pro (P)	CAU His (H)	CGU Arg (R)	U
	CUC Leu (L)	CCC Pro (P)	CAC His (H)	CGC Arg (R)	C
	CUA Leu (L)	CCA Pro (P)	CAA Gln (Q)	CGA Arg (R)	A
	CUG Leu (L)	CCG Pro (P)	CAG Gln (Q)	CGG Arg (R)	G
A	AUU Ile (I)	ACU Thr (T)	AAU Asn (N)	AGU Ser (S)	U
	AUC Ile (I)	ACC Thr (T)	AAC Asn (N)	AGC Ser (S)	C
	AUA Ile (I)	ACA Thr (T)	AAA Lys (K)	AGA Arg (R)	A
	AUG Met (M)	ACG Thr (T)	AAG Lys (K)	AGG Arg (R)	G
G	GUU Val (V)	GCU Ala (A)	GAU Asp (D)	GGU Gly (G)	U
	GUC Val (V)	GCC Ala (A)	GAC Asp (D)	GGC Gly (G)	C
	GUA Val (V)	GCA Ala (A)	GAA Glu (E)	GGA Gly (G)	A
	GUG Val (V)	GCG Ala (A)	GAG Glu (E)	GGG Gly (G)	G

and amino acids. Therefore, it is likely that translation originated in organisms in which RNAs played a central role and in which a mechanism of accurate replication of relatively long RNA sequences had already evolved.

The key to translation is the genetic code. This is the set of assignments between the 64 possible codons (three-letter sequences composed of U, C, A, and G nucleotides) in RNA and the 20 possible amino acids in proteins. The genetic code is decoded by complementary base pairing between the codon sequence in the messenger RNA and the anticodon sequence on the transfer RNAs (readers requiring more information should consult any biology textbook). The canonical genetic code shown in Table 4.3 is shared by the three domains of life (archaea, bacteria, and eukaryotes). Therefore, it evolved prior to the last universal common ancestor (LUCA). Understanding how it arose is a fundamental question in evolutionary biology.

As prebiotic synthesis of the early group of amino acids seems to be possible in a number of environments, it is likely that these early amino acids were available to RNA-based organisms. Prior to the evolution of the genetic code, amino acids could have played a role in metabolism, and it is possible that polypeptides might also have been synthesized if peptide bond formation was catalysed by a ribozyme.

For more details of possible types of chemistry that could have existed in this period, see Pascal *et al.* (2005). However, before the genetic code evolved, any peptides would have had stochastic amino acid compositions. When the genetic code arose, it became possible to synthesize specific proteins where the amino acid sequence was predetermined by the RNA sequence. Specific proteins can have specific structures and hence specific functions; therefore, they are much more useful to an organism than are stochastic peptides. The invention of the genetic code is a stroke of evolutionary brilliance which couples the hereditary mechanism inherent in the template-directed replication of nucleic acids to the catalytic possibilities of proteins. Prior to the genetic code, proteins could not undergo evolution.

Early versions of the code probably used a smaller repertoire of amino acids, each with a larger number of codons. The codon table was divided up into progressively smaller blocks as successive amino acids were added. This idea goes as far back as Crick (1968). Each addition would have opened up a whole new world of protein possibilities. Thus, there is a selective drive for adding new amino acids in the early stages of code development. This brings us back to question of the order of addition of amino acids to the genetic code. We have shown above that the early group of amino acids can be identified based on their appearance in meteorites, the Miller–Urey experiment, and a variety of other chemical syntheses designed to simulate prebiotic conditions. The strong correlation between the ranking, R_{obs}, that we derive and the free energy of formation, ΔG_{surf}, suggests that this ranking is a meaningful prediction of the relative frequencies of abiotically synthesized amino acids that would have been available to the first organisms. Essentially the same amino acids are placed early in R_{code} (Trifonov, 2004) as well.

We therefore envisage an organism in the RNA world that learned to use the available amino acids to synthesize proteins in a prescribed way and found this to be extremely useful. Such an organism would have begun to evolve metabolic pathways to synthesize these amino acids itself from other chemicals so that it was no longer reliant on the prebiotic supply. As this process proceeded, pathways developed for synthesis of the more complex, late group of amino acids that were never present in appreciable quantities prebiotically. At some point along the line, proteins became so useful and so essential that the RNA-based organisms were replaced by modern DNA/protein-based organisms. This scenario has long been advocated by Wong (1975) as part of what is known as the coevolution theory for the origin of the code. In an updated version of this theory, Wong (2005) proposed the same set of early and late amino acids as us, based on synthesis from atmospheric gases. We have shown above that this set of early amino acids seems likely, even if the place of origin is different.

4.7 How was the genetic code optimized?

There is considerable evidence that the canonical code has evolved to minimize the effect of errors (Freeland *et al.*, 2003). Mutations create errors in the gene sequence that are passed to the protein sequences translated from this gene. Translational errors due to codon–anticodon mispairing or mischarging of a tRNA will introduce occasional non-heritable errors into proteins. These errors are likely to involve only one of the three bases in the codon most of the time. Errors can be minimized by arranging for codons that differ by only one base to code for the same amino acid or amino acids with similar physical properties. Thus, when an error leads to an amino acid replacement, it is usually replaced by a similar amino acid, and the deleterious effect on the protein is minimized. The canonical code has been compared with large numbers of random codes created by reshuffling the amino acid positions in the table of codons. The fraction of random codes for which the average effect of an error is less than in the canonical code is as small as 10^{-6} (Freeland and Hurst, 1998), and can be even smaller than this if the effect of unequal amino acid frequencies is also accounted for (Gilis *et al.*, 2001).

It was originally thought that the canonical code was shared by all organisms and that it was frozen and unable to change. However, although the majority of organisms use the canonical code, there are now a large number of cases known where small changes to the code have occurred in specific lineages (Knight *et al.*, 2001). There are several different mechanisms by which codon reassignment can occur despite the negative selective effects that are present during the changeover period (Sengupta and Higgs, 2005, Sengupta *et al.*, 2007). The variant codes have all arisen since the establishment of the canonical code. In this chapter, we are more concerned with the origin of the canonical code itself.

One remarkable pattern in the code is that the five highest ranking amino acids, Gly, Ala, Asp, Glu, and Val, all have codons of the form GNN (codons with first position G are on the bottom row of Table 4.3). It is possible that only these GNN codons were assigned in the first code. Trifonov (1987) argued that the first position G is important for maintaining the correct reading frame during translation. Eigen and Winkler-Oswatitsch (1981) proposed an initial code with only the four GNC codons assigned (one codon each for Gly, Ala, Asp, and Val). However, such a system has the disadvantage that a large fraction of codons are unassigned, and that most mutations in a gene sequence would create untranslatable codons. It would be advantageous to rapidly progress to a stage where all codons were assigned, as this reduces the mutational load. A possibility that we consider likely is that there was a stage in which the code was divided simply into four columns: all NUN codons were Val, all NCN codons were Ala, all NAN codons were Asp (or a mixture of Asp and Glu), and all NGN codons were Gly. In this pattern, there is still a shift of three bases between successive codons, but only the second position base carries information. This means that the code can become more specific later by assigning significance to first and third positions without destroying the information contained at the second position (Crick, 1968). Later, prefix and suffix codons may have arisen (Wu *et al.*, 2005), where either the first and second or the second and third bases were significant. This would have broken up the four columns into smaller divisions.

This proposal is consistent with the observation that the code is error-minimizing. Selection would have acted to ensure each new amino acid was added in a favourable position. This would mean that amino acids would be added into positions that used to be occupied by an earlier amino acid with similar properties. Work by H. Goodarzi (personal communication) also makes this point. In the first column Phe, Leu, Ile, and Met are similar to Val, and in the second column Ser, Pro, and Thr are similar to Ala. A general property of the code is that amino acids in the same column are much more similar than those in the same row (Urbina *et al.*, 2006). This may be a relic of an early four-column stage of the code where all codons in the same column coded for the same amino acid.

Wong (1975, 2005) also supposes that large blocks of codons were divided up into smaller ones as new amino acids were added to the code; however, there are important differences to our argument above. From the pathways of biosynthesis of amino acids in modern organisms, it is possible to define precursor–product pairs. Wong argues that later amino acids were assigned to codons that were formerly occupied by their biochemical precursors. The earliest code would have contained only the amino acids that are at the start of the biochemical pathways. Di Giulio and Medugno (1999) showed the complex pattern of assignments that would have existed in the early code if every current amino acid were replaced by its earliest precursor. This pattern seems unlikely to us because it would require a very complex molecular recognition process for tRNA charging. The initial four-column pattern that we propose would require straightforward molecular recognition of the second base in the anticodon. The precursor–product argument predicts that precursor–product pairs in the canonical code should occupy neighbouring codons. The statistical significance of this depends on the details of how precursor–product pairs are counted (Ronneberg *et al.*, 2000) and is not as high as was previously claimed. Our argument is that later amino acids were added to positions that were formerly occupied by amino acids with similar properties, irrespective of whether the earlier amino acid was a precursor of the later. This predicts that amino acids on neighbouring codons should have similar properties, i.e., the code should be optimized in the sense of Freeland and Hurst (1998). The statistical evidence for this is much more robust, as discussed above.

In contrast to both these theories, it has also been proposed that there were specific interactions between amino acids and their codons or anticodons. Evidence for this comes from selection experiments on random RNA pools (Yarus *et al.*, 2005). Although this is intriguing, to us it is the error-minimizing properties of the code that most demand an explanation. An error-minimizing code can be reached by a pathway in which natural selection acts each time a new amino acid is added. Precursor–product relationships and RNA–amino acid interactions might influence which codons were most likely to be 'tried out' for a new amino acid, but selection would eliminate those trial codes in which an amino acid was added in 'the wrong place', and would favour the addition of a new amino acid in a place that was consistent with its physical properties. Thus our principle for amino acid addition explains how the code became optimized.

4.8 Protein evolution

The previous section envisages early proteins composed of a small set of amino acids. Several studies have considered whether such proteins could be functional. Using only three amino acids (Gln, Leu, and Arg), proteins with strong helical structure were found (Davidson *et al.*, 1995). Doi *et al.* (2005) found that random sequences of the five most abundant early amino acids according to our ranking, Gly, Ala, Asp, Glu, and Val, were more soluble than random sequences of 20 amino acids or Gln–Leu–Arg proteins. Riddle *et al.* (1997) took a 57-residue structural domain of a naturally occurring protein and found that almost all residues could be replaced by an amino acid from a simple alphabet of Ile–Lys–Glu–Ala–Gly. The smallest amino acids, Gly and Ala, were essential parts of this alphabet. The large aromatic amino acids, Trp, Phe, and Tyr, which are late according to our theory, were not able to be replaced by amino acids from the simple alphabet. Babajide *et al.* (1997) used inverse-folding programmes to locate sequences that are likely to fold to a specified natural protein structure. Suitable sequences could be found for some very small alphabets (including Ala–Asp–Leu–Gly, which is very similar to our proposed original set) but not others (including Gln–Leu–Arg). Taken together, these studies suggest that proteins composed of the early group of amino acids could well have formed similar structures to modern day proteins and could have performed useful functions in early organisms.

Another unusual proposal that is worth mentioning is the proteins-first hypothesis of Ikehara (2005). The same amino acids are again first in this theory, but proteins composed of Gly–Ala–Asp–Val evolve replication and metabolism prior to the origin of RNA. It is not clear to us how genetic information could be passed on by proteins alone, however.

We considered above the role that thermodynamics has in prebiotic synthesis of amino acids. It also seems likely that thermodynamics effects the evolution of modern proteins. Akashi and Gojobori (2002) showed that organisms appear to be sensitive to the ATP cost of amino acid synthesis (listed in Table 4.2). Specifically, the most highly expressed proteins had the highest frequency of the low-cost amino acids. Seligmann (2003) also found considerable evidence that amino acid usage in proteins is affected by cost minimization. Gutierrez *et al.* (1996) observed an increase in frequency of the base G in the first codon position in highly expressed genes, which can be interpreted as selection for low-cost amino acids Gly–Ala–Asp–Glu–Val.

It is possible that modern day proteins may still contain a signal of which amino acids were abundant in the early stages of evolution. Phylogenetic methods can be used to reconstruct estimates of ancestral sequences and hence to determine the frequencies of amino acids in those sequences. Brooks *et al.* (2002) estimated frequencies of amino acids in the LUCA, and found them to be noticeably different from current proteins. In Table 4.2, $\Delta p = p_{\mathrm{LUCA}} - p_{\mathrm{current}}$. Jordan *et al.* (2005) used a method that involved comparing sequences from triplets of related species. The earliest diverging species in each triplet can be used to give a direction to the amino acid substitutions occurring on the branches leading to the other two species. It is found that forward and backward substitution rates between pairs of amino acids are not equal, and that amino acids frequencies show an increasing or decreasing trend. $\mathrm{d}p/\mathrm{d}t$ is the estimate of the rate of change of the amino acid frequencies.

There are some limited points of agreement between $\mathrm{d}p/\mathrm{d}t$ and Δp. The most rapidly decreasing amino acid is Ala in both cases, and the three most frequent early amino acids, Gly, Ala, and Asp, are all decreasing. Beyond that, there is little agreement: only seven amino acids have the same sign in both studies. We would expect early amino acids to be decreasing and late amino acids to be increasing. Thirteen amino acids have the right sign according to Δp, and 14 have the right sign according to $\mathrm{d}p/\mathrm{d}t$. This is suggestive but not fully convincing. McDonald (2006) has cast doubt on the Jordan *et al.* (2005) study on methodological grounds, and a second study of Brooks *et al.* (2004) produced results that differ from their previous study (Brooks *et al.*, 2002) as to which amino acids are increasing and which decreasing. Our position is that trends in amino acid frequencies like this would have occurred over time and it makes sense to look for them. However, it is a long time since the LUCA and it may not be possible to see convincing trends above the noise.

One observation of Brooks *et al.* (2004) is relevant to the extremophile theme of this book. They found that the estimated frequencies of amino acids in the LUCA were closer to those of hyperthermophiles than mesophiles, which might suggest that LUCA was a hyperthermophile. It has also been argued that the genetic code was established at high temperature (Di Giulio, 2000) and high pressure (Di Giulio, 2005) based on comparison of amino acid frequencies in extremophiles and mesophiles. The hyperthermophile nature of LUCA is still controversial, and some cold water is poured on this idea in Chapter 8.

4.9 Summary

In this chapter, we have done our best to combine information from astrophysics, organic chemistry, molecular evolution, and bioinformatics. It became apparent to us that the strands of knowledge that link these diverse disciplines are becoming ever stronger and, indeed, are required to understand how life may have arisen in our Solar System and perhaps in others. The generality of planet formation and the universality of the organic chemistry (that is just starting to be explored) in protoplanetary disks is very encouraging. It suggests that the astrophysics of protoplanetary disks could play a major role in understanding how one goes from dust and gas to watery worlds equipped with the biomolecules that may have served as the first building blocks of life. Beyond this, we have also shown that the abundance and distribution of early amino acids may have played an important role in the origin of the earliest genetic code and how it may have subsequently evolved.

Acknowledgements

We thank the many excellent speakers at the Origins Institute's 2005 conference on 'Astrobiology and the Origins of Life' for the inspiration for this chapter and book.

References

Akashi, H., and Gojobori, T. (2002) Metabolic efficiency and amino acid composition in the proteomes of *Escherichia coli* and *Bacillis subtilis*. *Proc. Nat. Acad. Sci. USA*, **99**, 3695–3700.

Amend, J. P., and Shock, E. L. (1998) Energetics of amino acid synthesis in hydrothermal ecosystems. *Science*, **281**, 1659–1662.

Babjide, A., Hofacker, I. L., Sippl, M. J., and Stadler, P. F. (1997) Neutral networks in protein space: a computational study based on knowledge-based potentials of mean force. *Fold Des.*, **2**, 261–269.

Bar-Nun, A., Nar-Nun, N., Bauer, S. H., and Sagan, C. (1970) Shock synthesis of amino acids in simulated primitive environments. *Science*, **168**, 470–473.

Bergin, E. A., Aikawa, Y., Blake, G. A., and van Dishoeck, E. F. (2006). The chemical evolution of protoplanetary disks, in *Protostars and Planets V*, eds. B. Reipurth, D. Jewitt, and K. Keil. Tucson: University of Arizona Press, in press.

Bernstein, M. P., Dworkin, J. P., Sandford, S. A., Cooper, G. W., and Allamandola, L. J. (2002). Racemic amino acids from the ultraviolet photolysis of interstellar ice analogues. *Nature*, **416**, 401–403.

Brooks, D. J., Fresco, J. R., Lesk, A. M., and Singh, M. (2002). Evolution of amino acid frequencies in proteins over deep time: Inferred order of introduction of amino acids into the genetic code. *Mol. Biol. Evol.*, **19**, 1645–1655.

Brooks, D. J., Fresco, J. R., and Singh, M. (2004). A novel method for estimating ancestral amino acid composition and its application to proteins of the last universal ancestor. *Bioinformatics*, **20**, 2251–2257.

Chiang, E. I., and Goldreich, P. (1997). Spectral energy distributions of T Tauri stars with passive circumstellar disks. *Astrophys. J.*, **490**, 368–376.

Chakrabarti, S., and Chakrabarti, S. K. (2000). Can DNA bases be produced during molecular cloud collapse? *Astron. Astrophys.*, **354**, L6–L8.

Chyba, C. F., and Sagan, C. (1992). Endogenous production, exogenous delivery and impact-shock synthesis of organic molecules: an inventory for the origins of life. *Nature*, **355**, 125–132.

Crick, F. H. C. (1968). The origin of the genetic code. *J. Mol. Biol.*, **38**, 367–379.

Cronin, C. R., and Moore, C. B. (1971). Amino acid analyses of the Murchison, Murray and Allende carbonaceous chondrites. *Science*, **172**, 1327–1329.

D'Alessio, P., Canto, J., Calvet, N., and Lizano, S. (1998). Accretion disks around young objects I: the detailed vertical structure. *Astrophys. J.* **500**, 411–427.

Davidson, A. R., Lumb, K. J., and Sauer, R. T. (1995). Cooperatively folded proteins in random sequence libraries. *Nat. Struct. Biol.*, **2**, 856–864.

Di Giulio, M. (2000). The late stage of genetic code structuring took place at high temperatures. *Gene*, **261**, 189–195.

Di Giulio, M. (2005). The ocean abysses witnessed the origin of the genetic code. *Gene*, **346**, 7–12.

Di Giulio, M., and Medugno, M. (1999). Physicochemical optimization in the genetic code origin as the number of codified amino acids increases. *J. Mol. Evol.*, **49**, 1–10.

Doi, N., Kakukawa, K., Oishi, Y., and Yanagawa, H. (2005). High solubility of random sequence proteins consisting of five kinds of primitive amino acids. *Prot. Eng. Design and Selection*, **18**, 279–284.

Dubruelle, C. P., Hollenbach, D., Kamp, I., and D'Alessio, P. (2006) Models of the structure and evolution of protoplanetary disks, in *Protostars and Planets V*, eds. B. Reipurth, D. Jewitt, and K. Keil. Tuscon: University of Arizona Press, in press.

Dutrey, A., Guilloteau, S., and Guelin, M. (1997). Chemistry of protosolar-like nebulae: the molecular content of the DM Tau and GG Tau disks. *Astron. Astrophys.*, **317**, L55–L58.

Dutrey, A., Guilloteau, S., and Ho, P. (2006). Interferometric spectro-imaging of molecular gas in proto-planetary disks, in *Protostars and Planets V*, eds. B. Reipurth, D. Jewitt, and K. Keil. Tuscon: University of Arizona Press, in press.

Eigen, E., and Winkler-Oswatitsch, R. (1981). Transfer RNA, an early gene? *Naturwiss.*, **68**, 282–292.

Engel, M. H., and Nagy, B. (1982) Distribution and enantiomeric composition of amino acids in the Murchison meteorite. *Nature*, **296**, 837–840.

Freeland, S. J., and Hurst, L. D. (1998). The genetic code is one in a million. *J. Mol. Evol.*, **47**, 238–248.

Freeland, S. J., Wu, T., and Keulmann, N. (2003) The case for an error minimizing standard genetic code. *Origins Life Evol. Biosph.*, **33**, 457–477.

Gaidos, E., and Selsis, F. (2006). From protoplanets to protolife: the emergence and maintenance of life, in *Protostars and Planets V*, eds. B. Reipurth, D. Jewitt, and K. Keil. Tuscon: University of Arizona Press, in press.

Gilis, D., Massar, S., Cerf, N. J., and Rooman, M. (2001). Optimality of the genetic code with respect to protein stability and amino acid frequencies. *Genome Biol.*, **2(11)**, 49.1–49.12.

Gutierrez, G., Marquez, L., and Marin, A. (1996). Preference for guanosine at first position in highly expressed *Escherichia coli* genes. A relationship with translational efficiency. *Nucl. Acids. Res.*, **24**, 2525–2527.

Hennet, R. J. C., Holm, N. G., and Engel, M. H. (1992). Abiotic synthesis of amino acids under hydrothermal conditions and the origin of life: a perpetual phenomenon. *Naturwiss.*, **79**, 361–365.

Ikehara, K. (2005). Possible steps to the emergence of life: The GADV-protein world hypothesis. *Chemical Record*, **5**, 107–118.

Jeffares, D. C., Poole, A. M., and Penny, D. (1998). Relics from the RNA world. *J. Mol. Evol.*, **46**, 18–36.

Jordan, I. K., Kondrashov, F. A., Adzhubel, I. A., *et al.* (2005). A universal trend of amino acid gain and loss in protein evolution. *Nature*, **433**, 633–637.

Joyce, G. F. (2002). The antiquity of RNA-based evolution. *Nature*, **418**, 214–221.

Kasting, J. F. (1993). Earth's early atmosphere. *Science*, **259**, 920–926.

Kokubo, E., and Ida, S. (2002) Formation of protoplanet systems and diversity of planetary systems. *Astrophys. J.*, **581**, 666–680.

Knight, R. D., Freeland, S. J., and Landweber, L. F. (2001). Rewiring the keyboard: evolvability of the genetic code. *Nature Rev. Genet.*, **2**, 49–58.

Kvenvolden, K. A., Lawless, J. G., and Ponnamperuma, C. (1971). Nonprotein amino acids in the Murchison meteorite. *Proc. Nat. Acad. Sci. USA*, **68**, 486–490.

Lazcano, A., and Miller, S. L. (1996). The origin and early evolution of life: prebiotic chemistry, the pre-RNA world, and time. *Cell*, **85**, 793–798.

Lin, D. N. C., and Papaloizou, J. C. B. (1993). On the tidal interaction between protostellar disks and companions. *Protostars and Planets III*, 749–835.

Lowe, C. U., Rees, M. W., and Markham, R. (1963). Synthesis of complex organic compounds from simple precursors: formation of amino acids, amino-acid polymers, fatty acids and purines from ammonium cyanide. *Nature*, **199**, 219–222.

MacLow, M.-M., and Klessen, R. S. (2004). Control of star formation by supersonic turbulence. *Rev. Mod. Phys.*, **76**, 125–195.

Marshall, W. L. (1994). Hydrothermal synthesis of amino acids. *Geochim. Cosmochim. Acta*, **58**, 2099–2106.

Matsumura, S., Pudritz, R. E., and Thommes, E. W. (2007). Saving planetary systems: dead zones and planetary migration. *Astrophys. J.*, **660**, 1609–1623.

Mayer, L., Quinn, T., Wadsley, J., and Stadel, J. (2002). Formation of giant planets by fragmentation of protoplanetary disks. *Science*, **298**, 1756–1759.

McCaughrean, M. J., and Stauffer, J. R. (1994). High resolution near-infrared imaging of the trapezium: a stellar census. *Astron. J.*, **108**, 1382–1397.

McDonald, J. H. (2006). Apparent trend of amino acid gain and loss in evolution due to nearly neutral variation. *Mol. Biol. Evol.*, **23**, 240–244.

Meyer, M. R., Backman, D. E., Weinberger, A. J., and Wyatt, M. C. (2006). Evolution of circumstellar disks around normal stars: placing our solar system in context, in *Protostars and Planets V*, eds. B. Reipurth, D. Jewitt, and K. Keil. Tuscon: University of Arizona Press, in press.

Miller, S. L., and Orgel, L. E. (1974). *The Origins of Life on the Earth*. Englewood Cliffs: Prentice-Hall.

Miyakawa, S., Yamanashi, H., Kobayashi, K., Cleaves, H. J., and Miller, S. L. (2002). Prebiotic synthesis from CO atmospheres: implications for the origins of life. *Proc. Nat. Acad. Sci. USA*, **99**, 14628–14631.

Moore, P. B., and Steitz, T. A. (2002). The involvement of RNA in ribosome function. *Nature*, **418**, 229–235.

Morbidelli, A., Chambers, J., Lunine, J. I., *et al.* (2000). Source regions and time scales for the delivery of water to Earth *Meteorit. Planet. Sci.*, **35**, 1309–1320.

Munoz-Caro, G. M., Meierhenrich, U. J., Schutte, W. A., *et al.* (2002). Amino acids from ultraviolet irradiation of interstellar ice analogues. *Nature*, **416**, 403–406.

Natta, A., Testi, L., Calvet, N., *et al.* (2006) Dust in proto-planetary disks: properties and evolution, in *Protostars and Planets V*, eds. B. Reipurth, D. Jewitt, and K. Keil. Tuscon: University of Arizona Press, in press.

Pascal, R., Boiteau, L., and Commeyras, A. (2005). From the prebiotic synthesis of α-amino acids towards a primitive translation apparatus for the synthesis of peptides. *Top. Curr. Chem.*, **259**, 69–122.

Pierazzo, E., and Chyba, C. F. (1999). Amino acid survival in large cometary impacts. *Meteoritics Planet. Sci.*, **34**, 909–918.

Pollack, J. B., Hubickyj, O., Bedenheimer, P., Lissauer, J. J., Podolak, M., and Greenzweig, Y. (1996). Formation of the giant planets by concurrent accretion of solids and gas. *Icarus*, **124**, 62–85.

Raymond, S. N., Quinn, T., and Lunine, J. I. (2004). Making other earths: dynamical simulations of terrestrial planet formation and water delivery. *Icarus*, **168**, 1–17.

Riddle, D. S., Santiago, J. V., Bray-Hall, S. T., *et al.* (1997). Functional rapidly folding proteins from simplified amino acid sequences. *Nature Struct. Biol.*, **4**, 805–809.

Ronneberg, T. A., Landweber, L. F., and Freeland, S. J. (2000). Testing a biosynthetic theory of the genetic code: fact or artifact? *Proc. Nat. Acad. Sci. USA*, **97**, 13690–13695.

Schreyer, K., Semenov, D., Henning, T., Pavlyuchenkov, Y., and Dullemond, C. (2005). The massive disk around the young B-star AFGL 490, in *IAU Symposium 231: Astrochemistry – Recent Successes and Current Challenges*. Cambridge: Cambridge University Press.

Seligmann, H. (2003). Cost-minimization of amino acid usage. *J. Mol. Evol.*, **56**, 151–161.

Sengupta, S., and Higgs, P. G. (2005). A unified model of codon reassignment in alternative genetic codes. *Genetics*, **170**, 831–840.

Sengupta, S., Yang, X., and Higgs, P. G. (2007). The mechanisms of codon reassignments in mitochondrial genetic codes. *J. Mol. Evol.*, **64**, 662–688.

Sephton, M. A. (2002). Organic compounds in carbonaceous meteorites. *Nat. Prod. Rep.*, **19**, 292–311.

Shapiro, R. (2006). Small molecule interactions were central to the origin of life. *Quart. Rev. Biol.*, **81**, 105–125.

Shimoyama, A., Ponnamperuma, C., and Yanai, K. (1979). Amino acids in the Yamato carbonaceous chondrite from Antarctica. *Nature*, **282**, 394–396.

Simon, M., Dutrey, A., and Guilloteau, S. (2000). Dynamical masses of T Tauri stars and calibration of pre-main-sequence evolution. *Astrophys. J.*, **545**, 1034–1043.

Tilley, D. A., and Pudritz, R. E. (2004). The formation of star clusters I: simulations of hydrodynamic turbulence. *MNRAS*, **353**, 769–788.

Trifonov, E. N. (1987). Translation framing code and frame-monitoring mechanism as suggested by the analysis of messenger RNA and 16S ribosomal RNA nucleotide sequences. *J. Mol. Biol.*, **194**, 643–652.

Trifonov, E. N. (2000). Consensus temporal order of amino acids and the evolution of the triplet code. *Gene*, **261**, 139–151.

Trifonov, E. N. (2004). The triplet code from first principles. *J. Biomol. Struct. Dynam.*, **22**, 1–11.

Urbina, D., Tang, B., and Higgs, P. G. (2006). The response of amino acid frequencies to directional mutation pressure in mitochondrial genome sequences is related to the physical properties of the amino acids and the structure of the genetic code. *J. Mol. Evol.*, **62**, 340–361.

van Zadelhoff, G.-J., Aikawa, Y., Hogerheijde, M. R., and van Dishoeck, E. F. (2003). Axi-symmetric models of UV radiative transfer with applications to circumstellar disk chemistry. *Astron. Astrophys.*, **397**, 789–802.

Weber, A. L. and Miller, S. L. (1981). Reasons for the occurrence of the twenty coded protein amino acids. *J. Mol. Evol.*, **17**, 273–284.

Whittet, D. C. B. (1997). Is extraterrestrial organic matter relevant to the origin of life on Earth? *Orig. Life Evol. Biosph.*, **27**, 249–262.

Wolf, S., and D'Angelo, G. (2005). On the observability of giant protoplanets in circumstellar disks. *Astrophys. J.*, **619**, 1114–1122.

Wong, J. T. (1975). A coevolution theory of the genetic code. *Proc. Nat. Acad. Sci. USA*, **72**, 1909–1912.

Wong, J. T. (2005). Coevolution theory of the genetic code at age thirty. *BioEssays*, **27**, 416–425.

Wu, H. L., Bagby, S., and van den Elsen, J. M. H. (2005). Evolution of the genetic code via two types of doublet codons. *J. Mol. Evol.*, **61**, 54–64.

Yarus, M., Caporaso, J. G., and Knight, R. (2005). Origins of the genetic code: the escaped triplet theory. *Annu. Rev. Biochem.*, **74**, 179–198.

Yoshino, D., Hayatso, R., and Anders, E. (1971). Origin of organic matter in the early solar system – III. Amino acids: catalytic synthesis. *Geochim. Cosmochim. Acta*, **35**, 927–938.

5

Emergent phenomena in biology: the origin of cellular life

David Deamer
University of California, Santa Cruz

5.1 Introduction

The main themes of this chapter concern the phenomenon of emergence, the origin of life as a primary example of emergence, and how evolution begins with the inception of cellular life. The physical properties of certain molecular species are relevant to life's origins, because these properties lead to the emergence of more complex structures by self-assembly. One such property is the capacity of amphiphilic molecules such as soap to form membranous boundary structures, familiar examples being soap bubbles and cell membranes. A second example is the chemical bonding that allows biopolymers such as nucleic acids and proteins to assemble into functional sequences. Self-assembly processes can produce complex supramolecular structures with certain properties of the living state. Such structures are able to capture energy available in the environment and initiate primitive reactions associated with metabolism, growth, and replication. At some point approximately 4 billion years ago, cellular compartments appeared that contained macromolecular systems capable of catalysed growth and replication. Because each cellular structure would be slightly different from all others, Darwinian evolution by natural selection could begin, with the primary selective factor being competition for energy and nutrients.

5.2 Defining emergence

Researchers increasingly use the term emergence to describe processes by which more complex systems arise from seemingly simpler systems, typically in an unpredictable fashion. This usage is just the opposite of reductionism, the belief that any phenomenon can be explained by understanding the parts of that system. Although reductionism has been a powerful tool in the sciences, the concept of emergence is also useful in referring to some of the most remarkable phenomena that we observe in nature and in the laboratory.

Planetary Systems and the Origins of Life, eds. Ralph E. Pudritz, Paul G. Higgs, and Jonathon R. Stone.
Published by Cambridge University Press.

One example is the way that orderly arrangements of molecules appear when a relatively simple system of molecules undergoes a transition to a much more complex system. For instance, if we add soap molecules to water, at first there is nothing present except the expected clear solution of individual molecules dissolved in the water. At a certain concentration, additional molecules no longer dissolve, but instead begin to associate into small aggregates called micelles. As the concentration increases further, the micelles begin to grow into membranous layers that cause the originally clear solution to become turbid. Finally, if air is blown through a straw into the solution, much larger structures appear at the surface, the soap bubbles that are familiar to everyone.

Such emergent phenomena are called self-assembly processes, or sometimes self-organization. Typically the process is spontaneous, but in other instances an input of energy is required to drive self-assembly. One of the defining characteristics of an emergent property is the element of unpredictability. If we did not know by observation that soap molecules have properties of self-assembly, we could not have predicted that micelles would be produced simply by increasing the concentration of soap molecules in solution. Even knowing that micelles form, there is no physical theory available that can predict exactly what concentration of soap is required to form the micelles. And if soap bubbles were not such a common phenomenon, we would be astonished by their appearance when a soap solution is disturbed by addition of free energy in the form of an air stream.

5.3 Emergence of life: a very brief history

The origin of life was strongly dependent on physical conditions of the early Earth environment. Imagine that we could somehow travel back in time to the prebiotic Earth 3.9 billion years ago. The first thing we would notice is that it is very hot, with temperatures near the boiling point of water. Impacting objects the size of comets and asteroids add water, carbon dioxide, silicate minerals, and small amounts of organic material to the planetary surface. They also produce craters on the Moon that remain as a permanent record of the bombardment. The atmosphere has no oxygen, but instead is a mixture of carbon dioxide and nitrogen. Land masses are present, but they are volcanic islands resembling Hawaii or Iceland, rather than continents. We are standing on one such island, on a beach composed of black lava rocks, with tide pools containing clear seawater. We can examine the seawater in the tide pools with a powerful microscope, but there is nothing to be seen, only a dilute solution of organic compounds and salts. If we could examine the mineral surfaces of the lava rocks, we would see that some of the organic compounds have formed a film adhering to the surface, while others assembled into aggregates that disperse in the seawater.

Now imagine that we return 300 million years later. Not much has changed. The primary land masses are still volcanic in origin, although small continents are beginning to appear as well. The meteorite impacts have dwindled, and the global temperature has cooled to around 70 °C. We are surprised when we examine the tide pools, in which a turbidity has appeared that was not apparent earlier, and the mineral surfaces are coated with layers of pigmented films. When we examine the water and films with our microscope, we discover immense numbers of bacteria present in the layers. Life has begun.

What happened in the intervening time that led to the origin of life? This is a fundamental question of biology, and the answer will change the way we think about ourselves, and our place in the universe, because if life could begin on the Earth, it could begin by similar processes on Earth-like planets circling other stars in our galaxy. The origin of life is the most extraordinary example of emergent phenomena, and the process by which life began must involve the same kinds of intermolecular forces and self-assembly processes that cause soap to form membranous vesicles and allow monomers like amino acids and nucleotides to form functional polymers. The origin of life must also have incorporated the reactions and products that are the result of energy flowing through a molecular system, thereby driving it toward ever more complex systems with emergent properties.

5.4 The first emergent phenomena: self-assembly processes on the early Earth

All cellular life today incorporates two processes we will refer to here as self-assembly and directed assembly. The latter process always involves the formation of covalent bonds by energy-dependent synthetic reactions, and requires that a coded sequence in one type of polymer in some way directs the sequence of monomer addition in a second polymeric species. The first process – spontaneous self-assembly – occurs when certain compounds interact through non-covalent hydrogen bonds, electrostatic forces, and non-polar interactions to form closed membrane-bounded microenvironments. The self-assembly of membranes is by far the simplest of the processes associated with the origin of cellular life. It occurs whenever lipid-like molecules are present in an aqueous environment, and does not require the covalent bond formation needed for the synthesis of polymers such as proteins and nucleic acids. For this reason, it is reasonable to assume that membrane formation preceded the appearance of catalytic and replicating polymers. Boundary structures and the compartments they produce make energy available in the form of ion gradients, and can also mediate a selective inward transport of nutrients. Furthermore, membranous compartments are, in principle, capable of containing unique systems of macromolecules. If a replicating system of polymers could be encapsulated within a membrane-bounded compartment, the components of the system would share the same microenvironment, and the result would be a major step toward cellularity, speciation, and true cellular function (Deamer and Oro, 1980; Morowitz *et al.*, 1986; Morowitz, 1992; Dyson, 1995).

We know very little about how this event might have occurred at the origin of cellular life, but recent advances have provided clues about possible sources of self-assembling molecules and processes by which large molecules can be captured in membrane-bounded microenvironments. Here we will describe the chemical and physical properties of such systems, and several experimental models that incorporate certain properties related to the origin of cellular life.

5.5 Sources of amphiphilic molecules

The most striking examples of self-assembling molecules are called amphiphiles, because they have both a hydrophilic ('water loving') group and a hydrophobic ('water hating') group on the same molecule. Amphiphilic molecules are among the simplest of life's molecular components, and are readily synthesized by non-biological processes. Virtually any hydrocarbon having ten or more carbons in its chain takes on amphiphilic properties if one end of the molecule incorporates a polar or ionic group. The simplest common amphiphiles are therefore molecules such as soaps, more technically referred to as monocarboxylic acids. A good example is decanoic acid, which is shown below with its ten carbon chain:

$$CH_3–CH_2–CH_2–CH_2–CH_2–CH_2–CH_2–CH_2–CH_2–COOH$$

McCollom *et al.* (1999) and Rushdi and Simoneit (2001) demonstrated that a series of alkanoic acids and alcohols in this size range can be produced from very simple organic compounds by Fischer–Tropsch reactions that simulate geothermal conditions on the early Earth. It has been found that such molecules readily form membranous vesicles, as shown in Figure 5.1 (Apel *et al.*, 2002). This fact will become important later when we discuss the emergence of cellular life.

Two possible sources of organic compounds on a primitive planetary surface are delivery during late accretion, followed by chemical evolution, and synthesis by geochemical processes in the early atmosphere or hydrosphere. Early investigations focused on chemical synthesis of monomers common to the primary macromolecules involved in living systems, with the goal of determining whether it was possible that biologically relevant compounds were available on the primitive Earth. Most of these studies emphasized water-soluble compounds such as amino acids, nucleobases, and simple carbohydrates. The classic experiments of Miller (1953) and Miller and Urey (1959) showed that amino acids such as glycine

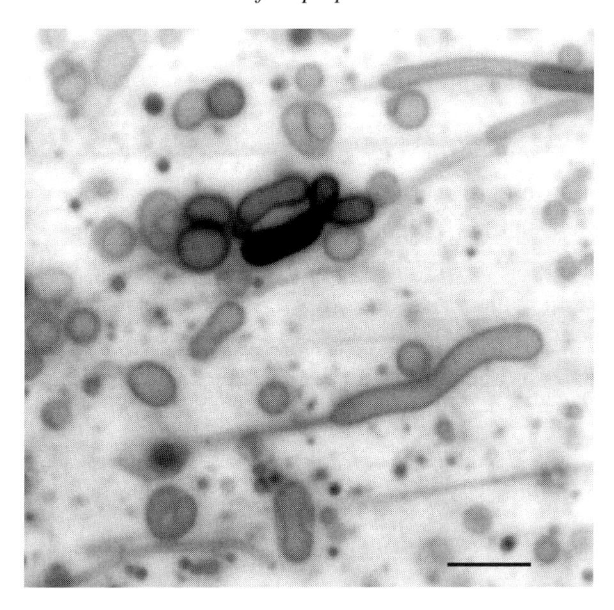

Fig. 5.1. Membranous vesicles are produced when simple amphiphilic molecules are dispersed in aqueous solutions at neutral pH ranges. The vesicles shown here are composed of a mixture of decanoic acid and decanol. Bar shows 20 μm.

and alanine could be obtained when a mixture of reduced gases was exposed to an electrical discharge. The mixture was assumed to be a simulation of the original terrestrial atmosphere which, by analogy with the outer planets, would have contained hydrogen, methane, ammonia, and water vapour. At sufficiently high energy fluxes, such mixtures of reducing gases generate hydrogen cyanide and formaldehyde, which in turn react to produce amino acids, purines, and a variety of simple sugars.

The possibility that organic compounds could be synthesized under prebiotic conditions was given additional weight when it was convincingly shown that carbonaceous meteorites contained amino acids, hydrocarbons, and even traces of purines (Kvenholden *et al.*, 1970; Lawless and Yuen, 1979; Cronin *et al.*, 1988). Such meteorites are produced by collisions in the asteroid belt between Mars and Jupiter. Asteroids range up to hundreds of kilometres in diameter, and are examples of planetesimals that happened to avoid accretion into the terrestrial planets. Asteroids and their meteoritic fragments therefore represent samples of the primitive Solar System in which the products of prebiotic chemical reactions have been preserved for over 4.5 billion years. It was reasonable to assume that similar reactions and products were likely to have occurred on the Earth's surface.

In the late 1970s it became increasingly clear that the archaean atmosphere was largely of volcanic origin and composed of carbon dioxide and nitrogen rather than the mixture of reducing gases assumed by the Millcr–Urey model (Holland, 1984; Kasting and Broun, 1998). This is consistent with the fact that approximately 65 atmosphere equivalents of carbon dioxide are present in the Earth's crust as carbonate minerals, all of which must have passed through the atmosphere at some point as a gaseous component. Carbon dioxide does not support synthetic pathways leading to chemical monomers, so interest was drawn to the second potential source of organic material, extraterrestrial infall in the form of micrometeorites and comets. This was first proposed by Oró (1961) and Delsemme (1984) and more recently extended by Anders (1989) and Chyba and Sagan (1992). The total organic carbon added by extraterrestrial infall over $\sim 10^8$ years of late accretion can be estimated to be in the rage of 10^{16}–10^{18} kg, which is several orders of magnitude greater than the 12.5×10^{14} kg total organic carbon in the biosphere (www.regensw.co.uk/technology/biomass-faq.asp). From such calculations it seems reasonable that extraterrestrial infall was a significant source of organic carbon in the prebiotic environment. Even today meteorites and interplanetary dust particles (IDPs) deliver organic materials to the modern Earth at a rate of $\sim 10^6$ kg/y^{-1} (Love and Brownlee, 1992; Maurette, 1998).

The discovery of biologically relevant compounds in meteorites also indicated that organic synthesis can occur in the interstellar medium, which immediately leads to the question of sources and synthetic pathways. The most important biogenic elements (C, N, O, S, and P) form in the interiors of stars, then are ejected into the surrounding interstellar medium (ISM) at the end of the star's lifetime during red giant, nova, and supernova phases. Following ejection, much of this material becomes concentrated into dense molecular clouds from which new stars and planetary systems are formed (Ehrenfreund and Charnley, 2000; Sandford, 1996). At the low temperatures in these dark molecular clouds, mixtures of molecules condense to form ice mantles on the surfaces of dust grains where they can participate in additional chemical reactions. Comparison of infrared spectra of low-temperature laboratory ices with absorption spectra of molecular clouds indicates that interstellar ices are mainly composed of H_2O mixed with CO, CO_2, CH_3OH, NH_3, and other components, the last ingredients generally comprising 5–15% of the total. The ices are exposed to ionizing radiation in the form of cosmic rays (and secondary radiation generated by their interaction with matter), and ultraviolet photons from stars forming within the cloud.

Laboratory experiments have shown that illuminating such ices with ultraviolet light leads to more complex molecular species (Greenberg and Mendoza-Gomez, 1993; Bernstein *et al.*, 1995; Gerakines *et al.*, 2000; Ehrenfreund *et al.*, 2001). Hundreds of new compounds are synthesized, even though the starting ices contain only a few simple common interstellar molecules. Many of the compounds formed in these experiments are also present in meteorites, comets, and 10 Ps, and some are relevant to the origin of life, including amino acids (Munoz-Caro *et al.*, 2002; Bernstein *et al.*, 2002; Bernstein *et al.*, 2001), and amphiphilic material (Dworkin *et al.*, 2001).

Although it is now clear that organic molecules are synthesized in dense molecular clouds, the molecules must be delivered to habitable planetary surfaces if they are to take part in the origin of life. This requires that they survive the transition from the dense cloud into a protostellar nebula and subsequent incorporation into planetesimals, followed by delivery to a planetary surface. During the late bombardment period, which lasted until about 4 billion years ago, the amount of extraterrestrial organic material brought to the prebiotic Earth was likely to have been several orders of magnitude greater than current rates of infall (Chyba and Sagan, 1992). Thus, the early Earth must have been seeded with organic matter created in the interstellar medium, protosolar nebula, and asteroidal/cometary parent bodies.

5.6 The emergence of primitive cells

Life on the Earth most likely arose from vast numbers of natural experiments in which various combinations of organic molecules were mixed and recombined to form complex interacting systems, then exposed to sources of energy such as light, heat, and oxidation–reduction potentials presented by donors and acceptors of electrons. This mixing and recombination probably did not occur in free solution, but rather in fluctuating environments at aqueous–mineral interfaces exposed to the atmosphere under conditions that would tend to concentrate the organic material so that reactions could occur. Through this process, incremental chemical evolution took place over a period of several hundred million years after the Earth had cooled sufficiently for water vapour to condense into oceans. At some point, membrane-bounded systems of molecules appeared that could grow and reproduce by using energy and nutrients from the environment. An observer seeing this endproduct would conclude that such systems were alive, but would be unable to pinpoint the exact time when the complex structures took on the property of life.

Here we will assume that the structures described above would be recognizable cells, and that cellular compartments were required for life to begin. This assumption differs from the conjectures of Kauffman (1993) and Wächtershäuser (1988) that life began as a series of reactions resembling metabolism. In their view, autocatalytic pathways were first established, perhaps on mineral surfaces. Over time the systems became increasingly complex to the point that self-reproducing polymers were synthesized, with cellular compartments appearing at a later stage. There is as yet little experimental evidence that allows a choice between the two perspectives of ‘compartments first’ or ‘metabolism first’. However, because membrane structures readily self-assemble from amphiphilic compounds known to be present in mixtures of organic compounds available on the prebiotic Earth, it is highly likely that membranous compartments were among the first biologically relevant structures to appear prior to the origin of life.

Evidence from phylogenetic analysis suggests that microorganisms resembling today's bacteria were the first form of cellular life. Traces of their existence can be found in the fossil record in Australian rocks at least 3.5 billion years old. Over the intervening years between life's beginnings and now, evolution has produced bacteria which are more advanced than the first cellular life. The machinery of life has become so advanced that when researchers began subtracting genes in one of the simplest known bacterial species, they reached a limit of approximately 265–350 genes that appears to be the absolute requirement for contemporary bacterial cells (Hutchinson *et al.*, 1999). Yet life did not spring into existence with a full complement of 300+ genes, ribosomes, membrane transport systems, metabolism, and the DNA → RNA → protein information transfer that dominates all life today. There must have been something simpler, a kind of scaffold life that was left behind in the evolutionary rubble.

Can we reproduce that scaffold? One possible approach was suggested by the RNA World concept that arose from the discovery of ribozymes, which are RNA structures having enzyme-like catalytic activities. The idea was greatly strengthened when it was discovered that the catalytic core of ribosomes is not composed of protein at the active site, but instead is composed of RNA machinery. This remarkable finding offers convincing evidence that RNA likely came first, and then was overlaid by more complex and efficient protein machinery (Hoang *et al.*, 2004).

Another approach to discovering a scaffold is to incorporate one or a few genes into microscopic artificial vesicles to produce molecular systems that display certain features associated with life. The properties of such systems may then provide clues to the process by which life began in a natural setting of the early Earth. What would such a system do? We can answer this question by listing the steps that would be required for a microorganism to emerge as the first cellular life on the early Earth:

- Boundary membranes self-assemble from amphiphilic molecules.
- Energy is captured either from light and a pigment system, or from chemical energy, or both.
- Ion concentration gradients are produced and maintained across the membrane by an energy-dependent process. The gradient is a source of energy to drive metabolism and synthetic reactions.
- The energy is coupled to the synthesis of activated monomers, which in turn are used to make polymers.
- Macromolecules are encapsulated, yet smaller nutrient molecules can cross the membrane barrier.

- The macromolecules grow by polymerizing the nutrient molecules, using the energy of metabolism.
- Macromolecular catalysts speed the metabolic reactions and growth processes.
- The macromolecular catalysts themselves are reproduced during growth.
- Information is contained in the sequence of monomers in one set of polymers, and this set is duplicated during growth. The information is used to direct the growth of catalytic polymers.
- The membrane-bounded system of macromolecules can divide into smaller structures.
- Genetic information is passed between generations by duplicating the sequences and sharing them between daughter cells.
- Occasional errors (mutations) are made during replication or transmission of information so that a population of primitive living organisms can evolve through selection.

Looking down this list, one is struck by the complexity of even the simplest form of life. This is why it has been so difficult to 'define' life in the usual sense of a definition, i.e., boiled down to a few sentences in a dictionary. Life is a complex system that cannot be captured in a few sentences, so perhaps a list of its observed properties is the best we can ever hope to do. Given the list, one is also struck by the fact that *all but one of the functions* – self-reproducing polymers – have now been reconstituted individually in the laboratory. For instance, it was shown 40 years ago that lipid vesicles self-assemble into bilayer membranes that maintain ion gradients (Bangham *et al.*, 1965). If bacteriorhodopsin is in the bilayer, light energy can be captured in the form of a proton gradient (Oesterhelt and Stoeckenius, 1973). If an ATP synthase is in the membrane, the photosynthetic system can make ATP by coupling the proton gradient to form a pyrophosphate bond between ADP and phosphate (Racker and Stoeckenius, 1974). Macromolecules such as proteins and nucleic acids can be easily encapsulated and can function in the vesicles as catalysts (Chakrabarti and Deamer, 1994) and as membrane transport agents. The system can grow by addition of lipids, and can even be made to divide by imposing shear forces, after which the vesicles grow again (Hanczyc *et al.*, 2003). Macromolecules like RNA can grow by polymerization in the vesicle, driven by catalytic proteins (Monnard and Deamer, 2006). And finally, samples of cytoplasm from a living cell like *E. coli* are easily captured, including ribosomes. A micrograph of such vesicles is shown in Figure 5.2.

The ability to capture structures as large as ribosomes has led to attempts to demonstrate translation in closed vesicles. This was first achieved by Yu *et al.* (2001) and later by Nomura *et al.* (2003), who captured samples of bacterial cytoplasm from *E. coli* in large liposomes. The samples included ribosomes, tRNAs and the hundred or so other components required for protein synthesis. The mRNA containing the gene for green fluorescent protein (GFP) was also included in the mix, which permitted facile detection of protein synthesis. The encapsulated translation

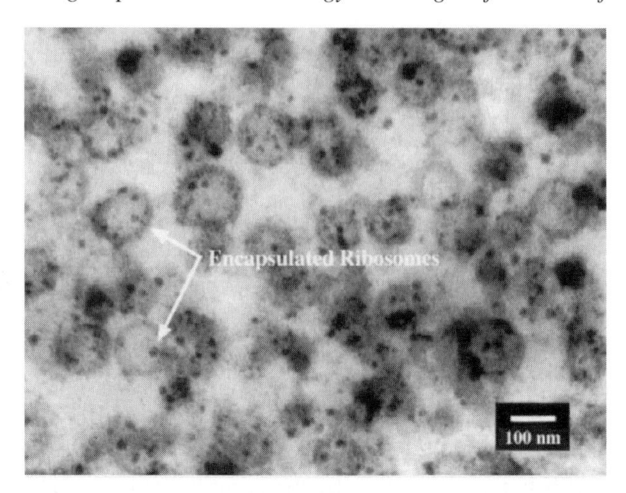

Fig. 5.2. Ribosomes from *E. coli* encapsulated in phospholipid vesicles. The vesicles were reconstituted from a detergent–lipid solution in the presence of cytoplasmic extracts from the bacteria. Several ribosomes were present in each vesicle. Micrograph courtesy of Z. Martinez.

systems worked, but only a few molecules of GFP were synthesized in each vesicle, because the only amino acids available were those captured within the vesicle. This limitation was resolved by Noireaux and Libchaber (2004), who included not only the mRNA for GFP in the mix, but also a second mRNA coding for the pore-forming protein alpha hemolysin. The hemolysin produced a channel in the lipid bilayer that allowed externally added 'nutrients' in the form of amino acids and ATP to cross the membrane barrier and supply the translation process with energy and monomers (Figure 5.3). The system worked well, and GFP synthesis continued for as long as four days.

These advances permit us to consider whether it might be possible to fabricate a kind of artificial life which is reconstituted from a complete system of components isolated from microorganisms. The system would by definition overcome the one limitation described above, the lack of a self-reproducing set of polymers, because everything in the system would grow and reproduce, including the catalytic macromolecules themselves and the lipid components of the boundary membrane. However, such a system requires that the genes for the translation system (ribosomes, tRNA, and so on) for DNA replication and transcription, and for lipid synthesis must all be present in a strand of synthetic DNA. When one adds up the number of essential genes, the total is over a hundred. This might at first seem like a daunting task, but in fact it is within the realm of possibility, in that the complete set of genes required to synthesize the polio virus has been assembled in a strand of synthetic DNA (Cello *et al.*, 2002).

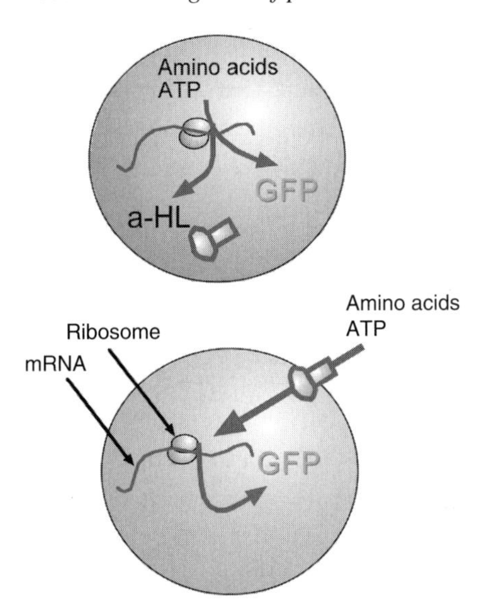

Fig. 5.3. Translation in a microenvironment. In the upper figure, amino acids and ATP encapsulated in the vesicle volume are used to make small amounts of GFP and alpha hemolysin (a-HL). The membrane prevents access to external amino acids and ATP, so the system will quickly exhaust the 'nutrients' trapped in the vesicle. However, because the a-HL is a pore forming protein, it migrates to the membrane and assembles into a heptamer with a pore large enough to allow amino acids and ATP to pass into the vesicle (lower figure). The system can now synthesize significant amounts of new protein, and the vesicle begins to show fluorescence from the accumulation of GFP (Noireaux and Libchaber, 2004).

This thought experiment clearly points out the limitations of our current understanding of the origin of cellular life. It does make clear that life could not begin as a complex molecular system of a hundred or more genes required to fabricate the simplest possible artificial cell that uses DNA, RNA, and ribosomes for self-reproduction. Instead, as noted earlier, there must have been a kind of molecular scaffold that was much simpler, yet had the capacity to evolve into the living systems we observe today. Is there any hope that we might discover such a system? One possible lead is to find a ribozyme that can grow by polymerization, in which the ribozyme copies a sequence of bases in its own structure (Johnston *et al.*, 2001). So far, the polymerization has only copied a string of 14 nucleotides, but this is a good start. If a ribozyme system can be found that catalyses its own complete synthesis using genetic information encoded in its structure, it could rightly be claimed to have the essential properties that are lacking so far in artificial cell models: reproduction of the catalyst itself. Instead of the hundred or so genes necessary for the translation system described above, the number is reduced to just a few genes that allow the ribozyme to control its own replication, synthesize catalysts related to primitive metabolism, including synthesis of membrane components, and perhaps a pore-forming molecule that allows nutrient transport. Given such a ribozyme, it is not difficult to imagine its incorporation into a system of lipid vesicles that would have the basic properties of the living state.

5.7 Self-assembly processes in prebiotic organic mixtures

We can now return to considering process that are likely to have occurred on the prebiotic Earth, ultimately leading to the beginning of life and initiating Darwinian evolution. We will first ask what physical properties are required if a molecule is to become incorporated into a stable cellular compartment. As discussed earlier, all membrane-forming molecules are amphiphiles, with a hydrophilic 'head' and a hydrophobic 'tail' on the same molecule. If amphiphilic molecules were present in the mixture of organic compounds available on the early Earth, it is not difficult to imagine that their self-assembly into molecular aggregates was a common process. Is this a plausible premise? In order to approach this question, we can assume that the mixture of organic compounds in carbonaceous meteorites such as the Murchison meteorite resembles components available on the early Earth through extraterrestrial infall. A series of organic acids represents the most abundant water-soluble fraction in carbonaceous meteorites (Cronin *et al.*, 1988). Samples of the Murchison meteorite have been extracted in an organic solvent commonly used to extract membrane lipids from biological sources (Deamer and Pashley, 1989). When this material was allowed to interact with aqueous phases, one class of compounds with acidic properties was clearly capable of forming membrane-bounded vesicles (Figure 5.4).

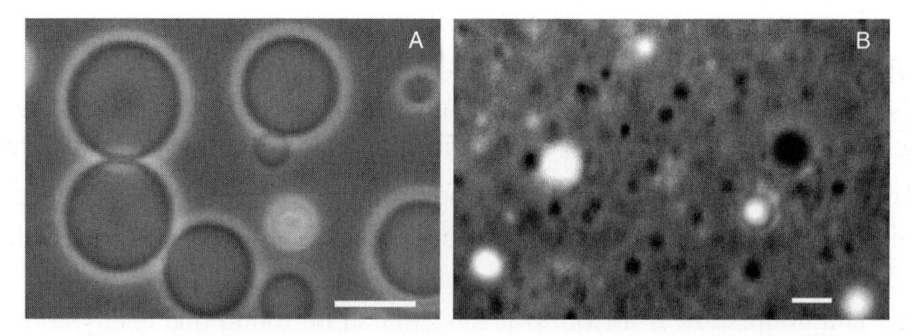

Fig. 5.4. Membranous compartments are produced by self-assembly of amphiphilic molecules extracted from the Murchison meteorite. The larger vesicles have a diameter of 2030 μm. The image on the right shows a fluorescent dye (1 mM pyranine) encapsulated within the vesicular volume. The fact that anionic dye molecules can be trapped shows that the vesicles are bounded by true membranes representing a permeability barrier. Bars show 20 μm.

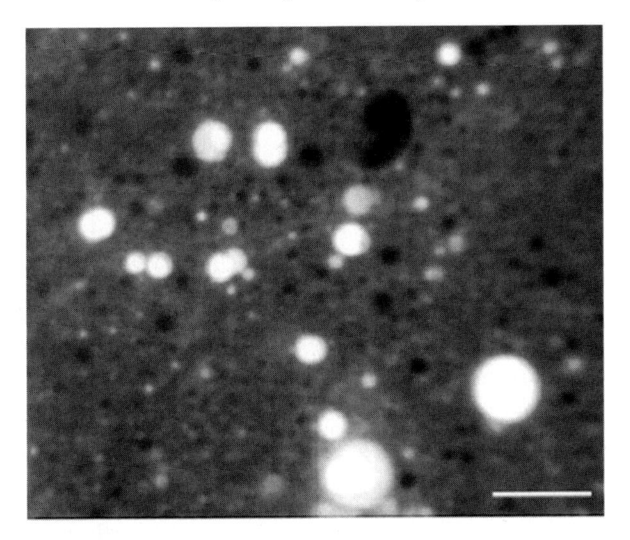

Fig. 5.5. Membranous compartments composed of very simple amphiphiles are capable of encapsulating macromolecules in stable vesicles. In the image above, DNA molecules have been trapped in decanoic acid vesicles. The DNA is stained with acridine orange so that it can be visualized by fluorescence microscopy. Bar shows 20 μm.

From these results, it is reasonable to conclude that a variety of simpler amphiphilic molecules were present on the early Earth that could participate in the formation of primitive membrane structures. Even if membranous vesicles were commonplace on the early Earth and had sufficient permeability to permit nutrient transport to occur, these structures would be virtually impermeable to larger polymeric molecules that were necessarily incorporated into molecular systems on the pathway to cellular life. The encapsulation of macromolecules in lipid vesicles has been demonstrated by hydration–dehydration cycles that simulate an evaporating lagoon (Shew and Deamer, 1985). Molecules as large as DNA can be captured by such processes. For instance, when a dispersion of DNA and fatty acid vesicles is dried, the vesicles fuse to form a multilamellar sandwich structure with DNA trapped between the layers. Upon rehydration, vesicles reform that contain highly concentrated DNA, a process that can be visualized by staining with a fluorescent dye (Figure 5.5).

5.8 Emergence of membrane functions

Membranes have many functions in addition to acting as a container for the macromolecular polymers of life. Three primary membrane functions associated with a protocell would include selective inward transport of nutrients from the environment, capture of the energy available in light or oxidation–reduction potentials, and coupling of that energy to some form of energy currency such as ATP in order to drive polymer synthesis.

The simplest of these functions is that of a permeability barrier which limits free diffusion of solutes between the cytoplasm and external environment. Although such barriers are essential for cellular life to exist, there must also be a mechanism by which selective permeation allows specific solutes to cross the membrane. In contemporary cells, such processes are carried out by transmembrane proteins that act as channels and transporters. Examples include the proteins that facilitate the transport of glucose and amino acids into the cell, channels that allow potassium and sodium ions to permeate the membrane, and active transport of ions by enzymes that use ATP as an energy source. It seems unlikely that the first living cellular systems evolved such highly specialized membrane transport systems, which brings up the question of how early cells overcame the membrane permeability barrier. One possibility is that simple diffusion across the bilayer may have been sufficient. To give a perspective on permeability and transport rates by diffusion, we can compare the fluxes of relatively permeable and relatively impermeable solutes across contemporary lipid bilayers. The measured permeability of lipid bilayers to small, uncharged molecules such as water, oxygen, and carbon dioxide is greater than the permeability to ions by a factor of $\sim 10^9$. Such values mean little by themselves, but make more sense when put in the context of time required for exchange across a bilayer: half the water in a liposome exchanges in milliseconds, while potassium ions have half-times of exchange measured in days.

We can now consider some typical nutrient solutes like amino acids and phosphates. Such molecules are ionized, which means that they would not readily cross the permeability barrier of a lipid bilayer. Permeability coefficients of liposome membranes to phosphate, amino acids, and nucleotides have been determined (Chakrabarti and Deamer, 1994; Monnard *et al.*, 2002) and were found to be similar to ionic solutes such as sodium and chloride ions. From these figures one can estimate that if a primitive microorganism depended on passive transport of phosphate across a lipid bilayer composed of a typical phospholipid, it would require several years to accumulate phosphate sufficient to double its DNA content, or pass through one cell cycle. In contrast, a modern bacterial cell can reproduce in as short a time as 20 minutes.

If solutes like amino acids and phosphate are so impermeant, how could primitive cells have had access to these essential nutrients? One clue may be that modern lipids are highly evolved products of several billion years of evolution, and typically contain hydrocarbon chains 16–18 carbons in length. These chains provide an interior 'oily' portion of the lipid bilayer that represents a nearly impermeable barrier to the free diffusion of ions such as sodium and potassium. The reason is related to the common observation that 'oil and water don't mix'. Using more technical language, ion permeation of the hydrophobic portion of a lipid bilayer faces a very high energy barrier which is associated with the difference in energy for an ion in water to 'dissolve' in the oily interior of a lipid bilayer composed of hydrocarbon chains. However, studies have shown that permeability is strongly dependent on chain length (Paula *et al.*, 1996). For instance, shortening phospholipid chains from 18 to 14 carbons increases permeability to ions by nearly three orders of magnitude. The reason is that thinner membranes have increasing numbers of transient defects that open and close on nanosecond timescales, so that ionic solutes can pass from one side of the membrane to the other without dissolving in the oily interior phase of the bilayer. Ionic solutes even as large as ATP can diffuse across a bilayer composed of a 14-carbon phospholipid (Monnard and Deamer, 2001). On the early Earth, shorter hydrocarbon chains would have been much more common than longer chain amphiphiles, suggesting that the first cell membranes were sufficiently leaky so that ionic and polar nutrients could enter, while still maintaining larger polymeric molecules in the encapsulated volume.

There are several implications of this conjecture. First, if a living cell is to take up nutrients by diffusion rather than active transport, the nutrients must be at a reasonably high concentration, perhaps in the millimolar range. It is doubtful that such high concentrations would be available in the bulk phase medium of a lake or sea, which suggests that early life needed to be in an environment such as a small pond undergoing periodic wet–dry cycles that would concentrate possible nutrients. On the other hand, that same cell also was likely to have metabolic waste products, so it could not be in a closed system or it would soon reach thermodynamic equilibrium in which the waste product accumulation inhibits forward reactions required for metabolism and synthesis. And finally, a leaky membrane may allow entry of nutrients, but it would also be useless for developing the ion gradients that are essential energy sources for all modern cells. All of these considerations should guide our thinking as we attempt to deduce conditions that would be conducive for the origin of cellular life.

5.9 Emergence of growth processes in primitive cells

Reports (Walde *et al.*, 1994) showed that vesicles composed of oleic acid can grow and 'reproduce' as oleoyl anhydride spontaneously hydrolyzed in the reaction mixture, thereby adding additional amphiphilic components (oleic acid) to the vesicle membranes. This approach has been extended by Hanczyc *et al.* (2003), who prepared myristoleic acid membranes under defined conditions of pH, temperature,

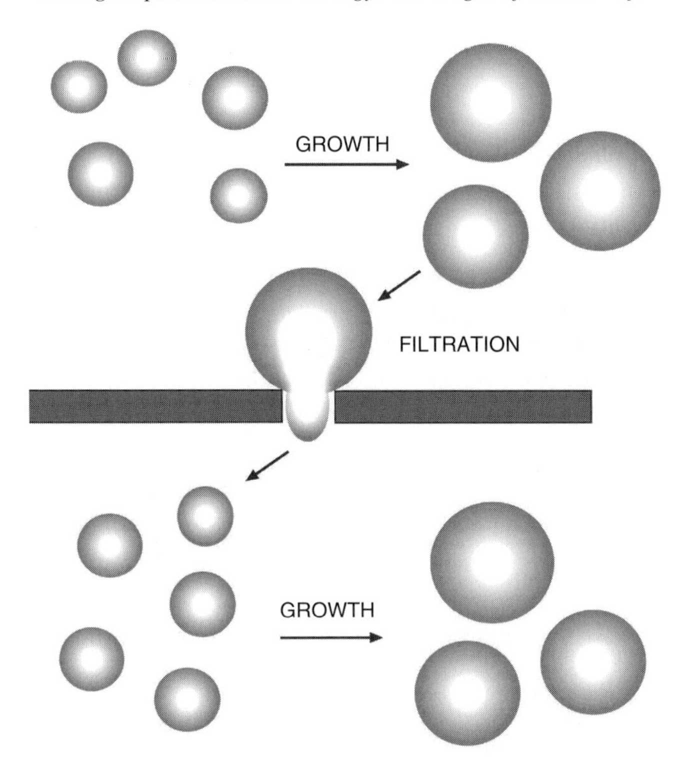

Fig. 5.6. Growth and division of lipid vesicles. (See text for details.)

and ionic strength. The process by which the vesicles formed from micellar solutions required several hours, apparently with a rate limiting step related to the assembly of 'nuclei' of bilayer structures. However, if a mineral surface in the form of clay particles was present, the surface in some way catalysed vesicle formation, reducing the time required from hours to a few minutes. The clay particles were spontaneously encapsulated in the vesicles. The authors further found that RNA bound to the clay was encapsulated as well. In a second series of experiments, Hanczyc and coworkers (Hanczyc *et al.*, 2003; Hanczyc and Szostak, 2004) showed that the myristoleic acid vesicles could be induced to grow by addition of fatty acid to the medium, presumably by incorporating fatty acid molecules into the membrane, rather than by fusion of vesicles. If the resulting suspension of large vesicles was then filtered through a polycarbonate filter having pores 0.2 μm in diameter, the larger vesicles underwent a kind of shear-induced division to produce smaller vesicles which could undergo further growth cycles (Figure 5.6). This remarkable series of experiments clearly demonstrated the relative simplicity by which complex system of lipid, genetic material and mineral catalysts can produce a model protocellular structure that can undergo a form of growth and division.

5.10 Environmental constraints on the first forms of life

Although self-assembly of amphiphilic molecules promotes the formation of complex molecular systems, the physical and chemical properties of an aqueous phase can significantly inhibit such processes, possibly constraining the environments in which cellular life first appeared. One such constraint is that temperature strongly influences the stability of vesicle membranes. It has been proposed that the last common ancestor, and even the first forms of life, were hyperthermophiles that developed in geothermal regions such as hydrothermal vents (Baross *et al.*, 1983) or deep subterranean hot aquifers (Pace, 1991). Such environments have the advantage of providing chemical energy in the form of redox potentials as well as abundant mineral surfaces to act as potential catalysts and adsorbants. However, because the intermolecular forces that stabilize self-assembled molecular systems are relatively weak, it is difficult to imagine how lipid bilayer membranes assembling from plausible prebiotic constituents would be stable under these conditions. All hyperthermophiles today have highly specialized lipid components, and it seems likely that these are the result of more recent adaptation than a molecular fossil of early life.

A second concern is related to the ionic composition of a marine environment. The high salt concentration of the present ocean has the potential to exert significant osmotic pressure on any closed membrane system. All marine organisms today have highly evolved membrane transport systems that allow them to maintain osmotic equilibrium against substantial salt gradients across their membranes. Furthermore, the concentrations of divalent cations, in particular Mg^{2+} and Ca^{2+}, were likely to exceed 10 mM in the early oceans, similar to their concentrations in seawater today. In the absence of oxygen, Fe^{2+} would also be present at millimolar concentrations, rather than the micromolar levels in contemporary seawater. At concentrations in the millimolar range, all such divalent cations have a strong tendency to bind to the anionic head groups of amphiphilic molecules. This causes aggregation and precipitation, strongly inhibiting the formation of stable membranes (Monnard *et al.*, 2002).

These considerations suggest that, from the perspective of membrane biophysics, the most plausible planetary environment for the origin of life would be at moderate temperature ranges ($<$ 60 °C), and the ionic content would correspond to low ionic strength and pH values near neutrality (pH 5–8) with divalent cations at submillimolar concentrations. This suggestion is in marked contrast to the view that life most likely began in a marine environment, perhaps even the extreme environment of a hydrothermal vent. One argument favouring a marine site for life's beginning is that fresh water would be rare on the early Earth. Even with today's extensive continental crust, fresh water only represents $\sim$1% of the contemporary

Earth's reservoir of liquid water. Another concern about a fresh water origin of life is that the lifetime of fresh water bodies tends to be short on a geological time scale. On the other hand, if sea water, with its high content of sodium chloride and divalent ions, markedly inhibits self-assembly processes and reactions that are essential to the emergence of cellular life, we may need to reconsider the assumption that life inevitably began in a marine environment. A more plausible site for the origin of cellular life may be a low ionic strength lacustrine environment such as a pond or lake. After the first form of cellular life was able to establish itself in a relatively benign environment, it would rapidly begin to adapt through Darwinian selection to more rigorous environments, including the extreme temperatures, salt concentrations and pH ranges that we associate with the limits of life on the Earth.

Acknowledgements

Portions of this chapter were adapted from a previous review (Deamer *et al.*, 2002).

References

Anders, E. (1989). Pre-biotic organic matter from comets and asteroids. *Nature*, **342**, 255–257.

Apel, C. L., Deamer, D. W., and Mautner, M. (2002). Self-assembled vesicles of monocarboxylic acids and alcohols: conditions for stability and for the encapsulation of biopolymers. *Biochim. Biophys. Acta*, **1559**, 1–10.

Bangham, A. D., Standish, M. M., and Watkins, J. C. (1965). Diffusion of univalent ions across the lamellae of swollen phospholids. *J. Mol. Biol.*, **13**, 238–252.

Baross, J. A., and Hoffman, S. E. (1983). Submarine hydrothermal vents and associated gradient environments as sites for the origin and evolution of life. *Origins of Life*, **15**, 327–335.

Bernstein, M. P., Sandford, S. A., Allamandola, L. J., Chang, S., and Scharberg, M. A. (1995). Organic compounds produced by photolysis of realistic interstellar and cometary ice analogs containing methanol. *Astrophys. J.*, **454**, 327–344.

Bernstein, M. P., Dworkin, J. P., Sandford, S. A., and Allamandola, L. J. (2001). Ultraviolet irradiation of naphthalene in H2O ice: Implications for meteorites and biogenesis. *Meteorit. Planet. Sci.*, **36**, 351.

Bernstein, M. P., Dworkin, J. P., Sandford, S. A., Cooper, G. W., and Allamandola, L. J. (2002). The formation of racemic amino acids by ultraviolet photolysis of interstellar ice analogs. *Nature*, **416**, 401–403.

Cello, J., Paul, A. V., and Wimmer, E. (2002). Chemical synthesis of poliovirus cDNA: generation of infectious virus in the absence of natural template. *Science*, **297**, 1016–1018.

Chakrabarti, A., and Deamer, D. W. (1994). Permeation of membranes by the neutral from fo amino acids and peptides: relevance to the origin of peptide translocation. *J. Mol. Evol.*, **39**, 1–5.

Chakrabarti, A., Joyce, G. F., Breaker, R. R., and Deamer, D. W. (1994). RNA synthesis by a liposome-encapsulated polymerase. *J. Mol. Evol.*, **39**, 555–559.

Chyba, C. F., and Sagan, C. (1992). Endogenous production, exogenous delivery and impact-shock synthesis of organic molecules: An inventory for the origin of life. *Nature*, **355**, 125–130.

Cronin J. R., Pizzarello S., and Cruikshank, D. P. (1988). In *Meteorites and the Early Solar System*, eds. J. F. Kerridge and M. S. Matthews. Tuscon: University of Arizona Press, p. 819.

Deamer, D. W., and Barchfeld, G. L. (1982). Encapsulation of macromolecules by lipid vesicles under simulated prebiotic conditions. *J. Mol. Evol.*, **18**, 203–206.

Deamer, D. W., and Oro, J. (1980). Role of lipids in prebiotic structures. *Biosystems*, **12**, 167–175.

Deamer, D. W., and Pashley, R. M. (1989). Amphiphilic components of carbonaceous meteorites. *Origins Life Evol. B.*, **19**, 21–33.

Deamer, D. W, Dworkin, J. P., Sandford, S. A., Bernstein, M. P., and Allamandola, L. J. (2002). The first cell membranes. *Astrobiology*, **2**, 371–382.

Delsemme, A. (1984). The cometary connection with prebiotic chemistry. *Origins of Life*, **14**, 51–60.

Dworkin, J. P., Deamer, D. W., Sandford, S. A., and Allamandola, L. J. (2001). Self-assembling amphiphilic molecules: synthesis in simulated interstellar/precometary ices. *Proc. Nat. Acad. Sci. USA*, **98**, 815–819.

Dyson, F. (1999). *The Origins of Life*. Princeton: Princeton University Press.

Ehrenfreund, P., and Charnley, S. B. (2000). Organic molecules in the interstellar medium, comets, and meteorites: A voyage from dark clouds to the early Earth. *Annu. Rev. Astron. Astr.*, **38**, 427–483.

Ehrenfreund, P., d'Hendecourt, L., Charnley, S. B., and Ruiterkamp, R. (2001). Energetic and thermal processing of interstellar ices. *J. Geophys. Res.*, **106**, 33291–33302.

Gerakines, P. A., Moore, M. H., and Hudson, R. L. (2000). Energetic processing of laboratory ice analogs: UV photolysis versus ion bombardment. *J. Geophys. Res.*, **106**, 3338.

Greenberg, M., and Mendoza-Gomez, C. X. (1993). Interstellar dust evolution: a reservoir of prebiotic molecules, in *The Chemistry of Life's Origins*, ed. M. Greenberg, C. X. Mendoza-Gomez and V. Pironella. Dordrecht: Academic Press, p. 1.

Hanczyc, M. M., and Szostak, J. W. (2004). Replicating vesicles as models of primitive cell growth and division. *Curr. Opin. Chem. Biol.*, **28**, 660–664.

Hanczyc M. M., Fujikawa S. M., and Szostak, J.W. (2003). Experimental models of primitive cellular compartments: encapsulation, growth, and division. *Science*, **302**, 618–622.

Holland, H. D. (1984). *The Chemical Evolution of the Atmosphere and Oceans*. Princeton, NJ: Princeton University Press.

Hoang, L., Fredrick, K., and Noller H. F. (2004). Creating ribosomes with an all-RNA 30S subunit P site. *Proc. Nat. Acad. Sci. USA*, **101**, 12439–12443.

Hutchison, C., Peterson, S., Gill, S., *et al.* (1999). Global transposon mutagenesis and a minimal Mycoplasma genome. *Science*, **286**, 2165–2169.

Ishikawa, K., Sato, K., Shima, Y., Urabe, I., and Yomo, T. (2004). Expression of a cascading genetic network within liposomes. *FEBS Letters*, **576**, 387.

Johnston, W. K., Unrau, P. J., Lawrence, M. S., Glasner, M. E., and Bartel, D. L. (2001). RNA-catalyzed RNA polymerization: accurate and general RNA-templated primer extension. *Science*, **292**, 1319–1325.

Kasting J. F. and Brown L. L. (1998). In *The Molecular Origins of Life*, ed. A. Brack. Cambridge: Cambridge University Press, p. 35.

Kauffman, S. (1993). *The Origin of Order. Self-Organization and Selection in Evolution.* New York: Oxford University Press.

Knauth, L. P. (2005). Temperature and salinity history of the Precambran ocean: implications for the course of microbial evolution. *Palaeogeogr. Palaeocl.*, **219**, 53–69.

Kvenvolden, K. A., Lawless, J. G., Pering, K., *et al.* (1970). Evidence for extraterrestrial amino-acids and hydrocarbons in the Murchison meteorite. *Nature*, **228**, 923.

Lawless, J. G., and Yuen, G. U. (1979). Quantitation of monocarboxylic acids in the Murchison carbonaceous meteorite. *Nature*, **282**, 396–398.

Love, S. G., and Brownlee, D. E. (1993). A direct measurement of the terrestrial mass accretion rate of cosmic dust. *Science*, **262**, 550–553.

Luisi, P. L. (1998). About various definitions of life. *Origins Life Evol. B.*, **28**, 613–622.

Maurette M. (1998). In *The Molecular Origins of Life*, ed. A. Brack. Cambridge: Cambridge University Press, p. 147.

McCollom, T. M., Ritter, G., and Simoneit, B. R. T. (1999). Lipid synthesis under hydrothermal conditions by Fischer-Tropsch-type reactions. *Origins of Life Evol. B.*, **29**, 153–166.

Miller, S. L. (1953). Production of amino acids under possible primitive Earth conditions. *Science*, **117**, 528–529.

Miller, S. L., and Urey, H. C. (1959). Organic compound synthesis on the primitive Earth. *Science*, **130**, 245–251.

Monnard, P.-A., and Deamer, D. W. (2001). Loading of DMPC based liposomes with nucleotide triphosphates by passive diffusion: A plausible model for nutrient uptake by the protocell. *Origins Life Evol. B.*, **31**, 147–155.

Monnard, P.-A., and Deamer, D. W. (2006). Models of primitive cellular life: polymerases and templates in liposomes. *Philos. T. Roy. Soc. B.*

Monnard, P.-A., Apel, C. L., Kanavarioti, A. and Deamer, D. W. (2002). Influence of ionic solutes on self-assembly and polymerization processes related to early forms of life: Implications for a prebiotic aqueous medium. *Astrobiology*, **2**, 139–152.

Morowitz, H. J. (1992). *Beginnings of Cellular Life*. New Haven: Yale University Press.

Morowitz, H. J., Heinz, B., and Deamer, D. W. (1988). The chemical logic of a minimum protocell. *Origins Life Evol. B.*, **18**, 281–7.

Munoz-Caro, G. M., Meierhenrich, U. J., Schutte, W. A. *et al.* (2002). Amino acids from ultraviolet irradiation of interstellar ice analogues. *Nature*, **416**, 403–406.

Noireaux, V., and Libchaber, A. (2004). A vesicle bioreactor as a step toward an artificial cell assembly. *Proc. Nat. Acad. Sci. USA*, **2**, 139–152., 17669–17674.

Nomura, S., Tsumoto, K., Hamada, T., *et al.* (2003). Gene expression within cell-sized lipid vesicles. *Chembiochem*, **4**, 1172–1175.

Oesterhelt, D. and Stoeckenius, W. (1973). Functions of a new photoreceptor membrane. *Proc. Nat. Acad. Sci. USA*, **70**, 2853–2857.

Oró, J. (1961). Comets and the formation of biochemical compounds on the primitive Earth. *Nature*, **190**, 389–390.

Pace, N. R. (1991). Origin of life – facing up to the physical setting. *Cell*, **65**, 531.

Paula, S, Volkov, A. G., Van Hoek, A. N., Haines, T.H., and Deamer, D.W. (1996). Permeation of protons, potassium ions, and small polar molecules through phospholipid bilayers as a function of membrane thickness. *Biophys. J.*, **70**, 339.

Racker, E., and Stoeckenius W. (1974). Reconstitution of purple membrane vesicles catalyzing light-driven proton uptake and adenosine triphosphate formation. *J. Biol. Chem.*, **249**, 662–663.

Rasmussen, S., Chen, L., Deamer, D., *et al.* (2004). Evolution. Transitions from nonliving to living matter. *Science*, **303**, 963–966.

Rushdi, A. I. and B. Simoneit. (2001). Lipid formation by aqueous Fischer–Tropsch type synthesis over a temperature range of 100 to 400 °C. *Origins Life Evol. B.*, **31**, 103–118.

Sandford, S. A. (1996). The inventory of interstellar materials available for the formation of the Solar System. *Meteorit. Planet. Sci.*, **31**, 449–476.

Sandford, S. A., Bernstein, M. P., and Dworkin, J. P. (2001). Assessment of the interstellar processes leading to deuterium enrichment in meteoritic organics. *Meteorit. Planet. Sci.*, **36**, 1117–1133.

Shew, R., and Deamer, D. W. (1985). A novel method for encapsulation of macromolecules in liposomes. *Biochim. Biophys. Acta*, **816**, 1–8.

Singer, S. J., and Nicolson, G. L. (1972). The fluid mosaic model. *Science*, **175**, 720–727.

Szostak, J. W., Bartel, D. P., and Luisi, P. L. (2001). Synthesizing life. *Nature*, **409**, 387–390.

Wachtershäuser, G. (1988). Before enzymes and templates: Theory of surface metabolism. *Microbiol. Rev.*, **52**, 452–484.

Walde, P., Wick, R., Fresta, M., Mangone, A., and Luisi, P. L. (1994). Autopoietic self-reproduction of fatty acid vesicles. *J. Am. Chem. Soc.*, **116**, 11649.

Walde, P., Goto, A., Monnard, P.-A., Wessicken, M., and Luisi, P.-A. (1994). Oparin's reactions revisited: enzymatic synthesis of poly(adenylic acid) in micelles and self-reproducing vesicles. *J. Am. Chem. Soc.*, **116**, 7541–7547.

Yu, W., Sato, K., Wakabayashi, M., *et al.* (2001). Synthesis of functional protein in liposome. *J. Biosci. Bioeng.*, **92**, 590.

Part II

Life on Earth

6

Extremophiles: defining the envelope for the search for life in the universe

Lynn Rothschild
NASA Ames Research Center

6.1 Introduction

Of the three big questions astrobiology addresses, none has captured the public imagination as much as the question, 'Are we alone?' As knowledge increases about the physical environments available in the universe, organisms on the Earth are used to gauge the minimum envelope for life. Organisms that live on the physical and chemical (and perhaps, biological) boundaries are called 'extremophiles'. Members of the domain Archaea are undoubtedly the high-temperature champions on Earth, surviving at temperatures far above the boiling point of water. However, a wide phylogenetic variety of organisms are able to inhabit other extremes, from low temperature to high salinity, from desiccation to high levels of radiation. In some cases, the adaptation is as simple as keeping ions out of the cell (low pH), but in others more profound adaptations are required. Reactive oxygen species are highly toxic, and thus organisms that are aerobic are arguably extremophilic. Thus, extraterrestrial habitats previously thought to be uninhabitable have been shown to be, at least theoretically, habitable, thus informing the search for life elsewhere.

Extremophilic organisms, or 'extremophiles', are integral to understanding the three big questions in astrobiology: Where do we come from? Where are we going? Are we alone? We do not know the environmental conditions under which life arose, although speculation includes hydrothermal areas; nor do we know the environments in which the earliest life forms existed. The future study of life on this planet and elsewhere is likely to include extremophiles, as our nearest neighbours present environments that are, in many respects, extreme. Herein, I focus on the importance of extremophiles for defining the boundary conditions for life. The focus will involve exploring the diversity of extremophiles, mechanisms for adaptation, examples of extreme ecosystems, and space as an extreme environment.

Although this chapter contains examples of organisms that live in extreme environments on Earth that overlap with environmental conditions found elsewhere,

Planetary Systems and the Origins of Life, eds. Ralph E. Pudritz, Paul G. Higgs, and Jonathon R. Stone.
Published by Cambridge University Press.

this does not entail these organisms will be found elsewhere. In fact, if they were, then their presence may be considered *prima facie* evidence of contamination, as it is extraordinarily unlikely that identical organisms would be found on other celestial bodies any other way. The critical point is that by studying extremophiles on Earth, currently the only source of living organisms, we can define an envelope for life in the Universe. We must be aware that this provides the minimum envelope for life. The envelope is likely to be extended as we discover new organisms, and life on Earth may not be able to or have had the opportunity to test all possible environmental habitats.

6.2 What is an extremophile?

The word 'extremophile', coined by Bob MacElroy (1974) from the Latin 'extremus' and the Greek 'philos', literally means 'lover of extremes'. Unfortunately, even the definition is not as clear as one might think. Extreme is usually taken to refer to physical, chemical, or, rarely, biological conditions. But what is extreme?

There is an objective definition of extreme, owing to the fact that life on Earth is based on organic carbon with water as a solvent. Thus, *conditions that disrupt the integrity or function of aqueous solutions of organic compounds are extreme.*

Must extremophiles 'love' extreme conditions? Scientists tend to be inconsistent; when it comes to temperature, extremophiles thrive optimally at the extreme temperatures. However, for many extremes, from pH to radiation, the extremophile organisms grow well, or even grow better, under less-extreme conditions.

The ability to thrive under extreme conditions may be limited to particular life stages. For example, some species of frogs, turtles, and snakes can tolerate freezing of extracellular water, but only after the production of appropriate antifreeze molecules in the autumn (Storey and Storey, 1996).

Which taxa contain extremophiles? With attention focused on temperature and salt tolerance, biologists tend to equate Archaea with extremophiles. This is inaccurate, as many extremophiles are Eubacteria and Eukarya, and additionally an increasing number of non-extremophilic Archaea are being uncovered.

Reasons for studying extremophiles are listed in Table 6.1. Only the last five are primarily or exclusively the territory of astrobiologists: the future use of extremophiles in space – perhaps for life support, perhaps for terraforming; the biodiversity of life on Earth, mechanisms to survive extreme environments, and the limits for life in the universe.

It is interesting to note that extremophiles have been used to look for life in the Universe not as a recent construct of astrobiology, but rather as a nineteenth-century suggestion. Richard Proctor, a popular science writer for the public, was

Table 6.1. *Reasons to study extremophiles*

Reason to study extremophiles	Important to astrobiology
food preservation	no
basic research; model organisms for basic research	no
biological warfare	no
biotech potential	no
future use in space	yes
biodiversity on Earth	yes
origin of life	yes
mechanisms of extremophile survival	yes
limits for life in the universe	yes

probably the first to connect the study of life in extreme environments and life on other planets (Proctor, 1870, Chapter 1):

> If we range over the [E]arth, from the arctic regions to the torrid zone, we find that none of the peculiarities which mark the several regions of our globe suffice to banish life from its surface.

While this chapter will focus almost exclusively on microbes, even humans inhabit extreme environments, including outer space (for a fascinating review of human adaptations to extreme environments, see Ashcroft (2001)).

6.3 Categories of extremophiles

The challenges posed by extreme environments range from physical factors such as temperature and pressure to chemical extremes such as pH and oxygen tension (reviewed in Rothschild and Mancinelli (2001); Table 6.2).

6.4 Environmental extremes

6.4.1 Temperature

Why should temperature make a difference to organisms? Organisms rely on biomolecules, all of which will break down at some temperature. For example, chlorophyll degrades near 75 °C (Rothschild and Mancinelli, 2001). The melting temperature of nucleic acids depends on the base composition (A–T bonds are less thermostable than are G–C bonds), the length of the polymer, and the solvent including the concentrations of ions in solution, but even DNA normally will degrade well below 100 °C.

Table 6.2. *Examples of natural habitats with environmental extremes*

Environmental parameter	Type	Definition	Example ecosystems	Example organisms
temperature	hyper-thermophile	growth > 80 °C	geyser	*Pyrolobus fumarii*, 113 °C
	thermophile	growth 60–80 °C	hotspring	
	mesophile	15–60 °C		*Homo sapiens*
	psychrophile	<15 °C	ice, snow	*Psychrobacter*, some insects
pH	alkaliphile	pH > 9	soda lakes	*Natronobacterium*, *Bacillus firmus OF4*, *Spirulina spp.* (all pH 10.5)
	acidophile	low pH loving	acid mine drainoff, hotsprings	*Cyanidium caldarium*, *Ferroplasma sp.* (both pH 0)
salinity	halophile	salt loving (2–5 M NaCl)	salt lakes, salt mines	*Halobacteriacea*, *Dunaliella salina*
desiccation	xerophiles	anhydrobiotic	desert	*Artemia salina*; nematodes, microbes, fungi, lichens
pressure	barophile piezophile	weight loving, pressure loving	deep ocean	unknown, for microbe, 130 MPa
radiation		tolerates high levels of radiation	reactors, high solar exposure, e.g., at altitude	*Deinococcus radiodurans*
oxygen	aerobe	tolerates oxygen	most of Earth today	
gravity	hypergravity	> 1 *g*		none known
	hypogravity	< 1 *g*		none known
vacuum		tolerates vacuum (space devoid of matter)		tardigrades, insects, microbes, seeds
chemical extremes	gases, metals	can tolerate high concentrations of metal (metalotolerant)	mine drainage	*Cyanidium caldarium* (pure CO_2), *Ferroplasma acidarmanus* (Cu, As, Cd, Zn); *Ralstonia* sp. CH34 (Zn, Co, Cd, Hg, Pb)

As temperature increases, proteins denature, which results in loss of quaternary, tertiary, and even secondary structures. This rules out their use as structural compounds.

Denatured enzymes lose their catalytic activity. But even when active, enzymes have a minimum, a maximum and an optimum temperature for function. By definition, at temperatures above or below the maximum, enzymatic function slows.

The solubility of gases in a liquid, including water, varies depending on temperature. As anyone who has opened a cold bottle of soda knows, the solubility of gases in liquids decreases with increasing temperature. As aquatic organisms often utilize soluble gases, such as CO_2 or O_2, high temperature decreases the availability of gases.

At low temperatures, water freezes. During freezing, solutes concentrate creating high, perhaps toxic, concentrations of solutes. Once the water is frozen, the ice crystals may act like tiny swords, piercing the cell membranes, and thus destroying cellular integrity.

These effects of temperature on biological compounds strongly suggest that life should have a fairly narrow temperature range, from above the freezing point of water to well below its boiling point. Thus, it was surprising to Tom Brock when, in July 1964, he detected living organisms in the run-off channels of hot springs in Yellowstone National Park. Later in the summer he noticed pink, gelatinous mats in Pool A of Octopus Spring, about six miles south of Old Faithful (Figure 6.1). At the time 'high temperature' was considered to be 60 °C. In 1966, Brock's graduate student, Hudson Freeze, cultured YT-1 (later known as *Thermus aquaticus*) from Mushroom Spring. *T. aquaticus* is the source for Taq polymerase which, when used as part of the polymerase chain reaction, revolutionized biology in the late twentieth century. Inspired by reports of this work (Brock, 1967), as well as the work of Woese (e.g., Woese and Fox (1977)) revealing the presence of the Archaea, Karl Stetter became the leading figure in thermophile research (see Chapter 7).

In 1977, the exploratory submarine *Alvin* travelled through the Panama Canal for the first time. Geological work in the Galapagos Rift was completed during February and March. The major discovery of an abundance of exotic animal life on and in the immediate proximity of warm water vents prompted theories about the generation of life. Because no light can penetrate through the deep waters, scientists concluded that the animal chemistry there is based on chemosynthesis, not photosynthesis. Since then, *Alvin* has located more than 24 hydrothermal sites in the Atlantic and Pacific Oceans. It also has allowed researchers to find and record about 300 new species of animals, including bacteria, foot-long clams and mussels, tiny shrimp, arthropods, and red-tipped tube worms that can grow up to 3 m long in some vents.

Fig. 6.1. Sources of thermophiles from Yellowstone National Park. Outflow channel of Octopus Spring, approximately 6 miles south of Old Faithful. Inset is close-up of the gelatinous (pink) mats, which contain the hyperthermophilic bacterium, *Thermocrinis ruber* (Reysenbach *et al.*, 1994; Huber *et al.*, 1998). The mats are growing at approximately 81 °C. Note that Octopus is 7352 ft above sea-level, and therefore the boiling point of water is depressed. In the front of the photo is a section of the source, which is boiling, yet it too contains living hyperthermophiles. Photo captured 3 August 2006.

The ability to withstand fire is not normally considered an extreme-environment adaption, but it is worth mention here. Fire-triggered germination is widespread in the plant families Fabaceae, Rhamnaceae, Convolvulaceae, Malvaceae, Cistaceae, and Sterculiaceae. The trigger is either heat shock or the chemical products of combustion, rather than fire itself (Keeley and Fotheringham, 1998).

The ability of organisms to thrive in cold temperature environments has a much longer history. The great nineteenth-century microbiologist, Ferdinand Cohn, noticed algae that grow on snow, a phenomenon often referred to as 'watermelon snow' (Figure 6.2). Cold temperature organisms, or psychrophiles, were studied from the marine environment by Certes in the 1980s and Conn in the early twentieth century. ZoBell and his students – particularly Dick Morita – studied these organisms and were the first to get deep-sea samples from the 1957 *Galathea* expeditions to ocean trenches. They were the first to study these psychrophiles and barophiles. By the 1980s, multiple field campaigns were being staged to the Dry Valleys of the Antarctic to study the microbes in the permanently ice-covered lakes (Figure 6.3). These lakes were seen as a model system for

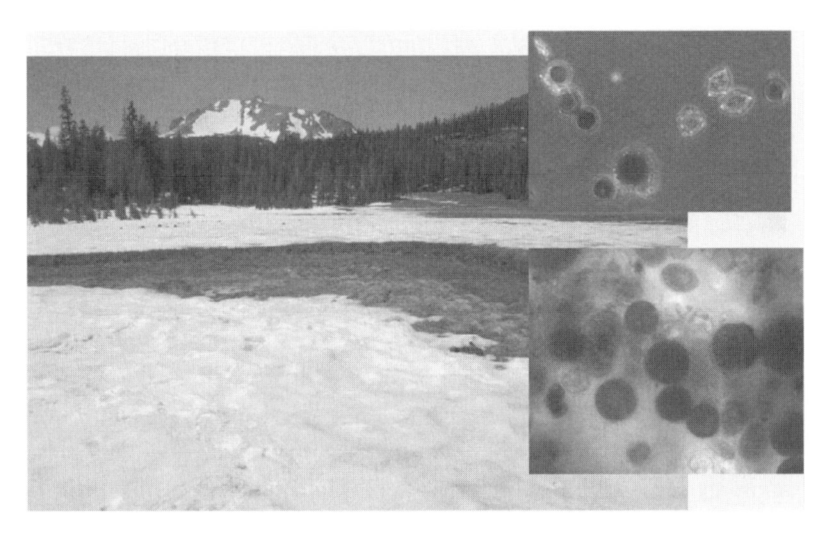

Fig. 6.2. Watermelon snow, Lassen National Park.

Fig. 6.3. Nymph Creek, an acidic creek several miles north of Norris Annex. At the time the photo was taken, the water was 41 °C. The water appears green due to *Cyanidium caldarium*, and an acidophilic *Chlorella* (Belly *et al.*, 1973). Photo captured 3 August 2006. Color photographs are available on the book website http://www.cambridge.org/9780521875486.

potential organisms living in rocks (e.g., Friedmann (1982), McKay and Friedmann (1988)), under ice-covered regions of ancient Mars, or even present-day Europa and now Enceladus (reviewed in Priscu *et al.* (1998), McKay *et al.* (2005); see Chapter 12).

6.4.2 pH limits for life

The acidity of a solution, referred to as pH, quantifies the activity of hydrogen ions. In dilute solutions, the activity is approximately equal to the numeric value of the concentration of the H^+ ion, denoted as $[H^+]$. Thus,

$$pH \sim -\log_{10} |[H^+]| \tag{6.1}$$

Seawater (currently) has a pH of 8.2. This means that it has a $[H^+]$ of $10^{-8.2}$ mol l^{-1}, or approximately 6.31×10^{-9} mol l^{-1}. While the pH scale is thought of as ranging from 0 to 14, a pH below 0 or above 14 is possible. For example, acid mine run-off can have a pH of -3.6, corresponding to a molar concentration of H^+ of 3981 M. The cytoplasm of most cells is near neutral, which is pH 7.

Life at pH extremes is difficult. At low pH, in the absence of a proton pump, hydrogen ions can rush into the cell, lowering the pH of the internal environment, thereby disrupting the internal biochemistry of the cell. At extremely low pH, proteins denature, which is the principle behind preservation at extremely low pH. Organisms that live at high pH and use proton pumps find it difficult to locate enough protons.

Yet, in spite of this, eukaryotic and prokaryotic microbes from low pH environments have been observed for several decades. The red alga, *Cyanidium caldarium* (Fig. 6.3), can live at pH $\sim$ 0 (Seckbach, 1999). The Archaea *Picrophilus oshimae* and *Picrophilus torridus* live in dry soils below pH 0.5 (Schleper *et al.*, 1995); and *Ferroplasma acidarmanus*, isolated from acid mine drainage, can grow at pH 0 (Edwards *et al.*, 2000). A diversity of organisms, from bacteria to cyanobacteria to rotifers, can live at pH 10.5 (e.g., Martins *et al.* (2001)). Microbial communities live at pH 12.9 in the soda lakes of Maqarin, Jordan (Pedersen *et al.*, 2004). It should be noted that soda lakes, in general, tend to be highly productive, with high concentrations of cyanobacteria because of the high concentration of dissolved CO_2 (Grant *et al.*, 1990; Rees *et al.*, 2004).

Work on life at low pH started because of the use of acidic conditions for food preservation (e.g., sauerkraut), but more recently low pH organisms have been useful as model systems for biochemistry research. In the late 1960s in particular (e.g., Seckbach *et al.* (1970)), but still to the present day, acidophiles have been used as model systems for a potential life form on Venus (e.g., Schulze-Makuch *et al.* (2004) and references therein).

The biotechnical potential also is enormous, as there are times in industrial processes where it is most economical to conduct the reactions at either high or low pH. An example of the economic significance of these organisms comes from the Kenya Wildlife Service (KWS), who planned to launch a multimillion dollar legal claim against Genencor and Procter & Gamble, alleging that a microbial cellulase enzyme illegally obtained from a soda lake in the country was used as an ingredient in the latter company's Tide laundry detergent (Sheridan, 2004).

6.4.3 Life in high salt

Red waters have been noted since biblical times, and Charles Darwin observed red ponds in South America during his voyage on HMS *Beagle*. But halophiles, organisms that live in 2–5 M salt, were first noted in the early twentieth century in conjunction with spoilage of food preserved in salt and yet red. In preparation for the Viking missions to Mars, halophiles became of importance in the search for life elsewhere. More recently, the discovery of evaporites on Mars and life in evaporites on Earth, and the presumed radiation resistance of halophilic organisms have also contributed to the astrobiological interest in halophiles (e.g., Rothschild (1990), Rothschild *et al.* (1994), Forsythe and Zimbelman (1995), McLennan *et al.* (2005)).

Halophiles are predominantly found among the Archaea and Bacteria. One eukaryote, the green alga *Dunaliella salina*, is well known from a variety of hypersaline environments. *D. salina* has been particularly useful commercially because the glycerol and β-carotene that it accumulates can be harvested. Halophiles are also desiccation resistant.

6.4.4 Desiccation (anhydrobiosis)

Desiccation is one of the oldest methods of food preservation. In ancient times, the sun and wind would have dried foods naturally, as they do with grapes on the vine producing raisins. There is evidence that the Middle Eastern cultures actively dried food in the sun as early as 12 000 BCE. The Romans in particular were fond of any dried fruit, but meats, especially pork, and cereal grains such as rice and wheat, have been dried or smoked for millennia. Desiccation preserves because all life on Earth relies heavily on the presence of liquid water, although it is possible that some lichens can survive on water vapour.

So, as with the other extremophiles that we have considered here, many organisms are extremely good at tolerating this extreme (Crowe *et al.*, 1992). Many microbes and plants use desiccation-resistant spores to weather dry periods and for dispersal. A variety of invertebrates, including nematodes and rotifers, are anhydrobiotic (Watanabe, 2006), and some, such as tardigrades, are known to survive desiccation for well over 100 years dried on herbarium samples.

Organisms that can survive extreme dehydration are also resistant to freezing to very low temperatures, elevated temperature for brief periods, and the effects of ionizing radiation (Crowe and Crowe, 1992).

6.4.5 Pressure

Organisms can live at extraordinarily high pressures, as evidenced by the lush microbial and invertebrate communities in the deep sea. Some of these organisms can also survive normal atmospheric pressure, while others cannot. Interestingly, the surface bacteria *Escherichia coli* and *Shewanella oneidensis* have been shown to be metabolically active up to 1600 MPa, substantially higher than ambient pressure (0.1 MPa) (Sharma *et al.*, 2002).

6.4.6 Radiation

Organisms are exposed to a variety of types of radiation on Earth. Some types of radiation have been constant for organisms over geological time, such as those from naturally occurring radioactive substances, while ultraviolet radiation received by organisms has varied as the Sun has aged and atmospheric oxygen has risen. Solar radiation exposure also varies seasonally and diurnally. Throughout the history of Earth, ultraviolet radiation below 200 nm has been blocked from reaching the surface by atmospheric CO_2. Since the rise of atmospheric oxygen and ozone over 2 billion years ago (see Chapter 11), much of the radiation below 300 nm has been attenuated as well. This is particularly important because the peak absorption for nucleic acids is $\sim$260 nm and that for proteins is, on average, $\sim$280 nm.

More recently, life has been exposed to radiation from man-made sources, such as nuclear reactors. And, of course, with the advent of space exploration, organisms from Earth have gone beyond its protective atmosphere to low orbit, the Moon, and, now, beyond the edge of our Solar System as hitchhikers on the Voyager spacecraft.

Radioactivity, a term coined by Marie Curie, was found to be hazardous to living organisms very quickly after the discovery of X-rays by Wilhelm Roentgen in 1895. Radiation damages nucleic acids, proteins, and lipids directly and indirectly through the production of reactive oxygen radicals. While the types of radiation vary, the types of damage are similar, ranging from modified bases to single-strand and double-strand breaks. With oxidative damage, oxidation of lipids and proteins can also occur. Thus, in 1956, when a can of γ radiation-sterilized meat surprisingly spoiled, Arthur W. Anderson at Oregon Agricultural Experiment Station in Corvallis, investigated and found the culprit: the aerobic bacterium *Deinococcus radiodurans*.

D. radiodurans is not just phenomenally radiation-resistant, but it is also desiccation-resistant. This has led to the hypothesis that the radiation resistance is incidental to desiccation resistance (Battista, 1997). Certainly organisms that are desiccated are not able to repair DNA damage; so, upon rehydration, there is pressure to repair the accumulated damage rapidly.

6.4.7 High oxygen

High oxygen seems an odd addition to a list of extreme parameters; are we not exquisitely adapted to living in 21% oxygen? In fact, oxygen levels on Earth are thought to have been as high as 35% during the Permo-Carboniferous, 248–354 million years ago (Berner *et al.*, 2003), deep into the age of multicellular life. Although we consider an aerobic lifestyle to be 'normal', I will argue that, in reality, oxygen presents one of the most severe challenges to organisms based on organic carbon, concluding that we, like all aerobes, are extremophiles.

Aerobic metabolism arose and became dominant because it is far more efficient than anaerobic metabolism, generating 18 times more ATP per molecule of glucose metabolized. However, during the reduction of oxygen to water, or the oxidation of water, a series of reactive oxygen species is produced, the worst of which, for a biological system, is the hydroxyl radical (·OH). These reactions, from oxygen to the superoxide anion (O_2^-), hydrogen peroxide (H_2O_2), the hydroxyl radical (·OH), and H_2O, are shown in Equation (6.2).

$$O_2 \leftrightarrow O_2^- \leftrightarrow H_2O_2 \leftrightarrow OH \leftrightarrow H_2O \tag{6.2}$$

Oxidative damage to DNA is well known and includes modification of bases to single-strand breaks to double-strand breaks. Reactive oxygen species also cause damage to lipids and proteins. Current research suggests that the increased production of reactive oxygen species is a feature of most, if not all, human disease, including cardiovascular disease and cancer (Jacob and Burri, 1996).

6.5 How do they do it?

Sometimes, evolution finds the route of the least number of changes, which I dub 'the laziness principle'. If the internal environment of a cell alters radically, there would have to be a massive reworking of the internal environment necessitating massive changes in the biomolecules and possibly pathways.

For these reasons, the simplest approach to living in extreme environments is to keep the external environment out. This is what we do by wearing clothes, mammals do by growing fur, and so on. Living under protective materials, such as sand or salt, is another strategy (Rothschild and Giver, 2003). A similar approach is taken to living at low pH. Acidophiles maintain a roughly neutral internal pH by evolving a strong proton pump or having low permeability to protons. Conversely, alkaliphiles (organisms that live at high pH) maintain an internal pH two or more units below the external medium by evolving an effective proton transport system.

As with pH, a variety of organisms can withstand anhydrobiosis, suggesting that this adaptation is relatively straightforward to evolve. Usually the synthesis of a disaccharide such as trehalose in microbes, animals, and fungi, or sucrose in plants, is sufficient to survive desiccation (Crowe *et al.*, 1992). Trehalose may provide effective protection against desiccation because it has superior biochemical and physicochemical properties for stabilizing membranes and biomolecules including proteins and lipids (Watanabe, 2006) by hydrogen bonding to polar residues in the dry macromolecular assemblages (Crowe *et al.*, 1998). However, both trehalose and sucrose form glasses in the dry state; so it has been suggested that glass formation (vitrification) is in itself sufficient to stabilize dry biomolecules (Sun *et al.*, 1998). Crowe *et al.* (1998) have shown that, while vitrification is required, it is not in itself sufficient to allow anhydrobiosis.

A further complication arises because anhydrobiosis occurs in the absence of trehalose or other disaccharides in bdelloid rotifers (Lapinski and Tunnacliffe, 2003). Thus, there may be alternative pathways to desiccation resistance.

Similarly, halophiles often are able to cope by the accumulation of a compatible solute. The internal osmotica include K^+ for Archaea, glycine betaine and organic compounds in Bacteria, and glycerol in the eukaryote *Dunaliella salina*.

If the external environment cannot be kept out, sometimes small modifications can be made to the internal environment. For example, enzyme activities can adapt to lower temperatures, and the lipid composition of the membrane can be altered to avoid loss of fluidity at low temperatures. At temperatures below freezing, ice forms, which is doubly dangerous. First, as the ice forms, it removes water and increases the concentration of solutes in the remaining water. Second, ice crystals themselves can cause physical damage to cell membranes. Antifreeze molecule(s) can prevent this damage. For example, the accumulation of glycerol in frogs allows natural tolerance to freezing (Schmid, 1982), as do the antifreeze proteins in fish provide the same function (e.g., Davies and Hew (1990), Clarke (2003)).

High temperature is normally lethal for a number of reasons, ranging from lowered solubility of gases at high temperature, to denaturation of proteins and nucleic acids. High temperature can be avoided by use of cold water circulation over the exterior, such as in the pompei worm. However, hyperthermophiles are required to make multiple changes to their internal biochemistry for survival. Salt bridges have long been known to stabilize proteins in thermophiles (Perutz, 1978). With increasing temperature, prokaryotes tend to acquire adenine and lose cytosine, while keeping thymine and guanosine relatively constant (Lambros *et al.*, 2003). Because of the economic interest in hyperthermophiles, this is an active area of interest.

Resistance to radiation damage, and the related phenomenon of oxidative damage, is critical for life. Thus, multiple pathways exist in all organisms (reviewed extensively elsewhere; e.g., Yasui and McCready (1998), Petit and Sancar (1999), Smith (2004)). These include photorepair, excision repair, and homologous recombinational repair, among others. However, the key to the remarkable ability of *D. radiodurans* to resist radiation might be the fact that the DNA is held in a ring-like structure, thus enabling the cell to ligate the broken strands in the correct order (Levin-Zaidman *et al.*, 2003), although its ability to accumulate manganese while keeping a low intracellular level of iron also may be a contributing factor to resistance (Daly *et al.*, 2004).

6.6 Examples of extreme ecosystems

Even excluding our nearly ubiquitous high oxygen terrestrial environment, there is a remarkable diversity of extreme habitats on Earth (Table 6.2; Figures 6.1–6.4). These range from the ice at the poles to the searing heat of hydrothermal vents. Paralana Spring, in the Flinders Ranges of Australia, bubbles radioactivity (Figure 6.4). As suggested above, many of these have multiple extreme environments simultaneously. Thus, while the source of Octopus Springs in Yellowstone National Park is at the boilina pH ~8.2, a pH similar to that of the ocean, Congress Springs also is boiling but with a pH of near 0. Laguna Colorada, a hypersaline lake in the Bolivian Altiplano, has a moderately alkaline pH of 8.4, but is hypersaline and, thus, populated by halophiles. At an altitude of 4364 m, and a location of 22 °15.836′ S; 67 °48.970′ W, the organisms are also subjected

Fig. 6.4. Paralana Spring, a natural spring containing high levels of radiation. Despite high levels of radon, radium, and uranium, not to mention a high ultraviolet flux, there is a lush microbial community growing in the source of Paralana Spring, shown here with bubbles of radon.

to exceptionally high levels of UVB radiation in the austral summer (Rothschild, unpublished).

An extreme environment that is poorly studied is that of the air. An air-borne organism must deal with the effects of desiccation, decreasing temperature, ultraviolet radiation, and a nutrient-poor environment. Of course, with increasing elevation, oxygen decreases as well. The great microbiologist Louis Pasteur did some of the early work on aerobiology, but current work is conducted primarily for military reasons, to investigate spread of air-borne pathogens, and to preserve monuments.

While many extreme environments occur naturally, there are new, human-made environments as well. In particular, nuclear reactors (high radiation) and acid mine drainage (high metals, low pH) can provide novel challenges.

6.7 Space: new categories of extreme environments

Assuming that all organisms on Earth are indigenous to this planet, outer space provides new extreme environments. Life beyond Earth is difficult for at least six reasons.

(i) Atmospheric composition is likely to be grossly different.
(ii) Gravity can range from nearly zero to lower than on Earth on other rocky planets (although higher on the gas giants).
(iii) Space itself is almost a vacuum.
(iv) Temperature extremes occur even in our planetary neighbourhood.
(v) Nutrient sources are different and may well be non-existent.
(vi) The radiation regimes are grossly different.

Cosmic and solar particle radiation are non-existent on Earth, but quickly add to the woes of increased ultraviolet flux beyond Earth's atmosphere and magnetosphere.

A series of experiments has been conducted, primarily by the Europeans under the aegis of the ESA, to test the ability of terrestrial organisms to withstand the rigours of space. The first studies were done on balloons and rockets, and, later, on spacecraft in low Earth orbit and outside the Earth's magnetic field (Apollo missions 16 and 17). The low Earth orbit experiments have been conducted on a variety of platforms including Spacelab 1, Spacelab D2, the Exobiology Radiation Assembly (ERA) on the European Retrievable Carrier (EURECA), NASA's Long Duration Exposure-Facility (LDEF), and Biopan on the Russian satellite Foton.

From 1994 to 2006, 25 exobiology experiments were performed on Biopan, an ESA space-exposure platform on a satellite, allowing two-week flight opportunities (Schulte *et al.*, 2006). Samples were exposed to the rigours of outer space vacuum and the resulting desiccation and radiation regime. Plans are finalized for a similar exposure facility to be attached to the outside of the International Space Station in 2007. The plan is that EXPOSE will allow one-year experiments. Isolated short duration experiments continue (e.g., Saffary *et al.* (2002)).

Early on Horneck showed that the spores of the bacterium *Bacillus subtilis* could survive such exposure (e.g., Horneck *et al.* (1984), Horneck (1993, 1999)). Since then, bacterial spores mixed with Mars soil analogues, permafrost soil samples with their embedded bacterial spores, and viruses also have survived. But, whereas spores are expected to be particularly resistant to environmental extremes (e.g., Nicholson *et al.* (2000)), Mancinelli *et al.* (1998) were the first to show survival of cells in the vegetative state. Since then, lichens and yeast have also been shown to survive space travel.

6.8 Life in the Solar System?

While planet Earth is the only body in the Solar System, indeed our Universe, that we know to be infested with life, there are several other possible habitats. Assuming that liquid water is a necessity (see Benner *et al.* (2004) and Shulze-Makuch and Irwin (2004) for a discussion of this and alternatives), there are two planets (Venus and Mars), and several moons (e.g., Europa and Enceladus) that currently are the most-likely candidates to harbour life (see Chapter 5). If one assumes that an organic solvent, such as methane alone or in combination with water, could be a solvent for life, then the search extends to Titan.

6.8.1 Venus

While Proctor and others have discussed the possibility of an extant Venusian biosphere, most people believe Venus to be sterile today. The surface of Venus is the hottest in the Solar System at 477 °C, so there is no liquid water. This was caused by a run-away greenhouse effect where the carbon-dioxide-rich atmosphere (96% CO_2 with 3% N_2) trapped incident sunlight raising temperatures. Additionally, clouds of sulphuric acid abound and the atmospheric pressure is over 90 times that of Earth at sea level. The information we have on the surface of Venus was obtained from the Russian Venera landers from the 1980s, which survived on the surface for about 60 minutes. Additional information has been gained from NASA's Magellan spacecraft, the Russian Venera orbiters and the ESA's Venus Express mission, which entered Venusian orbit on 11 April 2006.

In many ways, Venus began as Earth's close sister, perhaps even 'twin'. Although Venus is approximately 30% closer to the Sun than the Earth, it is similar in size, mass, density, and volume. Venus coalesced out of the same materials as the Earth and should have a similar bombardment history, so it is likely to have been similar when the planets formed. Since life arose on Earth by 3.8 billion years ago, and quite possibly much earlier, it is reasonable to assume that life could have arisen on Venus during that time as well. And, if life is not indigenous to Venus, transport from Earth or Mars during the time may have been possible (Gladman *et al.*, 2005). Unfortunately, reworking of the Venusian surface means that it is unlikely that we will ever know for certain.

First Sagan (1961) and more recently Cockell (1999), Grinspoon and Schulze-Makuch and others (e.g., Schulze-Makuch *et al.* (2004)) have envisioned a scenario where life could still be present on Venus. While the surface pressure and concentration of CO_2, though high, does not prohibit life, the high temperature does. However, in the lower and middle cloud layers of Venus, the temperatures are lower and water vapour is available, and thus these clouds could provide a refuge for life. Perhaps acidophilic sulphate-reducing chemoautotrophs suspended as aerosols could survive there even today. Ultraviolet radiation flux would be high, but Schulze-Makuch *et al.* (2004) have suggested a way that the flux could be attenuated allowing life to survive.

6.8.2 Mars

Mars is half again as far from the Sun as the Earth. Its mass is 0.12 that of the Earth, so there is one-sixth the gravity on the surface. As is the case for Venus, early Mars was similar to early Earth, and therefore if life is inevitable, or even likely, under early Earth conditions, it ought to have arisen on Mars as well. Unlike Venus, or even Earth, preservation of early fossils is possible, and, thus, Mars is a tantalizing target for palaeontologists.

The interest in extant life on Mars continued well into the twentieth century, inspired in part by the discovery of canali on Mars in 1877 by Giovanni Schiaparelli in Milan (see Zahnle (2001) for the history of nineteenth- and twentieth-century views of the habitability of Mars). It was NASA's Mariner IV mission, which gave the first close-up glimpse of the Martian surface, that finally destroyed all hope of a civilization on Mars or even extensive vegetation. Mariner IV arrived at Mars on 14 July 1965 and returned photos showing a heavily cratered barren landscape. In addition, the thinness of the atmosphere was realized.

The 1960s ended on a different note with Mariner 9's superior photos revealing gigantic volcanoes and a grand canyon stretching 4800 km across its surface. More surprisingly, the relics of ancient riverbeds were carved in the landscape of this seemingly dry and dusty planet. Thus, attention turned to the possibility of microbial life on Mars, even extant life. NASA's Viking landers arrived at Mars in 1976 with a variety of life detection experiments: a camera and a gas chromatograph/mass spectrometer (GC-MS) to detect organic compounds, and three metabolic experiments: labelled release, pyrolytic release, and gas exchange. While the approach was clever and the performance was excellent, unambiguous signs of life were not detected. Worse, organic carbon was not detected by the GC-MS. Lack of surface water and strong surface oxidants were suspected to be the culprits. Attention in the exobiology community turned to the possibility of finding signs of extinct life.

More recently, the possibility of extant microbial life has been revived. And, whether the Mars meteorite ALH84001 indeed harbours an extinct Martian biota (McKay *et al.*, 1996) or has only whetted our appetite for further studies, it remains one of the most exciting science stories of the twentieth century. The life forms would be found in the polar ice caps, in subsurface communities, in hydrothermal systems or in endoliths or evaporites (Rothschild, 1990). Terrestrial analogues exist for all of these potential life forms, and all circumvent the high flux of ultraviolet radiation on the surface. Primarily for these reasons, and the likelihood of eventual human settlement, Mars remains a prime target for missions, if not specifically to look for life, then to better understand the environment of the planet. By understanding the environment, habitability can be gauged. These include NASA's Mars Pathfinder, Mars Global Surveyor, Mars Odyssey, the Mars Exploration Rovers, and ESA's Mars Express. Sadly the British-designed Beagle 2 lander which was intended specifically to look for life, failed to communicate with Mars Express after its release in December 2003, and is now presumed 'dead'.

6.8.3 Europa

If water is the *sine qua non* for life, any body known to have liquid water instantly becomes a prime target in the search for life. For years it was thought that any body far from the Sun would not be able to harbour liquid water, but today there are several moons beyond Mars thought to contain liquid water beneath an icy covering. The best known of these is Europa.

Europa was discovered by Galileo Galilei in 1610 along with Jupiter's other three large moons, Io, Ganymede, and Callisto. In 1951, using Earth-based telescopes, H. Jeffreys proposed that Callisto might be partially or totally composed of water in the form of ice. This was suggested by Callisto's very low density and also its albedo. The presence of frost or ice on Europa, Ganymede and Callisto was confirmed by using infrared reflectivity observed from the solar telescope at Kitt Peak National Observatory. John Lewis (1971) and, later, others suggested that there could be a liquid water ocean beneath an icy crust. First the Voyager spacecraft, and, since the 1990s, NASA's Galileo spacecraft, have returned photos of ice-covered surfaces with coloured cracks (reviewed in Greenberg and Geissler (2002) and Greenberg (2005)). Current thinking focuses on the idea of a liquid water ocean beneath an icy crust that is under 50 million years old, so is actively resurfacing (Khurana *et al.*, 2002). The ocean is maintained by tidal resonance with Io and Ganymede and tidal friction with Jupiter.

As soon as it became known that there might be liquid water on Europa discussions began on the possibility that there might also be life there. Guy Consolmagno, who worked on the theoretical models of oceans on Europa, Ganymede, and Callisto with John Lewis at MIT, included an appendix in his 1975 Master's thesis 'Thermal history models of icy satellites', where he suggested that Europa could have the beginnings of organic chemistry if the rocky core is as rich in carbon as some of the primitive meteorites. He concluded by writing '... we stop short of postulating life forms in these mantles; we leave such to others more experienced than ourselves in such speculations.' Since then a dozen or so papers have appeared with similar speculations, including Gaidos *et al.* (1999), Chyba and Phillips (2002), and Greenberg (2005). While pictures of large invertebrates prowling the Europan ocean have entered the popular imagination, a more constrained microbial biosphere is widely discussed. Future missions, though long and expensive compared to Mars missions, are ardently desired.

6.8.4 Enceladus

Suddenly in March 2006 there was a new body in the Solar System in the ranks of suspected harbourers of life (Porco *et al.*, 2006). Enceladus, a tiny moon of Saturn, orbits in the outermost ring, the E ring. The Cassini mission had revealed a cloud of oxygen in the E ring, and Enceladus was suspected to be the source. Enceladus is thought to have an internal heat source and is geologically active, resulting in cold geysers that throw out water vapour and ice. The water decomposes into clouds of oxygen. Thus, there may be large pools of liquid water beneath the surface, and where there is water, the possibility for life exists.

6.9 Conclusions

Life 'as we know it' — and are likely to know it — is based on organic carbon. Furthermore, water is the solvent that supports life on Earth and, again, is likely to be used elsewhere, although the certainty is not quite as high as the argument for organic carbon. Based on these facts alone, environments that would be challenging to any life form exist.

The story of the so-called extremophiles is a conglomeration of many stories gathered for a multitude of purposes over the years. They reveal an envelope for life that is far beyond what we could have imagined 40 or even 10 years ago. For this reason, the potential habitats for life in the Universe have also expanded.

Some of the adaptations that organisms have made on Earth to these extreme environments are relatively simple from an evolutionary viewpoint, whereas others have required a suite of evolutionary changes. Particularly in the former case, examples of convergent evolution to solve the same problem the same way are known. Convergence suggests a relative simplicity to the adaptation, and thus increases the likelihood that it could occur elsewhere.

While it is impossible to predict whether extremophiles will be found elsewhere, two things are of near certainty. Terrestrial extremophiles will be among the first settlers of extraterrestrial bodies and the environmental limits for life presented here will someday be extended through further exploration of extreme environments on Earth.

Acknowledgements

The author wishes to thank the editors, whose patience can best be described as 'extreme'.

References

Ashcroft, F. (2001). *Life at the Extremes. The Science of Survival.* London: Flamingo Press.

Battista, J. R. (1997). Against all odds: the survival strategies of Deinococcus radiodurans. *Annu. Rev. Microbiol.*, **51**, 203–224.

Belly, R. T., Tansey, M. R., and Brock, T. D. (1973). Algal excretion of ^{14}C-labeled compounds and microbial interactions in Cyanidium caldarium mats. *J. Phycol.*, **9**, 123–127.

Benner, S. A., Ricardo, A., and Carrigan, M. A.. (2004). Is there a common chemical model for life in the universe? *Curr. Opin. Chem. Biol.*, **8**, 672–678.

Berner, R. A., Beerlind, D. J., Dudley, R., Robinson, J. M., and Wildman, R. A. J. (2003). Phanerozoic atmospheric oxygen. *Annu. Rev. Earth Pl. Sc.*, **31**, 105–134.

Brock, T. D. (1967). Life at high temperatures. *Science*, **158**, 1012–1019.

Chyba, C. F., and Phillips, C. B. (2002). Europa as an abode of life. *Origins Life Evol. B.*, **32**, 47–68.

Clarke, A. (2003). Evolution at low temperatures, in *Evolution on Planet Earth: The Impact of the Physical Environment*, eds. L. Rothschild and A. Lister. London: Academic Press.

Cockell, C. S. (1999). Life on Venus. *Planet Space Science*, **47**, 1487–1501.

Crowe, J. H., Hoekstra, F. A., and Crowe, L. M. (1992). Anhydrobiosis. *Annu. Rev. Physiol.*, **54**, 579–599.

Crowe, J. H., Carpenter, J. F., and Crowe, L. M. (1998). The role of vitrification in anhydrobiosis. *Annu. Rev. Physiol.*, **60**, 73–103.

Crowe L. M., and Crowe J. H. (1992). Anhydrobiosis: a strategy for survival. *Adv. Space Res.*, **12(4)**, 239–249.

Daly, M. J., Gaidamakova, E. K., Matrosova, V. Y., *et al.* (2004). Accumulation of Mn(II) in *Deinococcus radiodurans* facilitates gamma-radiation resistance. *Science*, **306**, 1025–1028.

Davies, P. L., and Hew, C. L. (1990). Biochemistry of fish antifreeze proteins. *FASEB J.*, **4**, 2460–2468.

Edwards, K. J., Bond, P. L., Gihring, T. M., and Banfield, J. F. (2000). An Archaeal iron-oxidizing extreme acidophile important in acid mine drainage. *Science*, **287**, 1796–1799.

Forsythe, R. D., and Zimbelman, J. R. (1995). A case for ancient evaporite basins on Mars. *J. Geophys. Res.*, **100(E3)**, 5553–5563.

Friedmann, E. I. (1982). Endolithic microorganisms in the Antarctic cold desert. *Science*, **215**, 1045–1053.

Gaidos, E. J., Nealson, K. H., and Kirschvink, J. L. (1999). Life in ice-covered oceans. *Science*, **284**, 1631–1633.

Gladman, B., Dones, L., Levison, H. F., and Burns, J. A.. (2005). Impact seeding and reseeding in the inner solar system. *Astrobiology*, **5**, 483–1496.

Grant, W. D., Mwatha, W. E., and Jones, B. E. (1990). Alkaliphiles, ecology, diversity and applications. *FEMS Microbiol. Rev.*, **75**, 25570.

Greenberg, R. (2005). *European – The Ocean Moon.* Chichester: Springer.

Greenberg, R. and Geissler, P. (2002). Europa's dynamic icy crust, An invited review. *Meteorit. Planet. Sci.*, **37**, 1685–1711.

Horneck, G. (1993). Responses of *Bacillus subtilis* spores to space environment: Results from experiments in space. *Origins Life Evol. B.*, **23**, 3752.

Horneck, G. (1999). European activities in exobiology in Earth orbit: results and perspectives. *Adv. Space Res.*, **23**, 38–386.

Horneck, G., Bücker, H., Reitz, G., *et al.* (1984). Microorganisms in the space environment. *Science*, **225**, 2268.

Huber, R., Eder, W., Heldwein, S., *et al.* (1998). *Thermocrinis ruber gen. nov., sp. nov.,* a pink-filament-forming hyperthermophilic bacterium isolated from Yellowstone National Park. *Appl. Environ. Microbiol.*, **64**, 357683.

Jacob, R. A. and Burri, B. J. (1996). Oxidative damage and defense. *Am. J. Clin. Nutrition*, **63**, 985S–990S.

Keeley, J. E., and Fotheringham, C. J. (1998). Smoke-induced seed germination in California chaparral. *Ecology*, **79**, 232036.

Khurana, K. K., Kivelson, M. G., and Russell, C. T. (2002). Searching for liquid water in Europa by using surface observatories. *Astrobiology*, **2**, 93–103.

Lambros, R. J., Mortimer, J. R., and Forsdyke, D. R. (2003). Optimum growth temperature and the base composition of open reading frames in prokaryotes. *Extremophiles*, **7**, 443–450.

Lapinski. J., and Tunnacliffe, A. (2003). Anhydrobiosis without trehalose in bdelloid rotifers. *FEBS Lett.*, **553(3)**, 387–390.

Lewis, J. S. (1971). Satellites of the outer planets: Their physical and chemical nature. *Icarus*, **15**, 174–185.

Levin-Zaidman, S., Englander, J., Shimoni, E., *et al.* (2003). Ringlike structure of the *Deinococcus radiodurans* genome: a key to radioresistance? *Science*, **299**, 254–256.

MacElroy, R. (1974). Some comments on the evolution of extremophiles. *Biosystems*, **6**, 74–75.

Mancinelli, R. L., White, M. R., and Rothschild, L. J. (1998). Biopan-survival I: exposure of the osmophiles *Synechococcus sp. (Nageli)* and *Haloarcula sp.* to the space environment. *Adv. Space Res.*, **22(3)**, 327–334.

Martins, R. F., Davids, W., Al-Sond, W. A., *et al.* (2001). Starch-hydrolyzing bacteria from Ethiopian soda lakes. *Extremophiles*, **5**, 135–144.

McKay, C. P., and Friedmann, E. I. (1988). The cryptoendolithic microbial environment in the Antarctic cold desert: temperature variations in nature. *Polar Biol.*, **4**, 19–25.

McKay, C. P., Andersen, D. T., Pollard, W. H., *et al.* (2005). Polar lakes, streams, and springs as analogs for the hydrological cycle on Mars, in *Advances in Astrobiology Biogeophysics*, pp. 219–233. Herdelberg: Springer-Verlag.

McKay, D. S., Gibson, E. K. Jr., Thomas-Keprta, K. L., *et al.* (1996). Search for past life on Mars: possible relic biogenic activity in Martian meteorite ALH84001. *Science*, **273**, 924–930.

McLennan, S. M., Bell, J. F. III, Calvin, W. M., *et al.* (2005). Provenance and diagenesis of the evaporite-bearing Burns formation, Meridiani Planum, Mars. *Earth Planet. Sc. Lett.*, **240**, 95121.

Nicholson, W. L., Munakata, N., Horneck, G., Melosh, H. J., and Setlow, P. (2000). Resistance of *Bacillus* endospores to extreme terrestrial and extraterrestrial environments. *Microb. Mol. Biol. Rev.*, **64**, 54872.

Pedersen, K., Nilsson, E., Arlinger, J., Hallbeck, L., and O'Neill, A. (2004). Distribution, diversity and activity of microorganisms in the hyper-alkaline spring waters of Maqarin in Jordan. *Extremophiles*, **8**, 151–164.

Perutz, M. F. (1978). Electrostatic effects in proteins. *Science*, **201**, 1187–1191.

Petit, C., and Sancar, A. (1999). Nucleotide excision repair: from *E. coli* to man. *Biochimie*, **81**, 15–25.

Porco, C. C., Helfenstein, P., Thomas, P. C., *et al.* (2006). Cassini observes the active South Pole of Enceladus. *Science*, **311**, 1393–1401.

Priscu, J. C., Fritsen, C. F., Adams, E. E., *et al.* (1998). Perennial Antarctic lake ice: an oasis for life in a polar desert. *Science*, **280**, 2095–2098.

Proctor, R. (1870). *Other Worlds Than Ours*. New York: Longmans.

Rees, H. C., Grant, W. D., Jones, B. E., and Heaphy, S. (2004). Diversity of Kenyan soda lake alkaliphiles assessed by molecular methods. *Extremophiles*, **8**, 63–71.

Reysenbach, A. L., Whickham, G. S., and Pace, N. R. (1994). Phylogenetic analysis of the hyperthermophilic pink filament community in Octopus Spring, Yellowstone National Park. *Appl. Environ. Microbiol.*, **60**, 21139.

Rothschild, L. J. (1990). Earth analogs for Martian life. Microbes in evaporites, a new model system for life on Mars. *Icarus*, **88**, 246–260.

Rothschild, L. J., and Giver, L. J. (2003). Photosynthesis below the surface in a cryptic microbial mat. *Intl. J. Astrobiol.*, **1**, 295–304.

Rothschild, L. J., and Mancinelli, R. L. (2001). Life in extreme environments. *Nature*, **409**, 1092–1101.

Rothschild, L. J., Giver, L. J., White, M. R., and Mancinelli, R. L. (1994). Metabolic activity of microorganisms in evaporites. *J. Phycol.*, **30**, 431–438.

Saffary, R., Nandakuma, R., Spence, D., *et al.* (2002). Microbial survival of space vacuum and extreme ultraviolet irradiation: strain isolation and analysis during a rocket flight. *FEMS Microbiol. Lett.*, **215**, 163–168.

Sagan, C. (1961). The planet Venus. *Science*, **133**, 849–858.

Schleper, C., Puehler, G., Holz, I. *et al.* (1995). *Picrophilus* gen. nov., fam. nov.: a novel aerobic, heterotrophic, thermoacidophilic genus and family comprising archaea capable of growth around pH 0. *J. Bacteriol*, **177**, 7050–7059.

Schmid, W. D. (1982). Survival of frogs in low temperature. *Science*, **215**, 697–698.

Schulte, W., Demets, R., Baglioni, P., *et al.* (2006). BIOPAN and ESPOSE: Space exposure platforms for exo/astrobiological research in earth orbit with relevance for Mars exploration. *Geophys. Res. Abs.*, **8**, 06643.

Seckbach, J. (1999). The Cyanidiophyceae: hot spring acidophilic algae, in *Enigmatic Microorganisms and Life in Extreme Environments*, ed. J. Seckbach. Dordrecht: Kluwer Academic Publishers, pp. 427–35.

Seckbach, J., Baker, F. A., and Shugarman, P. M. (1970). Algae survive under pure CO_2. *Nature*, **227**, 744–745.

Sharma, A., Scott, J. H., Cody, G. D., *et al.* (2002). Microbial activity at gigapascal pressures. *Science*, **295**, 1514–1516.

Sheridan, C. (2004). Kenyan dispute illuminates bioprospecting difficulties. *Nature Biotechnology*, http://www.nature.com/news/2004/041108/pf/nbt1104–1337_pf.html.

Shulze-Makuch, D., and Irwin, L. N.. (2004). *Life in the Universe*. Berlin: Springer.

Shulze-Makuch, D., Grinspoon, D. H., Abbas, O., Irwin, L. N., and Bullock, M. A. (2004). A sulfur-based survival strategy for putative phototrophic life in the Venusian atmosphere. *Astrobiology*, **4**, 11–17.

Smith, K. C. (2004). Recombinational DNA repair: the ignored repair systems. *BioEssays*, **26**, 1322–1326.

Storey, K. N., and Storey, J. M. (1996). Natural freezing survival in animals. *Annu. Rev. Ecol. Syst.*, **27**, 265–386.

Sun, W. Q., Davidson, P., and Chan, H. S. (1998). Protein stability in the amorphous carbohydrate matrix: relevance to anhydrobiosis. *Biochim. Biophys. Acta*, **1425**, 245–254.

Watanabe, M. (2006). Anhydrobiosis in invertebrates. *Appl. Entomol. Zool.*, **41**, 1531.

Woese, C. R., and Fox, G. E. (1977). Phylogenetic structure of the prokaryotic domain: The primary kingdoms. *Proc. Natl. Acad. Sci. USA*, **74**, 5088–5090.

Yasui, A., and McCready, S. J. (1998). Alternative repair pathways for UV-induced DNA damage. *BioEssays*, **20**, 291–297.

Zahnle, K. (2001). Decline and fall of the martian empire. *Nature*, **412**, 209–213.

7

Hyperthermophilic life on Earth – and on Mars?

Karl O. Stetter
Universität Regensburg

7.1 Introduction

Every living organism is adapted to a specific growth temperature. In the case of humans, this is 37 °C and an increase by 5 °C becomes fatal. In the world of microbes, the growth temperature range is much more diverse: heat lovers ('thermophiles') grow optimally (fastest) at temperatures up to 65 °C (Brock, 1978; Castenholz, 1979). Since the time of Pasteur, it had been assumed generally that growing (vegetative) cells of bacteria were killed quickly by temperatures of 80 °C and above. The Pasteurization technology is based on this observation. In contrast, during the past few decades, hyperthermophiles (HT; Stetter, 1992) that exhibit unprecedented optimal growth temperatures in excess of 80 °C have been isolated (Stetter *et al.*, 1981; Zillig *et al.*, 1981; Stetter, 1982). HT turned out to be very common in hot terrestrial and submarine environments. In comparing the growth requirements of these present-day HT with the conditions on ancient Earth, similar organisms could or even should have existed already by Early Archaean times. Propelled by impact energy, microbes could have spread in between the planets and moons of the early Solar System. Is there any evidence for the existence of microbes at that time? Most likely, yes. But the recognition of ancient microfossils on the basis of morphology turned out to be difficult, leading to controversy. Nevertheless, there are chemical traces of life within rocks from Precambrian deep sea vents (Schopf *et al.*, 1987; Brasier *et al.*, 2002; van Zullen *et al.*, 2002).

In my research, I search for HT, their environments, properties, and modes of living. So far, I have gathered a collection of more than 1500 strains of HT, among them are isolates that survive autoclaving at 121 °C. Whether such organisms are growing slowly or just surviving at 121 °C is unresolved (Blöchl *et al.*, 1997; Kashefi and Lovley, 2003; Cowan, 2004).

Planetary Systems and the Origins of Life, eds. Ralph E. Pudritz, Paul G. Higgs, and Jonathon R. Stone.
Published by Cambridge University Press.

7.2 Biotopes

HT form complex communities within water-containing geothermally and volcanically heated environments situated mainly along terrestrial and submarine tectonic spreading and subduction zones. Availability of liquid water is a fundamental prerequisite of life. At an increased boiling point of water (for example by elevated atmospheric, hydrostatic or osmotic pressure), several HT exhibit growth temperatures exceeding 100 °C. Due to the presence of reducing gasses like H_2S and the low solubility of O_2 at high temperatures, the biotopes of HT essentially are oxygen-free (anaerobic). HT have been isolated from terrestrial and submarine environments.

7.2.1 Terrestrial biotopes

Natural terrestrial biotopes of HT are mainly hot springs and sulfur-containing solfataric fields (named after Solfatara, Italy), with a wide range of pH values (pH 0–9) and usually low salinity (0.1–0.5%). Solfataric fields consist of soils, mud holes, and surface waters, heated by volcanic exhalations from magma chambers several kilometres below (Figure 7.1). Very often, solfataric fields are found close to active volcanoes and activity increases greatly during eruption phases. The chemical composition of solfataric fields is very variable and site-dependent. Many are rich in iron minerals, like ferric hydroxides and pyrite. Less usual compounds may be enriched at some sites, with magnetite and arsenic minerals like

Fig. 7.1. Solfataric field at Kafla, Iceland.

realgar and auripigment in Caldera Uzon, Kamchatka. Hydrogen may be formed either pyrolytically from water or chemically from FeS and H_2S (Drobner *et al.*, 1990). Steam within the solfataric exhalations is responsible mainly for the heat transfer. In addition, sites may contain carbon dioxide, variable amounts of H_2, methane, nitrogen, carbon monoxide, and traces of nitrate and ammonia. Active volcanoes may harbour hot-water lakes, which are heated by fumaroles. Usually, those abound in sulfur and are very acidic. Other suitable biotopes for HT are deep, subterranean, geothermally heated oil stratifications, approximately 3500 m below the bottom of the North Sea and the surface of the permafrost soil at the North Slope, Northern Alaska (Stetter *et al.*, 1993). The fluids produced in these regions contained up to 10^7 viable cells per litre of different species of HT. Artificial biotopes include smouldering coal refuse piles and hot outflows from geothermal and nuclear power plants.

7.2.2 Marine biotopes

Marine biotopes of HT consist of various hydrothermal systems situated at shallow to abyssal depths. Similar to ambient seawater, submarine hydrothermal systems usually contain high concentrations of NaCl and sulfate and exhibit a slightly acidic to alkaline pH (5.0–8.5). Otherwise, the major gases and life-supporting mineral nutrients can be similar to those in terrestrial thermal areas. Shallow submarine hydrothermal systems are found in many parts of the world, mainly on beaches with active volcanism, like at Vulcano Island, Italy, with temperatures of 80 to 105 °C.

Most impressive are the deep sea 'smoker' vents (Figure 7.2), where mineral-laden hydrothermal fluids with temperatures up to approximately 400 °C escape into the cold (2.8 °C), surrounding deep sea water and build up huge rock chimneys. Although these hot fluids are sterile, the surrounding porous smoker rock material appears to contain very steep temperature gradients, which provide zones of suitable growth temperatures for HT. Some smoker rocks are teeming with HT (for example 10^8 cells of *Methanopyrus* per gram of rock inside a Mid Atlantic Snake Pit hot vent chimney). Deep sea vents are located along submarine tectonic fracture zones (for example the 'TAG' and 'Snake Pit' sites situated at the Mid Atlantic Ridge in a depth of about 4000 m). Another type of submarine high temperature environment is provided by active sea mounts. Close to Tahiti, there is a huge abyssal volcano, Macdonald Seamount (28°58.7′ S, 140 °15.5′ W), the summit of which is situated approximately 40 meters below the sea surface. Samples taken during an active phase from the submarine eruption plume and rocks from the active crater contained high concentrations of viable HT (Huber *et al.*, 1990).

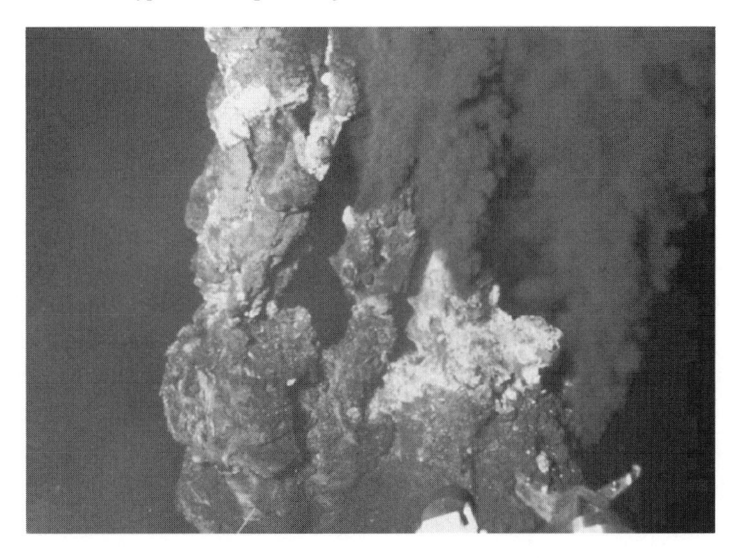

Fig. 7.2. Abyssal hot 'black smoker' chimneys at the East Pacific Rise, 21 °N. Depth: 2500 m, maximal fluid temperature: 365 °C.

7.3 Sampling and cultivation

Samples of possible HT containing hot waters, soils, rock and sediments may serve as primary material to set-up enrichment cultures in the laboratory. Special care has to be taken to avoid contamination of the sample by oxygen. In the presence of high temperatures of growth, it is toxic to anaerobic HT. In contrast, at low temperatures (for example: 4 °C) in the presence of oxygen, anaerobic HT may survive for years. Transportation and shipping can be performed at ambient temperature. In the lab, anaerobic samples (in tightly stoppered 100 ml storage bottles and stored at 4 °C) can be used for successful enrichment cultures at least for 10 years.

Enrichment cultures can be obtained by simulating the varying geochemical and geophysical composition of the environments. Various plausible electron donors and acceptors may be used under anaerobic, microaerophilic, or aerobic culture conditions. Depending on the (unknown) initial cell concentration and the doubling time of the organism, positive enrichment cultures of HT can be identified by microscopy within 1–7 days. For a deeper understanding of the organisms, the study of pure cultures is required. Due to the high incubation temperatures, the traditional manner of cloning by plating does not perform well in HT (even by using heat stable polymers like Gelrite). Therefore, we developed a new procedure for cloning single cells under the laser microscope, by employing optical tweezers (Ashkin and Dziedzic, 1987; Huber *et al.*, 1995). Large cell masses are required for biochemical and biophysical investigations. For mass culturing of HT, a new type of high temperature fermentor was invented in collaboration with an engineering

Fig. 7.3. Fermentation plant, University of Regensburg. Partial view, showing two 300 l fermenters and one 130 l fermenter for cultivation of HT.

company (Figure 7.3). Its steel casing is enamel-protected in order to resist the highly corrosive culture conditions. Sharp-edged parts like stirrers, gasing and sampling pipes and condensers are made of titanium. The cell yield of a 300 l fermentation may vary from approximately 3 g to 2 kg (wet weight), depending on the HT isolate.

7.4 Phylogenetic implications

Hot volcanic environments are among the oldest biotopes existing on Earth. What is known about the phylogenetic relationships of the organisms living there? Based on the pioneering work of Carl Woese, the small subunit ribosomal RNA (ss-rRNA) is widely used in phylogenetic studies of prokaryotes (Woese and Fox, 1977; Woese *et al.*, 1990). It consists of approximately 1500 nucleotides and is homologous to its eukaryotic counterpart. On the basis of sequence comparisons, a phylogenetic tree is now available (Figure 7.4). It shows a tripartite division of the living world into the bacterial, archaeal, and eukaryal domains. Within this tree, deep branches are evidence for early separation. The separation of the Bacteria from the stem shared by Archaea and Eukarya represents the deepest and earliest branching point. Short phylogenetic branches indicate a rather slow rate of evolution. In contrast to the eukaryal domain, the bacterial and archaeal domains within the phylogenetic tree exhibit some extremely short and deep branches. Surprisingly,

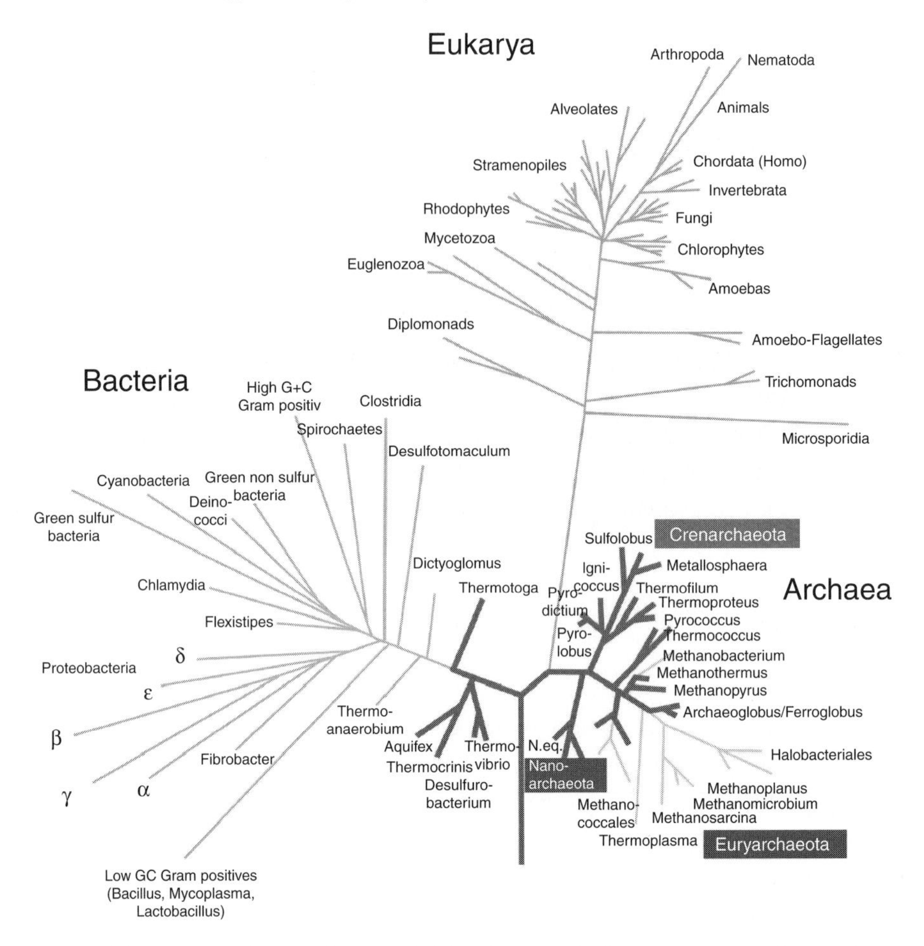

Fig. 7.4. Small subunit ribosomal RNA-based phylogenetic tree. The bulky lineages are representing HT.

those are covered exclusively by hyperthermophiles, which therefore form a cluster around the phylogenetic root (Figure 7.4, bulky lineages). The deepest and shortest phylogenetic branches are represented by the *Aquificales* and *Thermotogales* within the Bacteria and the *Nanoarchaeota*, *Pyrodictiaceae*, and *Methanopyraceae* within the Archaea. Long lineages represent mesophilic and moderately thermophilic Bacteria and Archaea (e.g., Gram-positives, Proteobacteria; *Halobacteriales*; *Methanosarcinaceae*) indicating their ss-rRNA had experienced a fast rate of evolution. Now, several total genome sequences are available. Phylogenetic trees based on genes involved in information management (for example, DNA replication, transcription, translation) parallel the ss-rRNA tree. Genes involved in metabolism, however, are prone to frequent lateral gene transfer, and a network rather than a tree might reflect their phylogenetic relations (Doolittle, 1999).

To date, approximately 90 species of HT Archaea and Bacteria, which had been isolated from different terrestrial and marine thermal areas in the world, are known. HT are very divergent, both in their phylogeny and physiological properties and are grouped into 34 genera and 10 orders (Boone and Castenholz, 2001).

7.5 Physiologic properties

7.5.1 Energy sources

Most species of HT exhibit a chemolithoautotrophic mode of nutrition. Inorganic redox reactions serve as energy sources (chemolithotrophic), and CO_2 is the only carbon source required to build up organic cell material (autotrophic). Therefore, these organisms fix CO_2 by chemosynthesis and are designated chemolithoautotrophs. The energy-yielding reactions in chemolithoautotrophic HT are those involved in anaerobic and aerobic respiration (Figure 7.5). Molecular hydrogen serves as an important electron donor. Other electron donors are sulfide, sulfur, and ferrous iron. As in mesophilic respiratory organisms, in some HT, oxygen may serve as an electron acceptor. In contrast, however, oxygen-respiring HT are usually microaerophilic and, therefore, grow only at reduced oxygen concentrations. Anaerobic respiration types are the nitrate-, sulfate-, sulfur- and carbon dioxide-types. While chemolithoautotrophic HT produce organic matter, there are some HT that depend on organic material as energy and carbon sources. They are designated as heterotrophs. Several chemolithoautotrophic HT are opportunistic heterotrophs. These are able to use organic material alternatively to inorganic nutrients whenever they are available from the environment (e.g., by

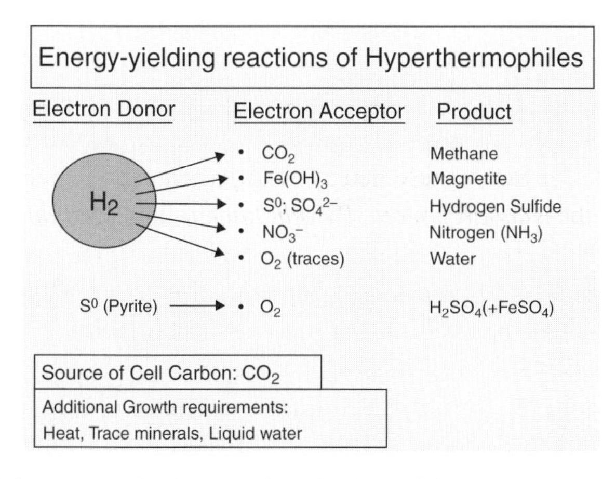

Fig. 7.5. Main energy-yielding reactions in chemolithoautotrophic HT (schematic drawing).

Table 7.1. *Growth conditions and morphology of hyperthermophiles*

Species	Min. Temp (°C)	Opt. Temp (°C)	Max. Temp (°C)	pH	Aerobic (ae) vs. anaerobic (an)	Morphology
Sulfolobus acidocaldarius	60	75	85	1–5	ae	lobed cocci
Metallosphaera sedula	50	75	80	1–4.5	ae	cocci
Acidianus infernus	60	88	95	1.5–5	ae/an	lobed cocci
Stygiolobus azoricus	57	80	89	1–5.5	an	lobed cocci
Thermoproteus tenax	70	88	97	2.5–6	an	regular rods
Pyrobaculum islandicum	74	100	103	5–7	an	regular rods
Pyrobaculum aerophilum	75	100	104	5.8–9	ae/an	regular rods
Thermofilum pendens	70	88	95	4–6.5	an	slender regular rods
Desulfurococcus mobilis	70	85	95	4.5–7	an	cocci
Thermosphaera aggregans	67	85	90	5–7	an	cocci in aggregates
Sulfophobococcus zilligii	70	85	95	6.5–8.5	an	cocci
Staphylothermus marinus	65	92	98	4.5–8.5	an	cocci in aggregates
Thermodiscus maritimus	75	88	98	5–7	an	disks
Aeropyrum pernix	70	90	100	5–9	ae	irregular cocci
Stetteria hydrogenophila	70	95	102	4.5–7	an	irregular disks
Ignicoccus islandicus	65	90	100	3.9–6.3	an	irregular cocci
Pyrodictium occultum	82	105	110	5–7	an	disks with cannulae
Hyperthermus butylicus	80	101	108	7	an	lobed cocci
Pyrolobus fumarii	90	106	113	4.0–6.5	ae/an	lobed cocci
Thermococcus celer	75	87	93	4–7	an	cocci
Pyrococcus furiosus	70	100	105	5–9	an	cocci

Archaeoglobus fulgidus	60	83	95	5.5–7.5	an	irregular cocci
Ferroglobus placidus	65	85	95	6–8.5	an	irregular cocci
Methanothermus sociabilis	65	88	97	5.5–7.5	an	rods in clusters
Methanopyrus kandleri	84	98	110	5.5–7	an	rods in chains
Methanococcus igneus	45	88	91	5–7.5	an	irregular cocci
Thermotoga maritima	55	80	90	5.5–9	an	rods with sheath
Aquifex pyrophilus	67	85	95	5.4–7.5	ae	rods

decaying cells). Heterotrophic HT gain energy either by aerobic or different types of anaerobic respiration, using organic material as electron donors, or by fermentation.

7.5.2 General physiologic properties

HT are adapted to environmental factors including composition of minerals and gases, pH, redox potential, salinity and temperature. Similar to mesophiles, they grow within a temperature range of about 25–30 °C between the minimal and maximal growth temperature (Table 7.1). Fastest growth is obtained at their optimal growth temperature, which may be up to 106 °C. Generally, HT do not propagate at 50 °C or below. Although unable to grow at the low ambient temperatures, they are able to survive for years. Based on their simple growth requirements, HT could grow on any hot, wet place, even on other planets and moons of our solar system, like Mars and Europa. Today, the surface of Mars is too cold and contains no liquid water and therefore is hostile to life as it is known on Earth. However, in a depth of a few kilometres below the permafrost layer, there may be hot liquid water and nutrients to support growth of HT. Life could have spread onto Mars via meteorites during the great bombardment, about 4 GA ago. At that time, the surface of Mars had been much hotter and contained liquid water, therefore being favourable to HT.

7.6 Examples of recent HT organisms

Within the Bacteria domain, the deepest phylogenetic branch is represented by the HT *Aquifex* (Huber *et al.*, 1992). Its type species *Aquifex pyrophilus* is a motile rod-shaped chemolithoautotroph (Figure 7.6). It is a facultative microaerophilic anaerobe. Under anaerobic conditions, *Aquifex pyrophilus* grows

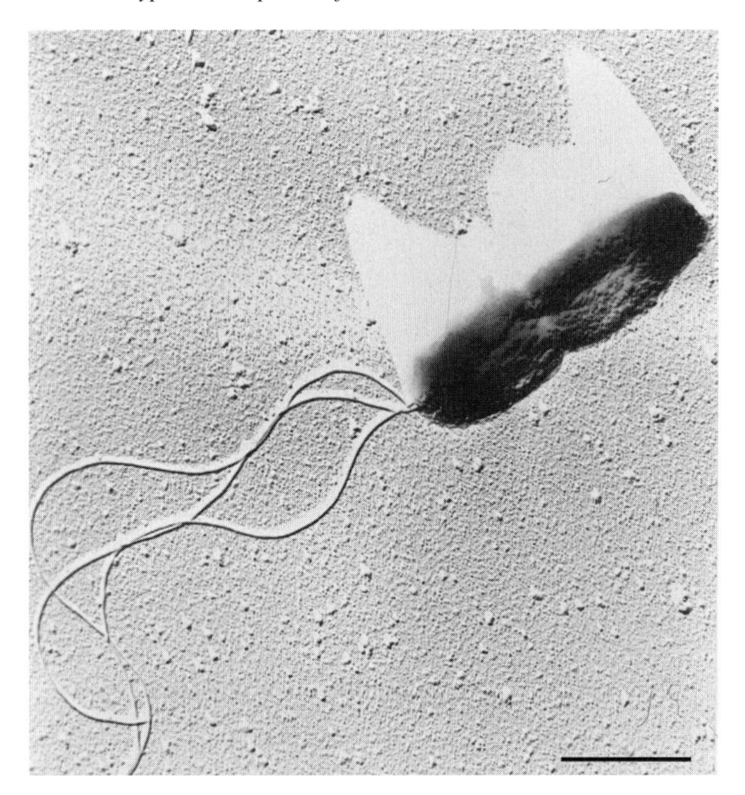

Fig. 7.6. *Aquifex pyrophilus*, dividing cell with a tuft of flagella. Pt-shadowing. Transmission electron micrograph. Scale bar, 1 μm.

by nitrate reduction with H_2 and S^0 as electron donors. Alternatively, at very low oxygen concentrations (up to 0.5%, after adaptation), it is able to gain energy by oxidation of H_2 and S^0, using oxygen as an electron acceptor. Members of *Aquifex* are found in shallow submarine vents. *Aquifex pyrophilus* grows up to 95 °C, the highest growth temperature observed within the Bacteria (Table 7.1).

From the walls of a black smoker at the Mid Atlantic Ridge, we had isolated the archaeon *Pyrolobus fumarii* (Blöchl *et al.*, 1997). Cells are lobed cocci, approximately 0.7–2.5 μm in diameter (Figure 7.7). The species *Pyrolobus fumarii* is adapted optimally to temperatures of superheated water, exhibiting an optimal growth temperature of 106 °C (Figure 7.8) and an upper temperature border of growth at 113 °C. It is so dependent to high temperatures that it is unable to grow below 90 °C. Cultures of *Pyrolobus fumarii*, similar to *Pyrodictium occultum* are able to survive autoclaving for one hour at 121 °C. A very closely related isolate ('strain 121') exhibits the same optimal growth temperature (106 °C), but has been reported to grow slowly even at 121 °C (Kashefi and Lovley, 2003). However, this result has not been confirmed.

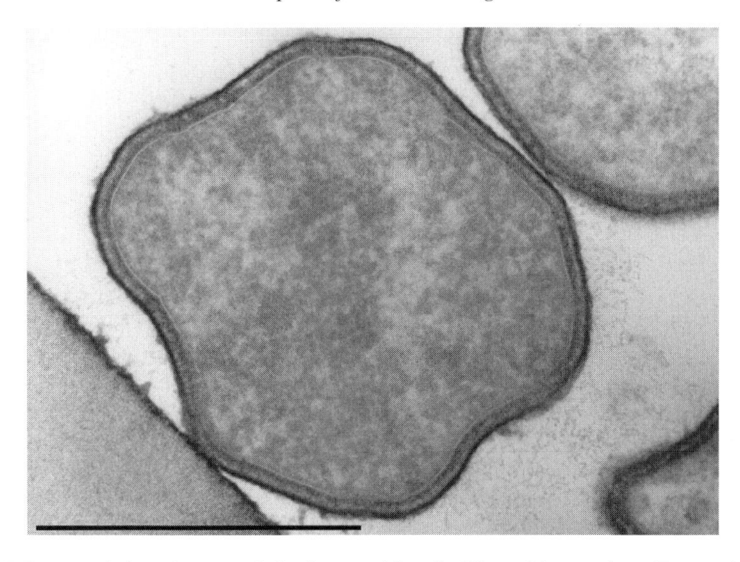

Fig. 7.7. *Pyrolobus fumarii*, lobed coccoid cell. Ultra thin section. Transmission electron micrograph. Scale bar, 0.5 μm.

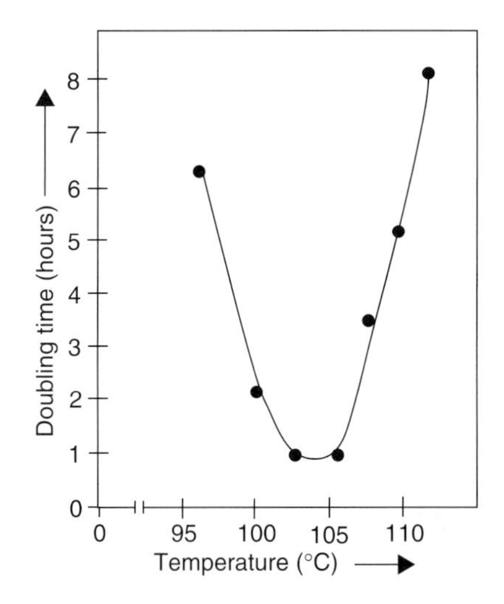

Fig. 7.8. *Pyrolobus fumarii*: Temperature dependence of its doubling time. Optimal growth (about 50 min. doubling time) occurs between 103 and 106 °C.

From a submarine hydrothermal system situated at the Kolbeinsey Ridge, north of Iceland, we were able to obtain our ultimate hyperthermophilic coccoid isolate *Nanoarchaeum equitans*, which represents a novel kingdom of Archaea (Huber *et al.*, 2002). With a cell diameter of only 400 nm, it is the smallest living organism known. Cells grow attached to the surface of a specific crenarchaeal

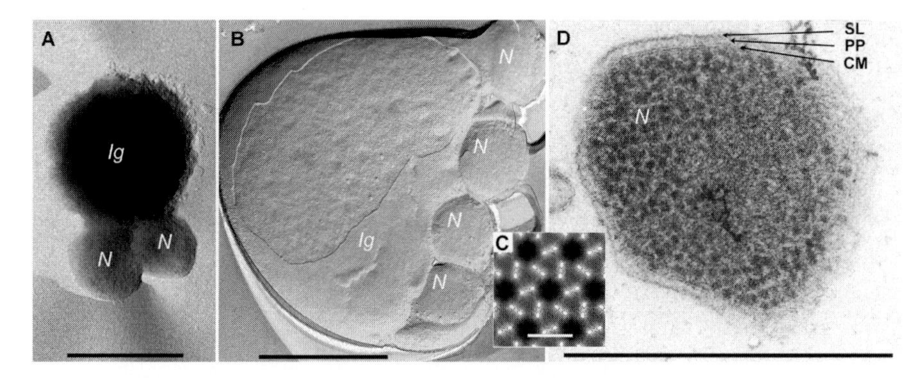

Fig. 7.9. *Nanoarchaeum equitans–Ignicoccus hospitalis*.: Transmission electron micrographs. (A) Two cells of *N. equitans* attached on the surface of the (central) *Ignicoccus* cell. Platinum shadowed. Scale bar, 1 μm. (B) Freeze-etched cell of *Ignicoccus hospitalis* (Ig) and attached cells of *N. equitans* (N) on the surface. Scale bar, 1 μm. (C) Surface relief reconstruction of *N. equitans*. Dark: cavity; Bright: elevation. Scale bar, 15 nm. (D) Ultrathin section of a *N. equitans* cell. Single cell. CM: cytoplasmic membrane; PP: periplasm; SL: S-layer. Scale bar: 0.5 μm.

host, a new member of the genus *Ignicoccus* (Figure 7.9). Owing to their unusual ss-rRNA sequence, members of *Nanoarchaeum equitans* remained undetectable by commonly used 'universal' primers in ecological studies based on the polymerase chain reaction. However, two different ss-rRNA genes could be detected in a coculture-derived DNA by Southern blot hybridization, by taking advantage of the generally high sequence homology of all ss-rRNA genes (about 75% similarity). After sequencing of the total genome, the phylogenetic relationships of *Nanoarchaeum equitans* could be investigated by concatenating and aligning the amino acid sequences of 35 ribosomal proteins (Waters *et al.*, 2003). The species *Nanoarchaeum equitans* was placed with high support at the most deeply branching position within the Archaea, suggesting that the *Nanoarchaeota* diverged early within the Archaea. With only 490 885 base pairs, the *Nanoarchaeum equitans* genome is the smallest microbial genome known to date and also the most compact, with 95% of the DNA predicted to encode proteins and stable RNAs. It harbours the complete machinery for information processing and repair, but lacks genes for lipid, cofactor, amino acid, and nucleotide biosynthesis. The limited biosynthetic and catabolic capacity suggests that the symbiotic relationship of *Nanoarchaeum equitans* to its *Ignicoccus hospitalis* host may be parasitic, making it the only known archaeal parasite. Unlike the small genomes of bacterial parasites, however, which are undergoing reductive evolution, the small genome of *Nanaoarchaeum equitans* has very few pseudogenes and a well-equipped DNA recombination system. Therefore, it may be a very ancient parasite. Alternatively, the *Ignicoccus hospitalis–Nanoarchaeum equitans* system could be remains from pre-Archaea communities, as postulated earlier (Kandler, 1994). At the molecular level, *Nanoarchaeum equitans* harbours further unexpected, most likely primitive properties, like separately encoded enzyme modules and – for the first time – t RNA gene fragments (Waters *et al.*, 2003; Randau *et al.*, 2005). Currently, we possess only rudimentary understanding of the *Nanoarchaeum equitans – Ignicoccus hospitalis* relationship. The *Nanoarchaeota* are distributed world-wide within hot environments and had been completely overlooked. Two ss-rRNA sequences from Uzon Caldeira (Kamchatka, Russia) and Yellowstone National Park (USA) exhibited 83% sequence similarity to *Nanoarchaeum equitans*, and, therefore, represent a distinct group within the Nanoarchaeota (Hohn *et al.*, 2002). Light microscopy and fluorescence *in situ* staining reveal that these novel *Nanoarchaeota* are again tiny cocci, approximately the size of *Nanoarchaeum equitans*, however, attached to a rod-shaped *Pyrobaculum*-like host. The discovery of the *Nanoarchaeota* suggests that further unrecognized major groups of microbes may remain undetected by current PCR primers and are waiting for isolation to tell us more about the origins and evolution of life.

References

Ashkin, A. and Dziedzic, J. M. (1987). Optical trapping and manipulation of viruses and bacteria. *Science*, **235**, 1517–1520.

Boone, D. R. and Castenholz, R. W. (2001). The Archaea and the deeply branching and phototrophic Bacteria, in *Bergey's Manual of Systematic Bacteriology*, vol. 1, 2nd edn., ed. G.M. Garrity. New York: Springer, pp. 169–387.

Blöchl, E., Rachel, R., Burggraf, S. *et al.* (1997). *Pyrolobus fumarii*, gen. and sp. nov., represents a novel group of archaea, extending the upper temperature limit for life to 113 °C. *Extremophiles*, **1**, 14–21.

Brasier, M. D., Green, O. R., Jephcoat, A. P. *et al.* (2002). Questioning the evidence for Earth's oldest fossils. *Nature*, **416**, 76–81.

Brock, T. D. (1978). Thermophilic microorganisms and life at high temperatures.

Castenholz, R. W. (1979). Evolution and ecology of thermophilic microorganisms. In *Strategies of Microbial Life in Extreme Environments*, ed. M. Shilo. Weinheim: Verlag Chemie, pp. 373–392.

Cowan, D. A. (2004). The upper limit of life–how far can we go. *Trends Microbiol.*, **12**, 58–60.

Doolittle, W. F. (1999). Phylogenetic classification and the universal tree. *Science*, **284**, 2124–2129.

Drobner, E., Huber, H., Wächtershäuser, G., Rose, D. and Stetter, K. O. (1990). Pyrite formation linked with hydrogen evolution under anaerobic conditions. *Nature*, **346**, 742–744.

Hohn, M. J., Hedlund, B. P. and Huber, H. (2002). Detection of 16S rDNA sequences representing the novel phylum 'Nanoarchaeota': indication for a broad distribution in high temperature. *Syst. Appl. Microbiol.*, **25**, 551–554.

Huber, H., Hohn, M. J., Rachel, R., Fuchs, T., V. C. Wimmer, V. C. and K. O. Stetter. (2002). A new phylum of Archaea represented by a nanosized hyperthermophilic symbiont. *Nature*, **417**, 63–67.

Huber, R, Stoffers, P., Cheminee, J. L., Richnow, H. H. and Stetter, K. O. (1990). Hyperthermophilic archaebacteria within the crater and open-sea plume of erupting Macdonald Seamount. *Nature*, **345**, 179–181.

Huber, R. Burggraf, S., Mayer, T., Barns, S. M., Rossnagel, P. and Stetter, K. O. (1995). Isolation of a hyperthermophilic archaeum predicted by in situ RNA analysis. *Nature*, **376**, 57–58.

Kandler, O. (1994). The early diversification of life. In *Early Life on Earth*, Nobel Symposium No. 84, ed. S. Bengtson. New York: Columbia University Press, pp. 152–160.

Kashefi, K. and Lovley, D. R. (2003). Extending the upper temperature limit of life. *Science*, **301**, 934.

Randau, L., Münch, R., Hohn, M. J. Jahn, D. and Söll, D. (2005). *Nanoarchaeum equitans* creates functional t-RNAs from separate genes for their 5'- and 3'- halves. *Nature*, **433**, 537–541.

Schopf, J. W., and Packer, B. M. (1987). Early Archean (3.3 billion to 3.5 billion-year-old) microfossils from Warrawoona Group, Australia. *Science*, **237**, 70–73.

Stetter, K. O., Thomm, M., Winter, J. *et al.* (1981). *Methanothermus fervidus*, sp. nov., a novel extremely thermophilic methanogen isolated from an Icelandic hot spring. *Zbl. Bakt. Hyg., I. Abt. Orig. C2*, 166–178.

Stetter, K. O. (1982). Ultrathin mycelia-forming organisms from submarine volcanic areas having an optimum growth temperature of 105 °C. *Nature*, **300**, 258–260.

Stetter, K. O. (1992). Life at the upper temperature border. In *Frontiers of Life*, eds. J. Tran Thanh Van, K. Tran Thanh Van, J. C. Mounolou, J. Schneider and C. McKay. Gif-sur-Yvette: Editions Frontieres, pp. 195–219.

Stetter, K. O., Huber, R., Blöchl, E. *et al.* (1993). Hyperthermophilic archaea are thriving in deep North Sea and Alaskan oil reservoirs. *Nature*, **365**, 743–745.

van Zullen, M. A., Lepland, A., and Arrhenius, G. (2002). Reassessing the evidence for the earliest traces of life. *Nature*, **418**, 627–630.

Waters, E., Hohn, M. J., Ahel, I. *et al.* (2003). The genome of *Nanoarchaeum equitans*: Insights into early archaeal evolution and derived parasitism. *Proc. Natl. Acad. Sci. USA*, **100**, 12984–12988.

Woese, C. R. and Fox, G. E. (1977). Phylogenetic structure of the prokaryotic domain: the primary kingdoms. *Proc. Natl. Acad. Sci. USA*, **74**, 5088–5090.

Woese, C. R., Kandler, O. and Wheelis, M. L. (1990). Towards a natural system of organisms: proposal for the domains Archaea, Bacteria and Eucarya. *Proc. Natl. Acad. Sci. USA*, **87**, 4576–4579.

Zillig, W., Stetter, K. O., Schäfer, W. *et al.* (1981). *Thermoproteales*: a novel type of extremely thermoacidophilic anaerobic archaebacteria isolated from Icelandic solfataras. *Zbl. Bakt. Hyg., I. Abt. Orig. C2*, 205–227.

8

Phylogenomics: how far back in the past can we go?

Henner Brinkmann, Denis Baurain, and Hervé Philippe
Université de Montréal

8.1 Introduction

The remnants of ancient life on Earth are extremely rare and difficult to interpret, since they have been modified by billions of years of subsequent geological history. Therefore, any understanding of the evolution of Earth's biosphere will heavily rely on the study of extant organisms for which complete genome sequences are already available or will eventually be established. A central point is the phylogenetic inference of the tree of life from such genomic data. Serving as a unifying framework, this tree will allow the disparate and often incomplete pieces of information gathered by various disciplines to be collected and structured. In the first half of this chapter, we will briefly explain the principles of phylogenetic inference and the major artefacts affecting phylogenetic reconstruction. Then we will introduce phylogenomics, starting with a theoretical presentation before illustrating the explained concepts through a case study centred on animal evolution. In the second half, we will review our present understanding of eukaryotic evolution and show how recent knowledge suggests that secondary simplification is an important mode of evolution that has been too long overlooked. We will also summarize the current views about the root and the shape of the tree of life. Finally, we will attempt to debunk two common hypotheses about the early evolution of life, i.e., a cell fusion at the origin of eukaryotes and a hyperthermophilic origin of life.

8.2 The principles of phylogenetic inference

8.2.1 The homology criterion

The reconstruction of the genealogical relationships of a set of species first requires the definition of a series of characters that are comparable across all studied organisms. This problem was solved in 1843 by the palaeontologist Richard Owen based on the principle of connectivity previously established by the anatomist

Planetary Systems and the Origins of Life, eds. Ralph E. Pudritz, Paul G. Higgs, and Jonathon R. Stone. Published by Cambridge University Press.

Etienne Geoffroy Saint-Hilaire: two organs, whatever their forms and functions, are homologous if and only if they are connected to the same structures. Owen simultaneously introduced an operational criterion and a theoretical distinction between mere analogy (similar function) and true homology (comparable structure across organisms, which would be explained a few years later by Charles Darwin as having been inherited from a common ancestor). Unfortunately, this important distinction is often ignored in modern biology, a common function being generally deduced from a crude homology assessment. At the molecular level (genes and proteins), the establishment of homology that corresponds to the alignment of nucleotide or amino acid sequences relies on the very same principle: two positions are homologous if they are connected to similar neighbouring nucleotides or amino acids. The alignment of molecular sequences is a complex step for which numerous methods have been developed (for a review see Wallace *et al.* (2005)). In particular, the homology within poorly conserved regions can be extremely difficult to assess, even if the surrounding sequences are indisputably homologous. This explains why it is difficult to decide whether the positions inside highly variable regions are indeed homologous, especially in the presence of sequence length heterogeneity. Because even the best tree reconstruction methods are sensitive to the 'garbage in, garbage out' principle, noisy areas can severely disturb the phylogenetic analysis when they are included in a dataset (Lake, 1991). Consequently, a conservative approach is to discard the regions of homologous sequences that are too divergent to be aligned with confidence (e.g., Castresana (2000)).

8.2.2 Cladistics and maximum parsimony

To explain how a phylogenetic tree can be inferred from a set of homologous characters, we will begin with the principles of cladistic analysis (Hennig, 1966) because they are easily understandable. The elementary task of phylogenetic inference is to group the species according to their relatedness. This is achieved by identifying positions that have the same character state (e.g., the same amino acid) in a given group of species, but a different state in the remaining species. To illustrate our point, we will refer to Figure 8.1, where the oldest events are on the left of the trees and the most recent on the right. Trees that depict the relative order of the speciation events are said to be rooted. In the following, we will assume that the character state of the root (i.e., the ancestral state on the far left) is known. On the tree of Figure 8.1A, a change of state is indicated on the internal branch of the group composed of taxa sp3 and sp4. This substitution actually occurred in the common ancestor shared by these species. Such a single change that characterizes a monophyletic group is termed a synapomorphy (i.e., shared derived character state). All the trees that group the species sharing the same

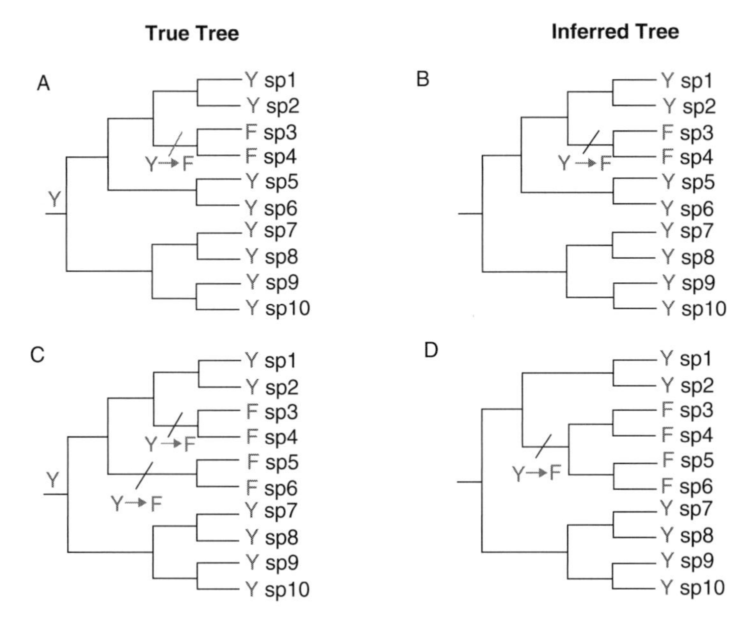

Fig. 8.1. Phylogenetic and non-phylogenetic signals. The true evolutionary histories of an amino acid position are depicted in A and C, while the inferred histories are depicted in B and D. In A, a single substitution occurred, which constitutes a phylogenetic signal, i.e., a synapomorphy allowing the recovery of the correct grouping (B). In C, two substitutions occurred and the same amino acid was independently acquired by convergence. This constitutes a non-phylogenetic signal that prevents the recovery of the correct tree (D).

character state will require a single substitution (Figure 8.1B) and will be preferred to the other trees that require two or more substitutions. The cladistic analysis is also known as maximum parsimony (MP) because it selects the tree(s) having the lowest number of substitutions (or steps). Problems crop up when the same character state is independently acquired by two unrelated lineages through convergence (Figure 8.1C). In such cases, the true phylogeny will require two substitutions, while erroneous trees grouping these unrelated lineages will require a single substitution (Figure 8.1D) and would be preferred. However, since not all characters support identical groups of species, cladistics tries to maximize the number of inferred synapomorphies by selecting the tree requiring the minimum number of steps, thereby following the principle of Occam's razor. The underlying hypothesis is that this approach should allow all synapomorphies and all convergences to be discovered. In practice, our assumption that the character state of the root is known is invalid. This issue is circumvented by applying the same parsimony criterion on unrooted trees. Then, the best tree is rooted using an outgroup, i.e., an externally recognized group of sequences that are a priori assumed to be located outside the group of interest. The rooting defines the polarization of the evolutionary history, which allows a chronological interpretation of the emergence order of the species and the character states in the ingroup.

8.3 Artefacts affecting phylogenetic reconstruction

8.3.1 Multiple substitutions and character saturation

In theory, the faster a character evolves the more phylogenetic signal (i.e., synapomorphies) it will contain. However, as soon as a single character undergoes multiple substitutions, it becomes difficult to extract its phylogenetic signal, because closely related species will not necessarily share the same character state (Figure 8.1C). Actually, for a fast evolving position, the number of required substitutions on two or more conflicting topologies can be so similar that the reconstruction method is unable to choose between the different alternatives. In practice, it is virtually impossible to extract the phylogenetic signal from very fast evolving characters, which are said to be mutationally saturated. When the level of saturation is high, inferred trees show a star-like topology, being totally unresolved, because the phylogenetic signal that would allow the different groups to be defined cannot be extracted.

8.3.2 Systematic biases and non-phylogenetic signals

Unfortunately, whenever many multiple substitutions accumulate in the data, the situation can be worse than a simple scenario based on the erosion of the phylogenetic signal. Indeed, in the presence of any kind of systematic bias, multiple substitutions will generate a non-phylogenetic signal potentially resulting in reconstruction artefacts, as we will illustrate in the case of the compositional bias of nucleotide sequences (Lockhart *et al.*, 1992). Let us assume that the content of guanine and cytosine (G+C) in a group is equal to 50%, except in the branches leading to species 1 and 3, for which the G+C content is 70% (Figure 8.2A). Because these two species have more often acquired a G or a C at the same position by convergence, while the remaining species have conserved the ancestral A or T, a compositional signal supporting the grouping of species 1 and 3 will be present. Furthermore, if the alignment used is so highly saturated that the phylogenetic signal for grouping species 1 and 2 has been seriously eroded, species 1 and 3 will be erroneously but strongly grouped together because of the non-phylogenetic signal due to the heterogeneous G+C content (Figure 8.2B).

The variation of evolutionary rates across lineages is another important cause of non-phylogenetic signal due to convergent evolution. It results in the long branch

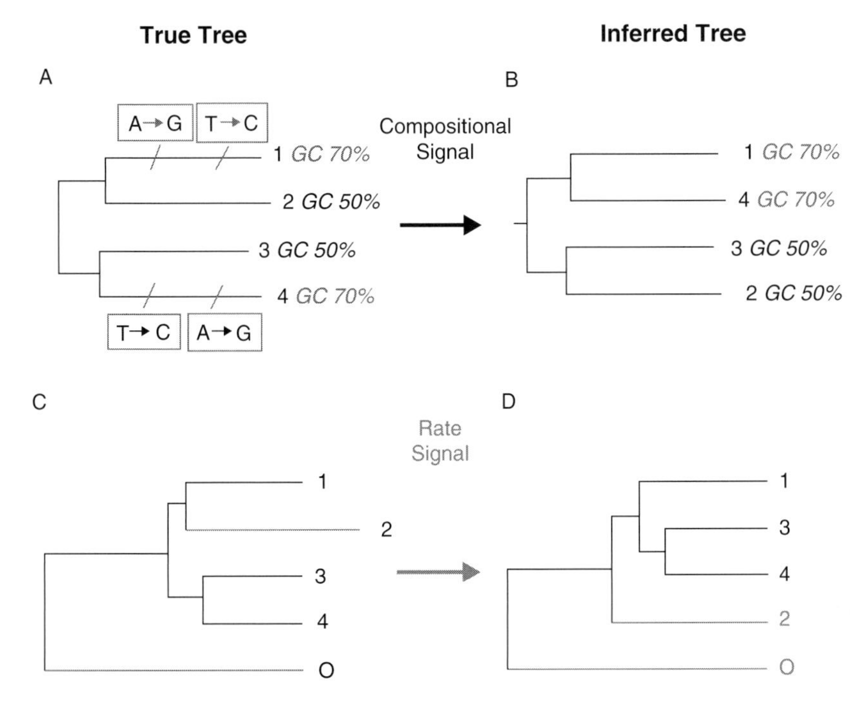

Fig. 8.2. Effects of non-phylogenetic signals on phylogenetic reconstruction. A. On the true tree, because of a bias of the substitutional process favouring the fixation of G and C versus A and T, the G+C content of lineages 1 and 3 independently increased from 50 to 70%. B. False tree resulting from the compositional signal. Lineages 1 and 3 are artefactually grouped because the convergence of their G+C content implies a high number of homoplastic positions, i.e., shared only by chance. C. On the true tree, after divergence from its common ancestor with lineage 1, lineage 2 was affected by an acceleration of its evolutionary rate. D. False tree inferred from the rate signal. Attracted by the long branch of the distant outgroup O, the long branch of lineage 2 artefactually emerges at the basis of the ingroup (species 1–4). This is the well-known LBA artefact.

attraction (LBA) artefact in which the two longest branches of a tree tend to be grouped even when not closely related (Felsenstein, 1978). A particular albeit frequent case may arise whenever a distant outgroup is included to root the tree, i.e., to allow the polarisation of the characters under study. Indeed, if a species of the studied group evolves significantly faster than the others (Figure 8.2C), it will artefactually emerge at the base of the group (Figure 8.2D), because its long branch is attracted by the long branch of the distant outgroup. This phenomenon wrongly suggests an early divergence of the fast species relative to the remaining species.

Early on, rate and compositional biases were identified as major sources of non-phylogenetic signals. An additional confounding bias that has drawn more attention is the heterotachous signal, which is due to a variation of the evolutionary rate of a given position throughout time (Kolaczkowski and Thornton, 2004; Lockhart *et al.*, 1996; Philippe and Germot, 2000) and remains to be thoroughly explored.

8.3.3 Gene duplication and horizontal gene transfer

For the sake of simplicity, we have so far assumed that it is sufficient to analyse homologous positions to infer phylogeny. Strictly speaking, this is correct as long as we are interested in gene phylogenies, but as soon as we want to deduce the species phylogenies, we must ensure that the genes under study are orthologous (Fitch, 1970). By definition, orthologous genes originate in speciation events, whereas paralogous genes originate in gene duplication events (Figure 8.3A).

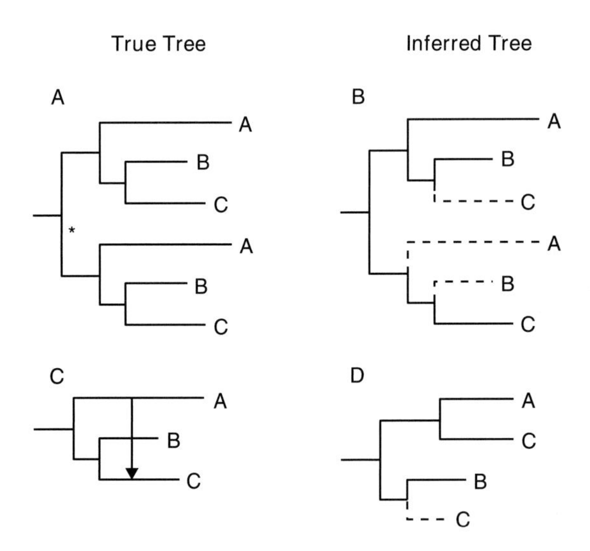

Fig. 8.3. Orthology, paralogy, and xenology and their consequence on phylogenetic inference. A. At some point in its history, a gene is duplicated and gives rise to two paralogous copies. The duplication event is indicated by a star. In the course of the subsequent speciation events, each copy evolves independently to generate a set of three orthologous genes. When a tree including both paralogues from each species (A, B, and C) is inferred, the true species phylogeny is recovered for each paralogue. B. In a tree inferred from different paralogues instead of orthologues, a wrong species phylogeny is recovered. The suboptimal gene sampling can be due to technical reasons (e.g., orthologous gene not yet sequenced) or to biological reasons (e.g., both copies have been differentially lost in the three lineages). C. True tree: during the evolution of lineages A, B, and C, a gene is horizontally transferred from lineage A to lineage C. D. False tree: because of the close similarity between the xenologue in C and the orthologue in A, a wrong organismal phylogeny grouping species A and C is recovered. As for paralogy above, the orthologue in C may be lacking for technical or biological reasons (e.g., the acquired xenologue has replaced the orthologous gene).

Phylogenetic analyses of species must be imperatively based on orthologous sequences, because trees based on cryptic paralogues can be extremely misleading (Figure 8.3B). Another problem is horizontal gene transfer (HGT), where a gene is transferred from a donor species to a possibly unrelated receiver species (Figure 8.3C). This alien copy, often called a xenologue, causes similar problems to paralogues, i.e., unrelated species are incorrectly grouped in phylogenetic trees (Figure 8.3D).

Since gene (and even genome) duplications are frequent in eukaryotes, while HGTs are frequent in prokaryotes, the number of genes that are perfectly orthologous when considering all comparable extant organisms is probably close to zero. For instance, only 14 genes were found to exist in one and only one copy in ten completely sequenced eukaryotic genomes (Philip *et al.*, 2005). Fortunately, if a given gene has undergone a single recent duplication event, the latter will be easy to detect and the gene will still be usable (Philippe *et al.*, 2005b) because it does not interfere with the species phylogeny. Although the same reasoning should hold for HGTs, it appears that the impact of transfer events on the phylogenetic structure is much more destructive (Philippe and Douady, 2003), even when only a few HGTs have occurred. While some researchers have suggested that because of transfer events species phylogenies cannot exist (nor can the tree representation be used) (Doolittle, 1999), the current consensus is that a core of genes (a few hundred within Bacteria) have experienced so few HGTs that they can be used to infer species phylogenies (for reviews see Brown (2003), Ochman *et al.* (2005), Philippe and Douady (2003)).

8.4 Strengths and limitations of phylogenomics

8.4.1 Stochastic error and the need for more data

Phylogenies based on a small number of characters (both morphological and molecular) are sensitive to stochastic (random or sampling) error. Consequently, the inferred trees are usually poorly resolved and often yield conflicting results, though differences are seldom statistically significant. By considering more characters and/or characters with more substitutions, the phylogenetic signal can be increased, since the stochastic error is due to the scarcity of substitutions that occurred along internal branches (the aforementioned synapomorphies). Starting from one or a few characters, as in the first classifications elaborated in the Middle Ages, to tens or a few hundred characters, as in most recent studies based on morphological characters, the rule ‘the more characters the better’ has always been applied. The advent of large-scale sequencing allowed a gain of about three orders of magnitude, resulting in an enormous improvement of the resolving power of phylogenetic inference. However, the switch from hundreds or a few thousand positions in single-gene phylogenies (e.g., rRNA tree (Woese, 1987)) to hundreds of thousand positions in phylogenomic studies based on complete genomes is quite recent. Phylogenies with a high statistical support for most nodes have been recently obtained for various groups, such as mammals (Madsen *et al.*, 2001; Murphy *et al.*, 2001), angiosperms (Qiu *et al.*, 1999), and eukaryotes (Rodriguez-Ezpeleta *et al.*, 2005).

8.4.2 Systematic error and the need for better reconstruction methods

While the use of many characters drastically reduces the stochastic error, it does not necessarily constitute a solution to the problem of tree reconstruction artefacts (Philippe *et al.*, 2005a). Indeed, the addition of more data increases both the phylogenetic and non-phylogenetic signals. Therefore, in the presence of a systematic bias, the latter will eventually become predominant, especially when the phylogenetic signal is rather weak (Jeffroy *et al.*, 2006). For example, a hyperthermophilic lifestyle gives rise to a systematically biased composition of all proteins (Kreil and Ouzounis, 2001), thus potentially leading to an artefactual albeit highly supported tree. Obviously, there is an urgent need to design reconstruction methods that are less sensitive to the systematic error induced by the use of large datasets. Hence, the research in this area is currently very active (Felsenstein, 2004). Although the previously presented MP method is intuitive, its improvement is difficult since it does not make any explicit assumptions about the underlying evolutionary process (see Steel and Penny (2000)). For example, the probability of a substitution event is implicitly assumed to be identical across all branches of the tree, whereas this assumption is clearly violated in cases where branch lengths are unequal, leading to the LBA artefact. In contrast, probabilistic methods such as maximum likelihood (ML) and Bayesian inference (BI) have been designed to take into account branch lengths and are therefore much less sensitive to the LBA artefact. More generally, in a probabilistic framework, the likelihood of a tree is computed using a model of sequence evolution able to handle numerous aspects of the underlying process of sequence evolution. This complex model enhances the extraction of the phylogenetic signal and greatly reduces the impact of non-phylogenetic signal because the probability that multiple substitutions have occurred is explicitly considered.

8.4.3 State of the art in evolutionary models

Currently, implemented models of sequence evolution have the following properties: (1) the various probabilities to substitute one character state by another are unequal (e.g., transitions are more frequent than transversions); (2) the stationary probabilities of the various character states can be unequal (e.g., A more frequent than T) and are generally estimated from observed frequencies; (3) evolutionary rates can be heterogeneous across sites (i.e., positions), this heterogeneity being usually modelled by a discrete gamma distribution (Yang, 1993). For nucleotides, the probabilities of the different substitutions are directly inferred from the data (GTR model (Lanave *et al.*, 1984)), whereas for amino acids, values previously computed from large alignments of closely related sequences are preferred (e.g., WAG substitution matrix (Whelan and Goldman, 2001)). Taking into account these evolutionary heterogeneities efficiently improves the extraction of the phylogenetic signal, as exemplified by the fact that the introduction of the Gamma distribution is often associated with changes of the tree topology (Yang, 1996). Other evolutionary features that have been modelled include: heterogeneity of the G+C content through a non-stationary model (Foster, 2004; Galtier and Gouy, 1995; Yang and Roberts, 1995), heterotachy through a covarion model (Galtier, 2001; Huelsenbeck, 2002), and heterogeneous probabilities of substitution types across sites using mixture models (Lartillot and Philippe, 2004; Pagel and Meade, 2004). Although newer models are complex and probabilistic methods have been shown to be robust to model violations, reconstruction artefacts can still interfere with sequence-based phylogenomic analyses, as we will illustrate later in a case study of animal evolution.

8.4.4 Current limits of phylogenomic performance

Profiting from the synergistic effects of the massive increase of sequence data and the improvement of tree reconstruction methods, the resolution of the tree of life is rapidly progressing (Delsuc *et al.*, 2005). However, a few important questions are expected to remain difficult to answer in the near future because the phylogenetic signal is either scarce or dominated by strong non-phylogenetic signals. Beside the afore-mentioned mutational saturation inherent in ancient events, the lack of phylogenetic signal can be due to the existence of short internal branches associated with numerous speciation events concentrated within a short time span (i.e., adaptive radiations). Furthermore, the resolution of ancient events is complicated by a dramatic reduction of the available data, because of the concomitant decrease in the number of orthologous genes (due to duplications and HGTs) and in the number of unambiguously aligned positions (due to considerable sequence divergence).

8.4.5 Corroboration from non-sequence-based phylogenomic methods

Formally defined as the inference of phylogenies from complete genomes, phylogenomics is not limited to primary sequence data. Instead, its principles can also be applied to virtually any heritable genomic feature such as gene content, gene order or intron positions (see Philippe *et al.* (2005a) for a review). Since the usual methods of tree reconstruction are used, strengths and limitations of phylogenies based on these other types of characters are very similar to those based on primary sequences. However, because these various characters are largely independent, they provide a major source of corroboration, which is of primary importance to validate historical studies (Miyamoto and Fitch, 1995). Indeed, the fact that phylogenies inferred from different character types converge to the same tree topology strongly suggests that the correct organismal tree has been reconstructed. Although such integrated approaches have rarely been applied so far, the first studies indicate a good congruence for Bacteria and Metazoa (for reviews see Delsuc *et al.* (2005), Philippe *et al.* (2005b)). Nevertheless, if the same systematic bias (e.g., rate acceleration) simultaneously affects all genomic features, the same reconstruction artefacts are likely to occur.

8.4.6 Case study: resolution of the metazoan evolution

Because of the general interest in animal evolution, we will present the resolution of this long-lasting problem as a case study to illustrate the theoretical concepts explained so far. Before 1997, metazoan taxonomy was essentially based on the presence or absence of true internal body cavities (coelom) (Adoutte *et al.*, 2000), with arthropods and vertebrates grouped, among others, into Coelomata to the exclusion of nematodes (Pseudocoelomata). Suspecting that the early emergence of the generally fast-evolving nematodes was the result of an LBA artefact (Philippe *et al.*, 1994), Alguinaldo *et al.* (1997) sequenced the SSU rRNA gene from dozens of nematodes, until they identified one slowly-evolving species, *Trichinella*. By using *Trichinella* they were able to overcome the LBA artefact, revolutionizing the picture of animal evolution by overruling the classical dichotomy between Coelomata and other animals. Instead, they found a new metazoan group named Ecdysozoa (Aguinaldo *et al.*, 1997). Including, among others, arthropods, nematodes, tardigrades, and onychophorans, these animals are characterized by a moult induced by a class of hormones known as the ecdysteroids. Nevertheless, several phylogenomic studies reject the Ecdysozoa hypothesis and find a significant support for the classical Coelomata hypothesis (Blair *et al.*, 2002; Dopazo *et al.*, 2004; Philip *et al.*, 2005; Wolf *et al.*, 2004), suggesting that the monophyly of Ecdysozoa represents an rRNA-specific anomaly. These analyses use a large number of characters but only a very limited number of species, i.e., the few completely sequenced model organisms. Furthermore, only species distantly related to animals (fungi, plants, or apicomplexans) are used as outgroups, thus increasing the probability of an LBA artefact.

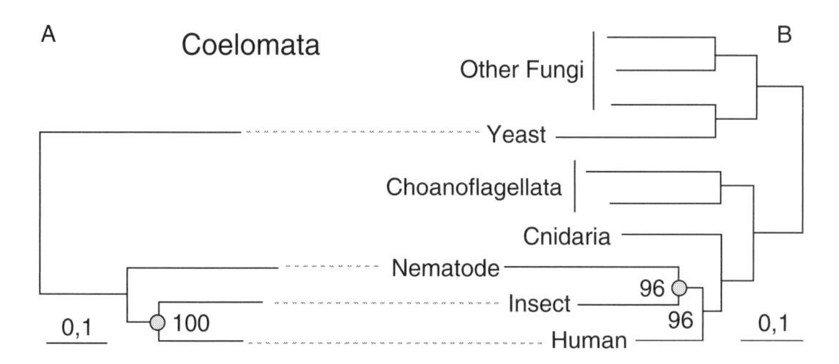

Fig. 8.4. Avoiding the LBA artefact through the use of a close outgroup. Phylogeny based on 146 genes inferred by the ML method. A. The long branch of the fast-evolving nematodes is attracted by the long branch of the yeast used as a distant outgroup. Because of this LBA artefact, the statistical support for the Coelomata (arthropods + chordates) hypothesis is maximal (100%). B. By breaking the long branch of the yeast, the addition of three basal animals as close outgroups allows the LBA artefact to be avoided and the true topology to be recovered. The statistical support for the Ecdysozoa (nematodes + arthropods) is now nearly maximal (96%). Redrawn from Delsuc *et al.* (2005).

To address this question, we assembled our own data set both species- and gene-rich (49 species, 146 proteins, and 35 371 amino acid positions) and our results strongly argue in favour of the group Ecdysozoa (Philippe *et al.*, 2005b). In agreement with previous studies, when a poor species sampling is used, a strong support for Coelomata is recovered (Figure 8.4A). In contrast, adding two choanoflagellates and a cnidarian (*Hydra*) has a dramatic effect since Ecdysozoa are now highly supported (Figure 8.4B). This topological change is not surprising because the longest branch of the tree (leading to the distant outgroup) has been broken (Delsuc *et al.*, 2005), which is known to reduce the impact of the artefact (Hendy and Penny, 1989). However, it is worth noting that this result is only achieved through the use of ML. Indeed, even in the presence of a close outgroup, the use of the LBA-sensitive MP still results in a similar support for Ecdysozoa and the artefactual Coelomata (data not shown).

Since reconstruction artefacts are primarily caused by multiple substitutions, eliminating the fastest-evolving characters should improve the quality of the phylogenetic inference, thereby reducing possible LBA artefacts. Actually, the removal of fast-evolving characters, performed using the SF method (Brinkmann and Philippe, 1999) produces exactly the same result as the addition of a close outgroup (Delsuc *et al.*, 2005), i.e., a topological shift from Coelomata to Ecdysozoa (Figure 8.5). Moreover, a similar result was independently obtained with a larger number of genes and two different data removal approaches (Dopazo and Dopazo, 2005; Philippe *et al.*, 2005b). Therefore, all these analyses demonstrate that

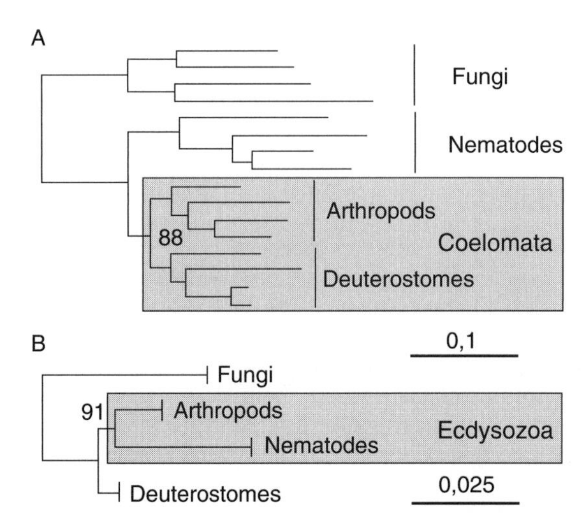

Fig. 8.5. Avoiding the LBA artefact through the elimination of fast-evolving positions. Phylogeny based on 146 genes inferred by the ML method. A. When all positions are considered, the fast-evolving nematodes are artefactually attracted by the long branch of fungi. B. When only the slowest evolving positions are used, nematodes are correctly located as the sister group of arthropods. Redrawn from Delsuc *et al.* (2005).

Ecdysozoa is a natural clade and that an LBA artefact is responsible for the high statistical support for Coelomata in species-poor studies.

While this case study shows that sophisticated reconstruction methods may still produce erroneous trees, there are major improvements due to the largely increased resolving power of phylogenomics. This is reflected in the fact that almost all other nodes in the metazoan phylogeny are well supported (Delsuc *et al.*, 2006; Philippe *et al.*, 2005b). In addition, reconstruction errors are drastically reduced as soon as a large number of species is considered, which is likely to become the rule in the next few years. Furthermore, starting from a large phylogenomic data set allows highly saturated characters to be removed, which can be very useful in cases where a close outgroup is not available (Brinkmann *et al.*, 2005; Burleigh and Mathews, 2004; Philippe *et al.*, 2005b).

8.5 The importance of secondary simplification

8.5.1 The rise and fall of Archezoa

As for Metazoa, advances in phylogenetic analyses have greatly modified our view of the eukaryotic tree. Originally, both morphological (Cavalier-Smith, 1987) and molecular studies (mainly based on rRNA) (Sogin, 1991) suggested that several simple lineages (e.g., devoid of mitochondrion, Golgi apparatus, or

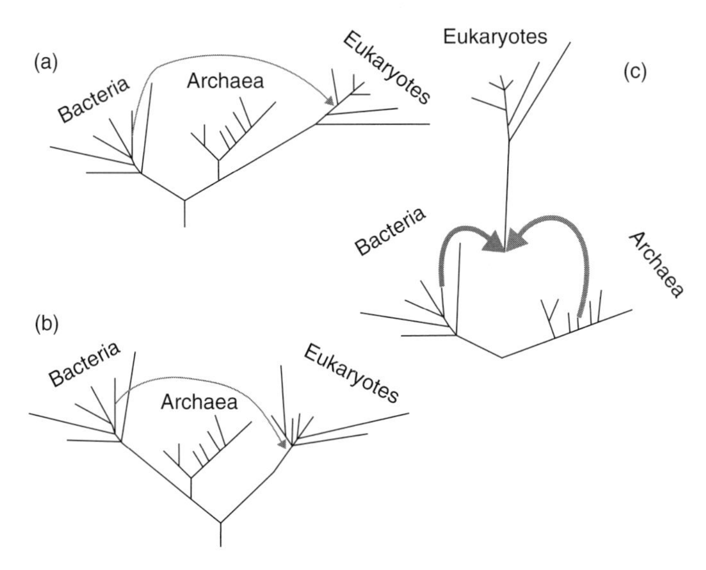

Fig. 8.6. Most common views of the universal tree of life. A. Schematic representation of Woese's paradigm. The root is located on the bacterial branch and the mitochondrial endosymbiosis occurred relatively late during the evolution of eukaryotes. B. Woese's tree corrected for LBA artefacts. Archaea and Bacteria are sister groups, rendering prokaryotes monophyletic. The diversification of extant eukaryotes occurred relatively recently, after the mitochondrial endosymbiosis. C. Genome fusion at the origin of eukaryotes.

peroxysome) emerged in a stepwise fashion at the base of eukaryotes, followed by an unresolved multifurcation containing all complex, often multicellular, eukaryotes. The common interpretation of the classical rRNA tree was that these simple groups (e.g., microsporidia, diplomonads, and trichomonads) were the direct descendants of genuinely primitive organisms and represented intermediate stages in the progressive complexification of eukaryotic cells. Hence, they were named Archezoa (Cavalier-Smith, 1987). This suggested that the endosymbiosis with an alpha-proteobacterium that gave rise to mitochondria (see below) had occurred relatively late in the course of eukaryotic evolution (Figure 8.6A). Therefore, the study of eukaryotes that were supposed never to have possessed mitochondria was regarded to be of prime importance for the understanding of early eukaryotic evolution (Sogin, 1991). The observation that hundreds of genes had been transferred from the protomitochondrion to the nucleus (Lang *et al.*, 1999) further underlined the dramatic modification of eukaryotic cells due to the mitochondrial endosymbiosis.

Unfortunately, a wealth of newly established data has been demonstrating more and more solidly that the Archezoa hypothesis is incorrect and that most likely none of the premitochondriate eukaryotes had survived (Embley and Hirt, 1998;

Philippe and Adoutte, 1998). First, the use of advanced reconstruction methods and/or of protein-encoding genes revealed that early rRNA trees (Sogin, 1991) were severely biased by LBA artefacts due to the non-phylogenetic rate signal (Edlind *et al.*, 1996; Philippe, 2000; Silberman *et al.*, 1999; Stiller and Hall, 1999). In fact, the lineages that emerge early in the rRNA tree are simply fast-evolving organisms that are erroneously attracted towards the base by the distant outgroup (Archaea). For example, microsporidia turned out to be highly derived fungi, while they had previously been thought to be genuinely 'primitive' early eukaryotes (Keeling and Fast, 2002). Actually, the correct placement of microsporidia is difficult to recover (Brinkmann *et al.*, 2005) because for most, but not all, of their proteins the non-phylogenetic signal (due to a high evolutionary rate) is stronger than the genuine phylogenetic signal.

8.5.2 Current view of mitochondrial origin and evolution

Since the new phylogenetic scheme places the former Archezoa back into the main eukaryotic radiation (known as the crown), it implies that the last common ancestor of extant eukaryotes (LCAEE) was much more complex than previously thought, i.e., the LCAEE probably presented all features that are typical for the crown group. Therefore, numerous independent secondary simplifications must have occurred to generate simple extant eukaryotes from such a complex common ancestor. For instance, the LCAEE likely contained a large number of spliceosomal introns, which were then massively and independently lost in most lineages (Roy and Gilbert, 2005). Similarly, ex-Archezoa were first interpreted as resulting from multiple mitochondrial losses. Accordingly, nuclear genes of mitochondrial origin found in these organisms would have been transferred to their nucleus before the loss of mitochondria (Clark and Roger, 1995; Germot *et al.*, 1997; Roger *et al.*, 1998). However, remnants of mitochondria, i.e., small double-membrane-bound organelles containing nuclear-encoded mitochondrial proteins, have subsequently been identified in all putatively 'amitochondriate' organisms investigated (Bui *et al.*, 1996; Tovar *et al.*, 1999; Tovar *et al.*, 2003; Williams *et al.*, 2002). This indicates that the previously called 'amitochondriate' organisms are simply lacking 'aerobic respiration', the most prominent function of mitochondria. Indeed, there are other functions performed by mitochondria that have persisted in all anaerobic eukaryotes, such as the synthesis of iron–sulphur clusters in diplomonads (Tovar *et al.*, 2003).

8.5.3 Current view of eukaryotic evolution

Following the rejection of the Archezoa hypothesis, some progress has been achieved in the resolution of deep eukaryotic phylogeny. Consequently, the division of all extant eukaryotes into six major groups has been proposed (Adl *et al.*, 2005; Keeling *et al.*, 2005; Patterson, 1999; Simpson and Roger, 2004). Some of these supergroups are solidly established, i.e., Opisthokonta (animals, choanoflagellates, and fungi) and Plantae (all three primary photosynthetic eukaryotes: green plants, red algae, and glaucophytes). Moreover, there is accumulating evidence for both Amoebozoa (containing a large part of amoebas, e.g., *Dictyostelium* and *Entamoeba*) and Chromalveolata (alveolates, stramenoplies, haptophytes, and cryptophytes); however, a final test based on the analysis of the phylogenomic data set containing representatives from all major lineages from this supergroup is still missing. The support for the monophyly of the two remaining groups (Excavata and Rhizaria) is much more tenuous and definitively requires more sequence data in the form of genome or expresseq sequence tag (EST) projects. Finally, it is noteworthy that anaerobic/amitochondriate species appear to be found in most of the six supergroups.

8.5.4 Philosophical grounds for the rejection of secondary simplification

Despite the aforementioned progress in eukaryotic phylogeny, the classical view, essentially rRNA-based, still largely prevails. Hence, we will try to explain why it is so difficult to obliterate, even in light of convincing evidence. The traditional taxonomy has been influenced by the assumption that the evolution of life was a steady rise to higher complexity, starting from 'primitive' or 'lower' (i.e., simple) organisms and ending with 'evolved' or 'higher' (i.e., complex) forms, especially humans. This conception is actually pre-Darwinian and can be traced back to Aristotle's *Scala Naturae*, the great chain of being (Nee, 2005). As a result, organisms having an apparently simple morphology were naturally located at the base of the tree of life. In contrast, all recent molecular phylogenetic studies demonstrate that secondary simplification constitutes a major evolutionary trend, encountered at all taxonomic levels. While this conclusion had already been drawn 60 years ago by André Lwoff (1943), its very slow acceptance by the scientific community was remarkably predicted in the very same book. Briefly, the idea of complexification is tightly linked to the concept of progress through the implicit equation 'progress = evolution towards more complexity'. Consequently, the simplification process has always been affected by a strong negative connotation tending to its denial (Lwoff, 1943). From a sociological point of view, it should be noted that the rediscovery of simplification occurred concomitantly with serious criticisms of the ideology of progress. Nevertheless, emphasizing simplification does not deny that complexification did occur, but rather means that both processes should be taken into account for the reconstruction of the evolutionary past (Forterre and Philippe, 1999; Gould, 1996).

8.6 The tree of life

8.6.1 Solid facts and open questions

In their long quest towards the resolution of the tree of life, phylogeneticists generally agree that several of its major branches can be currently considered as reliably inferred. For example, the monophyly of two of the three domains of life, Bacteria and Eukaryotes, is supported by numerous genes and does not seem to result from any known reconstruction artefact (Brown *et al.*, 2001; Philippe and Forterre, 1999). Of course, the monophyly of a domain depends as a last resort on the location of the root of the tree of life because falling within a domain would render it paraphyletic rather than monophyletic (see below). In contrast, the monophyly of Archaea, which is often taken for granted (Woese *et al.*, 1990), has never been significantly supported, even in multiple gene analyses (Brown *et al.*, 2001; Daubin *et al.*, 2002). Using a rate-invariant method applied to SSU rRNA, Lake (1988) proposed that Crenarchaeota are related to Eukaryotes and Euryarchaeota to Bacteria, thus rendering Archaea paraphyletic whatever the location of the root. A subsequent analysis of both the SSU rRNA and LSU rRNA, as well as of the concatenation of the two largest subunits of the RNA polymerase, led to the same conclusion (Tourasse and Gouy, 1999). Hence, the monophyly of the archaeal domain is not yet established and deserves further studies.

The ancient events that are by far the most strongly supported are the endosymbiotic origins of the mitochondrion and the plastid, respectively from an alphaproteobacterium and a cyanobacterium. Moreover, within the three domains, the monophyly of the major phyla (e.g., Proteobacteria, Spirochaetes, Cyanobacteria, Crenarchaeota, Animals, Ciliates) is consistently recovered (Brochier *et al.*, 2000; Daubin *et al.*, 2002; Philippe *et al.*, 2005b; Wolf *et al.*, 2001).

As we have already pointed out, there nevertheless exist a few situations where the phylogenetic signal may be genuinely weak (or at least difficult to recover). These include: (1) very rapid successions of speciation events that are often characteristic for an adaptive radiation; (2) the presence of a very strong non-phylogenetic signal (e.g., the rate signal in microsporidia); or (3) the absence of closely related outgroup species due to the extinction of related groups (e.g., coelacanth or *Amborella*). The lack of close outgroups is particularly worrisome because it concerns various important groups, among which are angiosperms, mammals, birds, tetrapods, and even the three domains of life: Archaea, Bacteria, and Eukaryotes. In all these cases, there exists an increased risk of tree reconstruction artefacts due to the long branch of the outgroup attracting any fast-evolving ingroup such as microsporidia or kinetoplastids (Brinkmann *et al.*, 2005).

Therefore, the most important open questions concerning the tree of life are the reliable positioning of its root (addressed in the next section), followed by the relative branching order of the major phyla within each of the three domains. Not surprisingly, both kinds of issues are undisputedly affected by LBA artefacts (Brochier and Philippe, 2002; Lopez *et al.*, 1999; Philippe *et al.*, 2000) because of the great divergence separating such ancient groups.

8.6.2 The current paradigm: a bacterial rooting of the tree of life and a sister-group relationship between Archaea and Eukaryotes

Nowadays, a majority of researchers in the field regard as an established fact the specific relationship between Archaea and Eukaryotes that is deduced from a bacterial rooting of the tree of life. Originally proposed in 1989 (Gogarten *et al.*, 1989; Iwabe *et al.*, 1989) on the basis of anciently duplicated genes, this scenario was explicitly formulated the year after by Carl Woese and his coworkers (Woese *et al.*, 1990). Because Woese *et al.*'s seminal paper led to the rapid and wide acceptance of these ideas, we will refer to the bacterial rooting of the tree of life and to the associated sister group relationship between Archaea and Eukaryotes as Woese's paradigm. So far, fewer than ten pairs of anciently duplicated genes, which already existed as two copies in the last universal common ancestor (LUCA) of extant life, have been identified. These universal paralogues can, in principle, be used to localize the root of the tree of life. Consistent with the paradigm, the analysis of all pairs but one (Lawson *et al.*, 1996) by standard phylogenetic methods results in the bacterial rooting of the tree of life. However, there are serious reasons to assume that this rooting is actually due to an LBA artefact (Brinkmann and Philippe, 1999; Forterre and Philippe, 1999; Lopez *et al.*, 1999; Philippe and Forterre, 1999).

Indeed, for nearly all molecular markers analysed for the rooting of the tree of life, the branches of Bacteria are the longest in each subtree, which potentially leads to their artefactual attraction by the very long branch of the outgroup (i.e., the paralogous group). Moreover, the long branch observed in Bacteria is also found not only for anciently duplicated genes, but also for most genes involved in translation and transcription. Such a long branch likely stems from an acceleration of the evolutionary rate. A possible explanation for this phenomenon could be a simplification process during which Bacteria were radically streamlining the informational system inherited from a more complex common ancestor shared with Archaea and Eukaryotes. According to this view, the multisubunit RNA polymerase and its TATA-box binding protein (the central element of the regulation of expression) found in Archaea and Eukaryotes would be ancestral, while the highly simplified bacterial polymerase (of the $\alpha 2\beta\beta'$ type) and its single sigma factor would be derived (Brinkmann and Philippe, 1999). Because this drastic change has likely been associated with an accelerated evolutionary rate of Bacteria, the higher similarity between Archaea and Eukaryotes, especially observed for non-metabolic (informational) genes, is best explained as the result of the shared conservation of ancestral features (symplesiomophies), which are not informative in terms of phylogeny.

Furthermore, an analysis of the SRP54–SRα pair (involved in protein secretion), the most similar pair of anciently duplicated genes, focusing on the most slowly evolving positions, recovers a eukaryotic rooting for the SRP54 part of the tree (Brinkmann and Philippe, 1999). More precisely, in the SRP54 subtree, eukaryotes are basal, while prokaryotes form a monophyletic group composed of paraphyletic Archaea and fast evolving Bacteria emerging from them. The reduced distance of the outgroup coupled with the use of the slowly evolving positions greatly enhances the phylogenetic–non-phylogenetic signal ratio, as shown for animals (Figures 8.4 and 8.5) and makes the eukaryotic rooting the most reliable hypothesis. Nevertheless, it should be noticed that this result is not statistically significant because too few positions remain available.

Certain authors interpret a shared insertion in the catalytic domain of the vacuolar ATPases (V-ATPases) from Archaea and Eukaryotes as a potential synapomorphy arguing for a sister group relationship of these two domains (Gogarten *et al.*, 1989; Zhaxybayeva *et al.*, 2005). However, the divergence observed between the V-ATPases and the bacterial F-ATPases is huge. Hence, there are at least three points to consider before drawing any conclusions based on V-ATPases. First, because producing a reliable alignment of the relevant part of the orthologues is not obvious, we do not feel able to position the shared insertion in an alignment of the two paralogues (which are even more distant). Second, the ATPase family has three paralogues with a wide distribution within Bacteria and it is almost impossible to clearly establish which of the three bacterial copies is orthologous to the vacuolar ATPase (Philippe and Forterre, 1999). Third, in addition to most Archaea, there is a wide diversity of bacteria that actually possess both subunits of the V-ATPase, thus making questionable the scenario of an archaeal origin followed by a lateral transfer to Bacteria (Philippe and Forterre, 1999; Zhaxybayeva *et al.*, 2005).

In conclusion, answering the question of the root of the tree of life and the related question of the basal groups within the three domains is of prime importance for our understanding of both LUCA and the early evolution of life on Earth. Indeed, in the standard scenario derived from early studies (Burggraf *et al.*, 1992; Gogarten *et al.*, 1989; Iwabe *et al.*, 1989), LUCA was a prokaryote-like hyperthermophilic organism (Figure 8.6A), while in our alternative scenario based on refined phylogenetic analyses (Brinkmann and Philippe, 1999; Brochier *et al.*, 2002), LUCA was an eukaryote-like mesophilic organism (Figure 8.6B). Other important arguments in favour of an eukaryotic rooting of the tree of life are based on the RNA world hypothesis and are not detailed here (see Poole *et al.* (1998, 1999)).

8.7 Frequent strong claims made with weak evidence in their favour

8.7.1 A genome fusion at the origin of eukaryotes

Starting a century ago with Mereschkowski's symbiogenesis (1905), several genome fusion or cell fusion scenarios have been proposed to explain the origin of eukaryotes (Lopez-Garcia and Moreira, 1999; Margulis, 1971). While these scenarios were originally grounded in cell biology, they gained acceptance when the rise of sequencing techniques suggested a chimerical nature for the eukaryotic cell, i.e., a greater similarity of metabolic genes to Bacteria and of informational genes to Archaea. The need to explain this chimerical nature has then triggered the development of a multitude of additional fusion scenarios beginning in the late 1980s (Zillig, 1987) and culminating in the 1990s (Gupta and Golding, 1996; Martin and Muller, 1998; Moreira and Lopez-Garcia, 1998). These scenarios can be separated into early and late variants depending on whether the fusion is supposed to have occurred before or after the divergence of the major prokaryotic phyla. Since they all assume that eukaryotes originated in a fusion of a bacterium and an archaeon, these scenarios actually imply that there were never any ancestral eukayotes prior to the fusion event, with Bacteria and Archaea being the only true ancestral lineages. Fusion scenarios are essentially compatible with Woese's paradigm (compare Figures 8.6A and 8.6C). An additional yet obvious corollary of fusion scenarios is that they are supposed to have happened only once in the past and are heavily inspired by contemporary organisms that thrive in the same habitat. Finally, fusion scenarios most likely require partners lacking cell walls, which would otherwise hinder the fusion.

As already mentioned, the fundamental basis of these fusion scenarios is grounded in the observation that metabolic (operational) genes from eukaryotes are predominantly more similar to those of Bacteria, whereas genes of the genetic machinery (replication, transcription, and translation) are more similar to those of Archaea. The similarity to bacterial metabolic genes can be straightforwardly explained by the massive gene transfer to the nucleus following both mitochondrial and plastid endosymbioses (Lang *et al.*, 1999). In contrast, the similarity of the genetic machinery between Archaea and eukaryotes can be interpreted in two radically different ways. The standard interpretation is that this similarity reflects a genuine common ancestry (synapomorphies), while the alternative is that it is due to sharing of ancestral states (symplesiomorphies), Bacteria being less similar because of an accelerated evolutionary rate. It is primordial to know approximately when this fusion event is proposed to have happened, because this will determine the expected relationships between nuclear genes of extant eukaryotes and genes found in the other two domains. Furthermore, one has to differentiate fusion hypotheses where the mitochondrial organelle has been simultaneously produced from those where the mitochondrial endosymbiosis is supposed to have happened later. For our purposes, we feel that a relatively cautious proposition should be the best starting point to verify deducible predictions. To our knowledge the most stringent formulation of a fusion scenario fitting the established scientific facts is the Hydrogen Hypothesis (Martin and Muller, 1998). The elegance of this scenario, which is essentially based on metabolic considerations, lies in the proposition that the bacterial partner (an alpha-protobacterium) will subsequently become the mitochondrion. In the related, yet more complex, Syntrophic Hypothesis (Moreira and Lopez-Garcia, 1998), the original bacterial partner is a delta-proteobacterium and the mitochondrial endosymbiosis occurs later as a separate event. Finally, the archaeal partner postulated by both scenarios is a methanogenic archaeon.

If we take a closer look at the bacterial part of the postulated fusion, there are many eukaryotic genes that are specifically related to alpha-proteobacteria, with most of them being encoded in the nucleus (Lang *et al.*, 1999). This relationship is recovered despite the usually faster rate of evolution of eukaryotic sequences of mitochondrial origin relative to their bacterial counterparts. An even more specific and solid relationship is found for the eukaryotic genes of cyanobacterial origin that were introduced by the plastid endosymbiosis. In addition, many genes located in the eukaryotic nucleus that are similar to bacterial genes cannot be associated with certainty to either of these two possibilities. This is not surprising for several reasons. First, the most likely is that eukaryotic sequences are so divergent that the genuine phylogenetic signal has been erased. Second, the corresponding gene could have been lost in the original bacterial lineage. Third, some eukaryotic genes could have been laterally acquired from other Bacteria (Doolittle, 1998). While we could conclude that the observations are in excellent agreement with the predictions, we should keep in mind that we are actually looking at the evidence of the solidly established mitochondrial endosymbiosis. Indeed, the latter observations are theoretically indistinguishable from the predictions for the bacterial part of the Hydrogen Hypothesis. Consequently, the only true phylogenetic test is to focus on the genetic machinery and to look for the sister group relationships between the genes of eukaryotes and those of methanogenic Archaea that are expected by both metabolic hypotheses (Martin and Muller, 1998; Moreira and Lopez-Garcia, 1998).

While multiple examples exist in which eukaryotic genes are more similar to their archaeal than to the bacterial counterpart, these usually involve distant relationships and do not point to any particular group of Archaea. This lack of specific affiliation is a major problem for the late fusion scenarios, because it suggests either that the sequences under study are too divergent or that the hypothetic methanogenic partner was not related to extant methanogens and has become extinct. To further address these two possibilities we will give first some additional information. In the analyses of the anciently duplicated genes that were used to infer the root of the tree of life, Archaea are usually the most slowly evolving group. Invariantly, any fusion scenario implies that the eukaryotic genes found in the nucleus of both bacterial and archaeal ancestry started to diverge from their prokaryotic counterparts at the same time. Therefore, it is difficult to explain the fact that the bacterial genes could be easily traced back to an alpha-proteobacteria, while the supposedly archaeal genes could not be affiliated to any extant methanogenic archaeon. In addition, the nuclear genes of archaeal ancestry have probably not been subject to the secondary simplification pressures that have been acting on mitochondrial genes of bacterial ancestry (and thus do not show an accelerated evolutionary rate). Taken together, these points rule out a large divergence as an explanation for the lack of a specific association between genes from eukaryotes and methanogenic Archaea comparable to the association observed between alpha-proteobacteria and mitochondrial genes. In order to examine the second possibility, i.e., that the hypothetic methanogenic partner was not related to extant methanogens and has become extinct, it is worth noting that methanogenesis most likely appeared only once within the domain Archaea (Slesarev *et al.*, 2002). Moreover, methanogens are limited to a monophyletic group located in a derived position within the subdomain Euryarchaeota (Bapteste *et al.*, 2005), which means that the ancestral metabolic type of this group was probably thermoacidophilic, as are the extant members of the subdomain Crenarchaeota. Finally, it seems that the split between these two archaeal subdomains is very deep (Brochier *et al.*, 2005) and likely older than the origin of the late arising alpha-Proteobacteria, because the latter only evolved after the separation of the epsilon, the delta, and the common ancestor of alpha and beta/gamma proteobacteria. Hence, the lack of any specific affinity to extant methanogenic Archaea is difficult to reconcile with the predictions of the two late fusion scenarios for the origin of eukaryotes. While phylogenetic data favour an extinct archaeal linaeage that would be basal to the known archaeal diversity, the assumption of contemporaneity inherent to any fusion scenario makes this hypothetical archaeal partner incompatible with the time frame of the bacterial partner. Indeed, this would suggest that the primary separation into Cren- and Euryarchaeota only occurred after the diversification of alpha-proteobacteria (Cavalier-Smith, 2002). Therefore, sequence similarities between eukaryotes and Archaea would be better explained in terms of common ancestry, as in Woese's original paradigm (Figure 8.6A) or symplesiomorphies (Figure 8.6B).

8.7.2 A hyperthermophilic origin of life

Claims about a hyperthermophilic origin of life are based on two different lines of arguments. The first involves the geological record, while the second results from early phylogenetic analyses of the SSU rRNA gene. In its distant past,

Earth was very hot due to the frequent impacts of large meteorites that were generating enormous amounts of energy. According to most palaeontologists, life started very early in the history of our planet (about 3 800 mya) and only shortly after the end of the massive meteorite bombardment. Therefore, an ancestral adaptation to extreme thermophilic conditions would make sense, at least at first sight. Formally, an organism is recognized as a hyperthermophile when its optimal growth temperature is above 80 °C. Such a property is not known for any extant eukaryote. On the other hand, hyperthermophily in Bacteria is found in two groups, Thermotogales and Aquificales, that were among the most basal lineages in early rRNA trees. However, analyses suggest that these groups are probably related and not the most basal lineages in Bacteria (Brochier and Philippe, 2002; Daubin *et al.*, 2002; but see Griffiths and Gupta, 2004). Actually, Thermotogales and Aquificales are only moderate hyperthermophiles, the most extreme hyperthermophiles being found within the Archaea. In that group, hyperthermophiles are broadly distributed, suggesting a hyperthermophilic origin of extant Archaea (Forterre *et al.*, 2002). Interestingly, the reverse gyrase (involved in DNA supercoiling) found in Thermotogales and Aquificales appears to be of independent archaeal origin. This indicates that at least some of the sophisticated mechanisms required to thrive at high temperature first evolved in the common ancestor of Archaea to be only secondarily horizontally transferred to bacterial lineages adapting to a hyperthermophilic lifestyle (Forterre *et al.*, 2000).

Nevertheless, the observation that there is no known hyperthermophilic eukaryote is often overlooked. The explanation most likely lies in the high instability of RNA at elevated temperatures, which implies that the half-life of messenger RNA (mRNA) molecules is drastically reduced in a hyperthermophilic environment. While this is not a major issue for prokaryotes, which have a coupled transcription/translation machinery, it certainly penalises eukaryotes. In eukaryotes, the premature mRNAs need to be processed (intron excision, addition of a cap and a polyA-tail) before transfer across the nuclear envelope, in order to overcome the physical separation of transcription (nucleus) and translation (cytosol). These facts argue strongly against the possibility that hyperthermophilic eukaryotes ever existed and limit possible scenarios for the relationships of the three domains of life. Indeed, under a bacterial rooting, where eukaryotes derive late from Archaea, a hyperthermophilic origin of life could be possible, whereas under an eukaryotic rooting, it would be excluded. However, the latter scenario is compatible with a secondary adaptation of the monophyletic prokaryotes to hyperthermophilic conditions, for example in the context of the Thermoreduction Hypothesis (Forterre, 1995).

Whatever scenario of the origin and early evolution of life is preferred, be it thermo-, meso-, or even psychrophilic, there is an almost general consensus about the existence of an intermediate phase in the history of Earth known as the 'RNA world'. According to this convincingly substantiated theory (Jeffares *et al.*, 1998; Poole *et al.*, 1998), many functions today catalysed by protein-based enzymes were originally performed by catalytic RNA molecules. Because of the thermolability of RNA, the RNA world hypothesis seems impossible to reconcile with the extreme thermal conditions associated with a hyperthermophilic origin of life. Besides, current studies of thermophilic organisms and thermoadaptive mechanisms indicate that life at very high temperatures relies on the establishment and maintenance of complicated devices that are specific to present-day hyperthermophiles. Taken together, these observations support the idea of a secondary adaptation of already evolved and quite complex cells to thermo- and later hyperthermophilic conditions, rather than a hyperthermophilic origin of life.

8.8 Conclusions

The advent of large-scale sequencing techniques has given rise to phylogenomics, a new discipline that attempts to infer phylogenetic relationships from complete genome data. Though not immune to the systematic biases present in some lineages, phylogenomics has largely benefited from the progress of probabilistic methods. Its resolving power has led to a drastic revision of the tree of life. One important lesson from our current understanding, especially in the eukaryotic domain, is that the equation 'simple equals primitive' rarely holds, i.e., many morphologically simple organisms have actually evolved from complex ancestors through secondary simplification. This does not mean that the eukaryotic cell was complex from its very beginning, but rather that all extant eukaryotes can be traced back to an already highly evolved common ancestor, which has logically phased out its inferior competitors. Similarly, we have provided arguments in favour of a relatively sophisticated LUCA for all extant life. In that context, prokaryotes would be the products of a streamlining process. While the emergence of the prokaryotic cell may reflect a secondary adaptation to thermophily (still largely present in Archaea), it is noteworthy that the RNA-world hypothesis rules out per se a hyperthermophilic origin of life. Finally, there seem to be no stringent constraints on the early evolution of life on Earth from a phylogenetic perspective. Hence, future research in astrobiology should not rely on the biased picture of evolution that was largely prevailing at the turn of the last century while looking for the manifestations of life on other planets.

Acknowledgements

This work was supported by operating funds from Genome Québec and NSERC. H. P. is a member of the Program in Evolutionary Biology of the Canadian Institute for Advanced Research (CIAR), whom we thank for interaction support, and is grateful to the Canada Research Chairs Program and the Canadian Foundation for Innovation (CFI) for salary and equipment support. D. B. is a postdoctoral researcher of the Fonds National de la Recherche Scientifique (FNRS) at the University of Liège (Belgium). D. B. is gratefully indebted to the FNRS for the financial support of his stay at the University of Montréal.

References

Adl, S. M., Simpson, A. G., Farmer, M. A., *et al.* (2005). The new higher level classification of eukaryotes with emphasis on the taxonomy of protists. *J. Eukaryot. Microbiol.*, **52**, 399–451.

Adoutte, A., Balavoine, G., Lartillot, N., *et al.* (2000). The new animal phylogeny: reliability and implications. *Proc. Natl. Acad. Sci. USA*, **97**, 4453–4456.

Aguinaldo, A. M., Turbeville, J. M., Linford, L. S., *et al.* (1997). Evidence for a clade of nematodes, arthropods and other moulting animals. *Nature*, **387**, 489–493.

Bapteste, E., Brochier, C., and Boucher, Y. (2005). Higher-level classification of the Archaea: evolution of methanogenesis and methanogens. *Archaea*, **1**, 353–363.

Blair, J. E., Ikeo, K., Gojobori, T., and Hedges, S. B. (2002). The evolutionary position of nematodes. *BMC Evol. Biol.*, **2**, 7.

Brinkmann, H., Giezen, M., Zhou, Y., Raucourt, G. P., and Philippe, H. (2005). An empirical assessment of long-branch attraction artefacts in deep eukaryotic phylogenomics. *Syst. Biol.*, **54**, 743–757.

Brinkmann, H., and Philippe, H. (1999). Archaea sister group of Bacteria? Indications from tree reconstruction artifacts in ancient phylogenies. *Mol. Biol. Evol.*, **16**, 817–825.

Brochier, C., and Philippe, H. (2002). Phylogeny: a non-hyperthermophilic ancestor for bacteria. *Nature*, **417**, 244.

Brochier, C., Bapteste, E., Moreira, D., and Philippe, H. (2002). Eubacterial phylogeny based on translational apparatus proteins. *Trends Genet.*, **18**, 1–5.

Brochier, C., Forterre, P., and Gribaldo, S. (2005). An emerging phylogenetic core of Archaea: phylogenies of transcription and translation machineries converge following addition of new genome sequences. *BMC Evol. Biol.*, **5**, 36.

Brochier, C., Philippe, H., and Moreira, D. (2000). The evolutionary history of ribosomal protein RpS14: horizontal gene transfer at the heart of the ribosome. *Trends Genet.*, **16**, 529–533.

Brown, J. R. (2003). Ancient horizontal gene transfer. *Nat. Rev. Genet.*, **4**, 121–132.

Brown, J. R., Douady, C. J., Italia, M. J., Marshall, W. E., and Stanhope, M. J. (2001). Universal trees based on large combined protein sequence data sets. *Nat. Genet.*, **28**, 281–285.

Bui, E. T., Bradley, P. J., and Johnson, P. J. (1996). A common evolutionary origin for mitochondria and hydrogenosomes. *Proc. Nat. Acad. Sci. USA*, **93**, 9651–9656.

Burggraf, S., Olsen, G. J., Stetter, K. O., and Woese, C. R. (1992). A phylogenetic analysis of Aquifex pyrophilus. *Syst. Appl. Microbiol.*, **15**, 352–356.

Burleigh, J. G., and Mathews, S. (2004). Phylogenetic signal in nucleotide data from seed plants: Implications for resolving the seed plant tree of life. *Am. J. Bot.*, **91**, 1599–1613.

Castresana, J. (2000). Selection of conserved blocks from multiple alignments for their use in phylogenetic analysis. *Mol. Biol. Evol.*, **17**, 540–552.

Cavalier-Smith, T. (1987). Eukaryotes with no mitochondria. *Nature*, **326**, 332–333.

Cavalier-Smith, T. (2002). The neomuran origin of archaebacteria, the negibacterial root of the universal tree and bacterial megaclassification. *Int. J. Syst. Evol. Microbiol.*, **52**, 7–76.

Clark, C. G., and Roger, A. J. (1995). Direct evidence for secondary loss of mitochondria in Entamoeba histolytica. *Proc. Natl. Acad. Sci. USA*, **92**, 6518–6521.

Daubin, V., Gouy, M., and Perriere, G. (2002). A phylogenomic approach to bacterial phylogeny: evidence of a core of genes sharing a common history. *Genome Res.*, **12**, 1080–1090.

Delsuc, F., Brinkmann H., Chourrout D., *et al.* (2006). Tunicates and not cephalochordates are the closest living relatives of vertebrates. *Nature*, **439**, 965–968.

Delsuc, F., Brinkmann, H., and Philippe, H. (2005). Phylogenomics and the reconstruction of the tree of life. *Nat. Rev. Genet.*, **6**, 361–375.

Doolittle, W. F. (1998). You are what you eat: a gene transfer ratchet could account for bacterial genes in eukaryotic nuclear genomes. *Trends Genet.*, **14**, 307–311.

Doolittle, W. F. (1999). Phylogenetic classification and the universal tree. *Science*, **284**, 2124–2129.

Dopazo, H., and Dopazo, J. (2005). Genome-scale evidence of the nematode-arthropod clade. *Genome Biol.*, **6**, R41.

Dopazo, H., Santoyo, J., and Dopazo, J. (2004). Phylogenomics and the number of characters required for obtaining an accurate phylogeny of eukaryote model species. *Bioinformatics*, **20**, i116–i121.

Edlind, T. D., Li, J., Visvesvara, G. S., *et al.* (1996). Phylogenetic analysis of beta-tubulin sequences from amitochondrial protozoa. *Mol. Phylogenet. Evol.*, **5**, 359–367.

Embley, T. M., and Hirt, R. P. (1998). Early branching eukaryotes? *Curr. Opin. Genet. Dev.*, **8**, 624–629.

Felsenstein, J. (1978). Cases in which parsimony or compatibility methods will be positively misleading. *Syst. Zool.*, **27**, 401–410.

Felsenstein, J. (2004). *Inferring Phylogenies*. Sunderland: Sinauer Associates, Inc.

Fitch, W. M. (1970). Distinguishing homologous from analogous proteins. *Syst. Zool.*, **19**, 99–113.

Forterre, P. (1995). Thermoreduction, a hypothesis for the origin of prokaryotes. *C R Acad. Sci. III*, **318**, 415–422.

Forterre, P., and Philippe, H. (1999). Where is the root of the universal tree of life? *BioEssays*, **21**, 871–879.

Forterre, P., Bouthier De La Tour, C., Philippe, H., and Duguet, M. (2000). Reverse gyrase from hyperthermophiles: probable transfer of a thermoadaptation trait from archaea to bacteria. *Trends Genet.*, **16**, 152–154.

Forterre, P., Brochier, C., and Philippe, H. (2002). Evolution of the Archaea. *Theor. Popul. Biol.*, **61**, 409–422.

Foster, P. G. (2004). Modeling compositional heterogeneity. *Syst. Biol.*, **53**, 485–495.

Galtier, N. (2001). Maximum-likelihood phylogenetic analysis under a covarion-like model. *Mol. Biol. Evol.*, **18**, 866–873.

Galtier, N., and Gouy, M. (1995). Inferring phylogenies from DNA sequences of unequal base compositions. *Proc. Natl. Acad. Sci. USA*, **92**, 11317–11321.

Germot, A., Philippe, H., and Le Guyader, H. (1997). Evidence for loss of mitochondria in Microsporidia from a mitochondrial-type HSP70 in Nosema locustae. *Mol. Biochem. Parasitol.*, **87**, 159–168.

Gogarten, J. P., Kibak, H., Dittrich, P., *et al.* (1989). Evolution of the vacuolar H+-ATPase: implications for the origin of eukaryotes. *Proc. Natl. Acad. Sci. USA*, **86**, 6661–6665.

Gould, S. J. (1996). *Full House: The Spread of Excellence From Plato to Darwin.* New York: Harmony Books.

Griffiths, E., and Gupta, R. S. (2004). Signature sequences in diverse proteins provide evidence for the late divergence of the Order Aquificales. *Int. Microbiol.*, **7**, 41–52.

Gupta, R. S., and Golding, G. B. (1996). The origin of the eukaryotic cell. *Trends Biochem. Sci.*, **21**, 166–71.

Hendy, M. D., and Penny, D. (1989). A framework for the quantitative study of evolutionary trees. *Syst. Zool.*, **38**, 297–309.

Hennig, W. (1966). *Phylogenetic Systematics.* Urbana: University of Illinois Press.

Huelsenbeck, J. P. (2002). Testing a covariotide model of DNA substitution. *Mol. Biol. Evol.*, **19**, 698–707.

Iwabe, N., Kuma, K., Hasegawa, M., Osawa, S. and Miyata, T. (1989). Evolutionary relationship of archaebacteria, eubacteria, and eukaryotes inferred from phylogenetic trees of duplicated genes. *Proc. Natl. Acad. Sci. USA*, **86**, 9355–9359.

Jeffares, D. C., Poole, A. M., and Penny, D. (1998). Relics from the RNA world. *J. Mol. Evol.*, **46**, 18–36.

Jeffroy, O., Brinkmann, H., Delsuc, F., *et al.* (2006). Phylogenomics: the beginning of incongruence? *Trends Genet.*, **22**, 225–231.

Keeling, P. J., and Fast, N. M. (2002). Microsporidia: biology and evolution of highly reduced intracellular parasites. *Annu. Rev. Microbiol.*, **56**, 93–116.

Keeling, P. J., Burger, G., Durnford, D. G., *et al.* (2005). The tree of eukaryotes. *Trends Ecol. Evol.*, **20**, 670–676.

Kolaczkowski, B., and Thornton, J. W. (2004). Performance of maximum parsimony and likelihood phylogenetics when evolution is heterogeneous. *Nature*, **431**, 980–984.

Kreil, D. P., and Ouzounis, C. A. (2001). Identification of thermophilic species by the amino acid compositions deduced from their genomes. *Nucleic Acids Res.*, **29**, 1608–1615.

Lake, J. A. (1988). Origin of the eukaryotic nucleus determined by rate-invariant analysis of rRNA sequences. *Nature*, **331**, 184–186.

Lake, J. A. (1991). The order of sequence alignment can bias the selection of tree topology. *Mol. Biol. Evol.*, **9**, 378–385.

Lanave, C., Preparata, G., Saccone, C., and Serio, G. (1984). A new method for calculating evolutionary substitution rates. *J. Mol. Evol.*, **20**, 86–93.

Lang, B. F., Gray, M. W., and Burger, G. (1999). Mitochondrial genome evolution and the origin of eukaryotes. *Annu. Rev. Genet.*, **33**, 351–397.

Lartillot, N., and Philippe, H. (2004). A Bayesian mixture model for across-site heterogeneities in the amino-acid replacement process. *Mol. Biol. Evol.*, **21**, 1095–1109.

Lawson, F. S., Charlebois, R. L., and Dillon, J. A. (1996). Phylogenetic analysis of carbamoylphosphate synthetase genes: complex evolutionary history includes an internal duplication within a gene which can root the tree of life. *Mol. Biol. Evol.*, **13**, 970–977.

Lockhart, P. J., Howe, C. J., Bryant, D. A., Beanland, T. J., and Larkum, A. W. (1992). Substitutional bias confounds inference of cyanelle origins from sequence data. *J. Mol. Evol.*, **34**, 153–162.

Lockhart, P. J., Larkum, A. W., Steel, M., Waddell, P. J., and Penny, D. (1996). Evolution of chlorophyll and bacteriochlorophyll: the problem of invariant sites in sequence analysis. *Proc. Natl. Acad. Sci. USA*, **93**, 1930–1934.

Lopez, P., Forterre, P., and Philippe, H. (1999). The root of the tree of life in the light of the covarion model. *J. Mol. Evol.*, **49**, 496–508.

Lopez-Garcia, P., and Moreira, D. (1999). Metabolic symbiosis at the origin of eukaryotes. *Trends Biochem. Sci.*, **24**, 88–93.

Lwoff, A. (1943). *L'Évolution Physiologique. Étude des Pertes de Fonctions Chez les Microorganismes.* Paris: Hermann et Cie.

Madsen, O., Scally, M., Douady, C. J., *et al.* (2001). Parallel adaptive radiations in two major clades of placental mammals. *Nature*, **409**, 610–614.

Margulis, L. (1971). Symbiosis and evolution. *Sci. Am.*, **225**, 48–57.

Martin, W., and Muller, M. (1998). The hydrogen hypothesis for the first eukaryote. *Nature*, **392**, 37–41.

Mereschkowski, C. (1905). Ueber Natur und Ursprung der Cromatophoren im Pflanzenreiche. *Biologisches Centralblatt*, **25**, 593–604.

Miyamoto, M. M., and Fitch, W. M. (1995). Testing species phylogenies and phylogenetic methods with congruence. *Syst. Biol.*, **44**, 64–76.

Moreira, D., and Lopez-Garcia, P. (1998). Symbiosis between methanogenic archaea and delta-Proteobacteria as the origin of eukaryotes: The syntrophic hypothesis. *J. Mol. Evol.*, **47**, 517–530.

Murphy, W. J., Eizirik, E., Johnson, W. E., *et al.* (2001). Molecular phylogenetics and the origins of placental mammals. *Nature*, **409**, 614–618.

Nee, S. (2005). The great chain of being. *Nature*, **435**, 429.

Ochman, H., Lerat, E., and Daubin, V. (2005). Examining bacterial species under the specter of gene transfer and exchange. *Proc. Natl. Acad. Sci. USA*, **102**, Suppl 1:6595–6599.

Pagel, M., and Meade, A. (2004). A phylogenetic mixture model for detecting pattern-heterogeneity in gene sequence or character-state data. *Syst. Biol.*, **53**, 571–581.

Patterson, D. J. (1999). The diversity of eukaryotes, *Am. Nat.*, **154**, S96–S124.

Philip, G. K., Creevey, C. J., and McInerney, J. O. (2005). The Opisthokonta and the Ecdysozoa may not be clades: stronger support for the grouping of plant and animal than for animal and fungi and stronger support for the Coelomata than Ecdysozoa. *Mol. Biol. Evol.*, **22**, 1175–1184.

Philippe, H. (2000). Long branch attraction and protist phylogeny. *Protist*, **51**, 307–316.

Philippe, H., and Adoutte, A. (1998). The molecular phylogeny of Eukaryota: solid facts and uncertainties, in *Evolutionary Relationships among Protozoa*, eds. G. Coombs, K. Vickerman, M. Sleigh, and A. Warren. Dordrecht: Kluwer Academic Publishers, pp. 25–56.

Philippe, H., and Douady, C. J. (2003). Horizontal gene transfer and phylogenetics. *Curr. Opin. Microbiol.*, **6**, 498–505.

Philippe, H., and Forterre, P. (1999). The rooting of the universal tree of life is not reliable. *J. Mol. Evol.*, **49**, 509–523.

Philippe, H., and Germot, A. (2000). Phylogeny of eukaryotes based on ribosomal RNA: long-branch attraction and models of sequence evolution. *Mol. Biol. Evol.*, **17**, 830–834.

Philippe, H., Chenuil, A., and Adoutte, A. (1994). Can the cambrian explosion be inferred through molecular phylogeny? *Development*, **120**, S15–S25.

Philippe, H., Delsuc, F., Brinkmann, H., and Lartillot, N. (2005a). Phylogenomics. *Annu. Rev. Ecol. Evol. Syst.*, **36**, 541–562.

Philippe, H., Lartillot, N., and Brinkmann, H. (2005b). Multigene analyses of bilaterian animals corroborate the monophyly of Ecdysozoa, Lophotrochozoa, and Protostomia. *Mol. Biol. Evol.*, **22**, 1246–1253.

Philippe, H., Lopez, P., Brinkmann, H., *et al.* (2000). Early-branching or fast-evolving eukaryotes? An answer based on slowly evolving positions. *Proc. R. Soc. Lond. BS*, **267**, 1213–1221.

Poole, A., Jeffares, D., and Penny, D. (1999). Early evolution: prokaryotes, the new kids on the block. *Bioessays*, **21**, 880–889.

Poole, A. M., Jeffares, D. C., and Penny, D. (1998). The path from the RNA world. *J. Mol. Evol.*, **46**, 1–17.

Qiu, Y. L., Lee, J., Bernasconi-Quadroni, F., *et al.* (1999). The earliest angiosperms: evidence from mitochondrial, plastid and nuclear genomes. *Nature*, **402**, 404–407.

Rodriguez-Ezpeleta, N., Brinkmann, H., Burey, S., *et al.* (2005). Monophyly of primary photosynthetic eukaryotes: green plants, red algae and glaucophytes. *Curr. Biol.*, **15**, 1325–1330.

Roger, A. J., Svard, S. G., Tovar, J., *et al.* (1998). A mitochondrial-like chaperonin 60 gene in Giardia lamblia: evidence that diplomonads once harbored an endosymbiont related to the progenitor of mitochondria. *Proc. Natl. Acad. Sci. USA*, **95**, 229–234.

Roy, S. W., and Gilbert, W. (2005). The pattern of intron loss. *Proc. Natl. Acad. Sci. USA*, **102**, 713–718.

Silberman, J. D., Clark, C. G., Diamond, L. S., and Sogin, M. L. (1999). Phylogeny of the genera Entamoeba and Endolimax as deduced from small-subunit ribosomal RNA sequences. *Mol. Biol. Evol.*, **16**, 1740–1751.

Simpson, A. G., and Roger, A. J. (2004). The real 'kingdoms' of eukaryotes. *Curr. Biol.*, **14**, R693–R696.

Slesarev, A. I., Mezhevaya, K. V., Makarova, K. S., *et al.* (2002). The complete genome of hyperthermophile Methanopyrus kandleri AV19 and monophyly of archaeal methanogens. *Proc. Natl. Acad. Sci. USA*, **99**, 4644–4649.

Sogin, M. L. (1991). Early evolution and the origin of eukaryotes. *Curr. Opin. Genet. Dev.*, **1**, 457–463.

Steel, M., and Penny, D. (2000). Parsimony, likelihood, and the role of models in molecular phylogenetics. *Mol. Biol. Evol.*, **17**, 839–850.

Stiller, J., and Hall, B. (1999). Long-branch attraction and the rDNA model of early eukaryotic evolution. *Mol. Biol. Evol.*, **16**, 1270–1279.

Tourasse, N. J., and Gouy, M. (1999). Accounting for evolutionary rate variation among sequence sites consistently changes universal phylogenies deduced from rRNA and protein-coding genes. *Mol. Phylogenet. Evol.*, **13**, 159–168.

Tovar, J., Fischer, A., and Clark, C. G. (1999). The mitosome, a novel organelle related to mitochondria in the amitochondrial parasite Entamoeba histolytica. *Mol. Microbiol.*, **32**, 1013–1021.

Tovar, J., Leon-Avila, G., Sanchez, L. B., *et al.* (2003). Mitochondrial remnant organelles of Giardia function in iron-sulphur protein maturation. *Nature*, **426**, 172–176.

Wallace, I. M., Blackshields, G., and Higgins, D. G. (2005). Multiple sequence alignments. *Curr. Opin. Struct. Biol.*, **15**, 261–266.

Whelan, S., and Goldman, N. (2001). A general empirical model of protein evolution derived from multiple protein families using a maximum-likelihood approach. *Mol. Biol. Evol.*, **18**, 691–699.

Williams, B. A., Hirt, R. P., Lucocq, J. M., and Embley, T. M. (2002). A mitochondrial remnant in the microsporidian Trachipleistophora hominis. *Nature*, **418**, 865–869.

Woese, C. R. (1987). Bacterial evolution. *Microbiol. Rev.*, **51**, 221–271.

Woese, C. R., Kandler, O., and Wheelis, M. L. (1990). Towards a natural system of organisms: proposal for the domains Archaea, Bacteria, and Eucarya. *Proc. Natl. Acad. Sci. USA*, **87**, 4576–4579.

Wolf, Y. I., Rogozin, I. B., Grishin, N. V., Tatusov, R. L., and Koonin, E. V. (2001). Genome trees constructed using five different approaches suggest new major bacterial clades. *BMC Evol. Biol.*, **1**, 8.

Wolf, Y. I., Rogozin, I. B., and Koonin, E. V. (2004). Coelomata and not ecdysozoa: evidence from genome-wide phylogenetic analysis. *Genome Res.*, **14**, 29–36.

Yang, Z. (1993). Maximum-likelihood estimation of phylogeny from DNA sequences when substitution rates differ over sites. *Mol. Biol. Evol.*, **10**, 1396–1401.

Yang, Z. (1996). Among-site rate variation and its impact on phylogenetic analyses. *Trends Ecol. Evol.*, **11**, 367–370.

Yang, Z., and Roberts, D. (1995). On the use of nucleic acid sequences to infer early branchings in the tree of life. *Mol. Biol. Evol.*, **12**, 451–458.

Zhaxybayeva, O., Lapierre, P., and Gogarten, J. P. (2005). Ancient gene duplications and the root(s) of the tree of life. *Protoplasma*, **227**, 53–64.

Zillig, W. (1987). Eukaryotic traits in Archaebacteria. Could the eukaryotic cytoplasm have arisen from archaebacterial origin? *Ann. N. Y. Acad. Sci.*, **503**, 78–82.

9

Horizontal gene transfer, gene histories, and the root of the tree of life

Olga Zhaxybayeva
Dalhousie University
and
J. Peter Gogarten
University of Connecticut

9.1 Introduction

Tracing organismal (species) histories on large evolutionary timescales remains a big challenge in evolutionary biology. Darwin metaphorically labelled these relationships the 'tree of life', but in his notebook he expressed unhappiness with this label, because in the 'tree of life' depicting species evolution only the tips of the branches are alive; this layer of living organisms rests on dead ancestors and extinct relatives. Darwin mused that therefore the term 'coral of life' would be more appropriate (B26 in Darwin (1987)); however, this alternative label did not gain popularity. Different markers have been utilized to elucidate the relationship among different lineages: from morphological characters to complete genomes. Since complete genomes are now available for organisms from all three domains of life, it is possible to use large amounts of data to attempt to decipher the relationships between all known organisms.

Many comparative genome analyses have shown that different genes in genomes often have different evolutionary histories (e.g., Hilario and Gogarten (1993), Nesbo *et al.* (2001), and Zhaxybayeva *et al.* (2004)), which implies that the tree of life metaphor (and a bifurcating tree as a model for evolutionary relationships in general) might be no longer adequate (Doolittle, 1999). The incongruence between gene histories can be attributed to many factors, one of which is horizontal gene transfer (HGT). Simulations based on coalescence have shown that HGT can affect not only the topology of an inferred phylogeny (and therefore inferences of last common ancestors), but also divergence times (Zhaxybayeva and Gogarten, 2004; Zhaxybayeva *et al.*, 2005). In this chapter we review challenges in multigene analyses in light of HGT and discuss the consequences for inferring the position of the root of the tree/coral of life.

Planetary Systems and the Origins of Life, eds. Ralph E. Pudritz, Paul G. Higgs, and Jonathon R. Stone. Published by Cambridge University Press.

9.2 How to analyse multigene data?

In addition to different models and methods for sequence analyses (see Chapter 8 for an overview), many methodologies exist for the analysis of multiple genes. One approach is to combine individual genes into a single dataset to extract phylogenetic information that might be distributed over many gene families; this so-called 'supermatrix approach' is often cited as a way to improve the resolution of the inferred phylogeny (Brown *et al.*, 2001; Delsuc *et al.*, 2005). In this approach, the individual gene alignments are concatenated into one superalignment, if the individual gene histories are found to have compatible (non-conflicting) phylogenetic histories. Then the superalignment is analysed as if it were one gene alignment, but with an advantage of containing a larger number of sites informative for phylogenetic analysis.

A problem with direct concatenation is the selection of data to include. This selection is complicated by the fact that the absence of evidence for transfer cannot be taken as evidence for the absence of transfer. If one applies a stringent measure for the detection of conflict, nearly all genes agree with each other within the limits of confidence. The amount of conflict detected depends on the chosen limits of confidence and on the extent of taxon sampling (Snel *et al.*, 2002; Daubin *et al.*, 2003; Mirkin *et al.*, 2003; Ge *et al.*, 2005; Kunin *et al.*, 2005).

Testing the compatibility between different trees and the alignments from which these trees were derived using the Shimodaira–Hasegawa (Shimodaira and Hasegawa, 1999) or approximately unbiased (AU) test (Shimodaira, 2002) has become the preferred tool for assessing potential conflict between individual gene families (e.g., Lerat *et al.* (2003)). In these tests, the fit of alternative topologies to an alignment is evaluated and the trees under which the data have a significantly worse probability are rejected and considered as incompatible with the data. (The probability of observing the data given a model of evolution and a phylogenetic tree is also known as the likelihood of a phylogenetic tree.) However, the failure to reject a tree should not be mistaken for evidence for congruence with a tree (Bapteste *et al.*, 2004). A gene might indeed have evolved with a different history, and this history might be different from the consensus phylogeny, but the individual gene family contains too little phylogenetically useful information to make the likelihood of the two phylogenies significantly different. This is analogous to the failure in detecting a significant correlation between fat intake and cancer (Prentice *et al.*, 2006): it does not prove that the correlation does not exist; it only says that the correlation was not significant in the dataset, possibly because too small a sample was studied.

Another challenge to inferences based on concatenation may come from hidden, or unrecognized, paralogy in lineages that went through frequent gene duplication and aneupolyploidization (i.e., having multiple, albeit each incomplete, sets of chromosomes). Especially in animals with metameric organization (i.e., with the body divided into a number of similar segments), gene and genome duplications have long been postulated to have created the regulatory complexity necessary for the different bodyplans (e.g., Nam and Nei (2005)). Multiple whole genome duplications were inferred for the early evolution of plants (Cui *et al.*, 2006) and vertebrates (Escriva *et al.*, 2002; Meyer and Van de Peer, 2005). These gene duplications have led to gene families with often astounding diversity (e.g., Foth *et al.* (2006)).

However, it is not the complexity of the gene families in itself that generates problems in phylogenetic reconstruction; rather, the frequent loss of one or the other paralogue (Hughes and Friedman, 2004; Nam and Nei, 2005) can lead to the inclusion of unrecognized paralogues in the datasets, with the result that some of the events in the genes' histories reflect gene duplication and not speciation events. For example, analysing the homeobox gene superfamily in 11 genomes, Nam and Nei (2005) inferred 88 homeobox genes to have been present in the ancestor of bilateral animals. Thirty–forty of these were completely lost in one of the 11 species analysed, and many of the ones still represented underwent frequent gene duplications, especially in vertebrates, where more than 200 homeobox genes are found per haploid genome. In a study of four animal genomes, Hughes and Friedman (2004) observe massive (19–20% of gene families present in common ancestor) parallel loss in *Caenorhabditis elegans* and *Drosophila melanogaster*.

An alternative to the supermatrix approaches (Delsuc *et al.*, 2005) is to analyse genes individually and to combine the resulting trees (or the bipartitions/embedded quartets constituting the trees) into a consensus signal (i.e., the phylogenetic signal supported by at least a plurality of genes) by using supertree methods (Beiko *et al.*, 2005), bipartition plotting (Zhaxybayeva *et al.*, 2004), or quartet decomposition (Zhaxybayeva *et al.*, 2006). As an example, Figure 9.1A shows bipartition analysis of 678 gene families present in ten cyanobacterial genomes (see Zhaxybayeva *et al.* (2004) for more details). A bipartition plot shows all bipartitions significantly supported by at least one gene family and allows us to extract quickly the plurality signal as well as gene families conflicting with it. Notably, for cyanobacteria, only three compatible bipartitions are supported by the plurality of genes, resulting in a not very resolved plurality consensus (Figure 9.1B). And even those three bipartitions are conflicted strongly by 13 gene families (i.e., these genes support a conflicting partition with >99% bootstrap support; see Figure 9.1 legend for the list of gene families). An advantage of the bipartition plotting approach is that only the signal retained in the plurality of datasets is used to synthesize the consensus; the

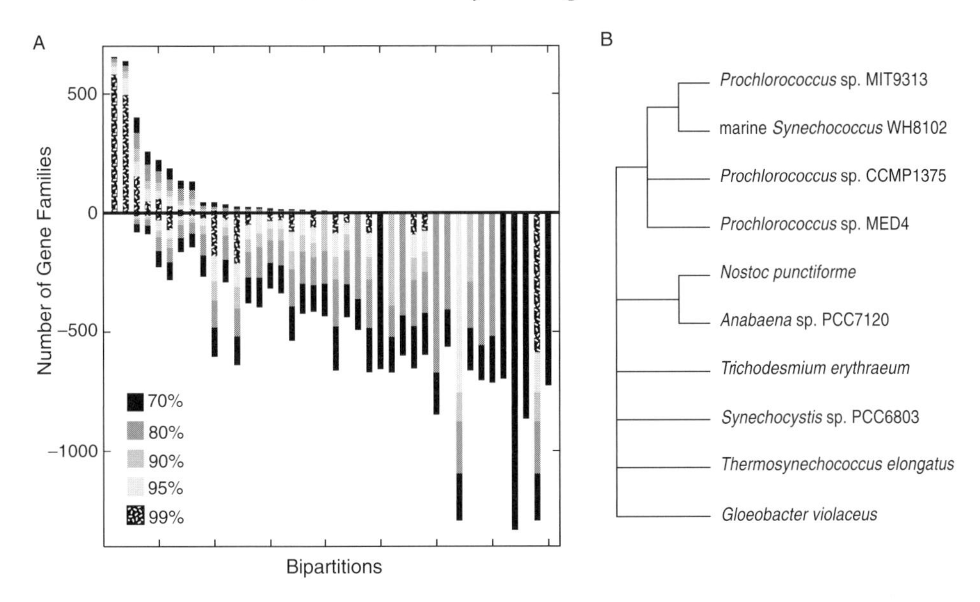

Fig. 9.1. Bipartition plot analysis of 678 gene families in ten cyanobacterial genomes. An unrooted phylogenetic tree can be represented as a set of bipartitions (or splits). If a branch of a tree is removed, the tree 'splits' into two sets of leaves. A bipartition of an unrooted phylogenetic tree is defined as a division of the tree into two mutually exclusive sets of leaves. A. Plot of bipartitions with at least 70% bootstrap support. Each column represents the number of gene families that support (columns that are pointing upwards) or conflict (columns that are pointing downwards) a bipartition. The columns are sorted by number of gene families supporting each bipartition. The level of bootstrap support is coded by different shades of gray. For details of phylogenetic analyses see Zhaxybayeva *et al.* (2004). B. Plurality consensus reconstructed from the three most supported partitions. The genes that are in conflict with the consensus at >99% bootstrap support encode ribulose bisphosphate carboxylase large subunit, cell division protein FtsH, translation initiation factor IF-2, ferredoxin, geranylgeranyl hydrogenase, amidophosphoribosyltransferase, photosystem II protein D2, photosystem II CP43 protein, photosystem II CP47 protein, photosystem I core protein A2, photosystem I core protein A1, photosystem II manganese-stabilizing protein, and 5′-methylthioadenosine phosphorylase.

disadvantage is that individual gene analyses often suffer from a lack of resolution due to an insufficient number of phylogenetically informative positions.

Regardless of the method used to arrive at the consensus, the question still remains: what does the signal inferred from multigene data mean? Does it serve as a proxy for an organismal phylogeny? Does it sometimes reflect grouping by ecotypes? In the next section we explore these questions by considering an example of four marine cyanobacteria.

9.3 What does the plurality consensus represent? Example of small marine cyanobacteria

At the time of the analysis, four genomes of small coccoid marine cyanobacteria of broadly similar lifestyles had been completely sequenced: *Prochlorococcus marinus* CCMP1375 (also known as SS120), *Prochlorococcus marinus* MED4, *Prochlorococcus marinus* MIT9313, and marine *Synechococcus* WH8102. Members of the *Prochlorococcus* genus have only been discovered quite recently due to their small size and anomalously low fluorescence (Chisholm *et al.*, 1988). Marine *Synechococcus* and *Prochlorococcus* are proposed to diverge from a common phycobilisome-containing ancestor (Ting *et al.*, 2002). The phycobilisome is a light-harvesting complex associated with photosystem II and is used as a light-harvesting antenna in most cyanobacteria. Phylobilisomes, in contrast to the light-harvesting complexes of higher plants, are not embedded in the photosynthetic membrane, and they contain phycobilins as pigments, which give the cyanobacteria their typical blue-green colour. While marine *Synechococcus* still uses phycobilisomes as light-harvesting antennae (an ancestral trait), members of the *Prochlorococcus* genus lack phycobilisomes and utilize a different chlorophyl-containing antenna complex (Pcb). *Prochlorococcus* also possess derivatives of chlorophyll a and b pigments that are unique to this genus (see Partensky *et al.* (1999) for a review). In addition, marine *Synechococcus* and *Prochlorococcus* are adapted to different ecological niches: marine *Synechococcus* is prevalent in coastal waters, while *Prochlorococcus* is ubiquitous in open subtropical and tropical ocean. Within *Prochlorococcus marinus* two 'ecotypes' are differentiated: low-light adapted and high-light adapted types (Rocap *et al.*, 2003). In the 16S rRNA tree, low-light adapted *Prochlorococcus* spp. form a paraphyletic clade with respect to high-light adapted *Prochlorococcus* spp. (Ting *et al.*, 2002). Beiko *et al.* (2005) reported more than 250 HGT events among these marine cyanobacteria. In their supertree as well as in our bipartition analyses (Zhaxybayeva *et al.*, 2004) the marine cyanobacteria cluster into two clades: *P. marinus* CCMP1375 ('low-light' adapted) groups with *P. marinus* MED4 ('high-light'), and *P. marinus* MIT9313 ('low-light') groups with marine *Synechococcus* WH8102 (Beiko *et al.* (2005) and Figure 9.1B). Interestingly, the relationship among these four genomes as captured by the supertree and bipartition analyses does not support the relationship inferred from phylogenetic analyses of 16S rRNA (e.g., Ting *et al.*, 2002), or the groupings based on proposed ecotypes (Ting *et al.*, 2002; Rocap *et al.*, 2003), or the many derived characteristics shared between the three *Prochlorococcus* species. These results are confirmed through independent analyses of cyanobacteria utilizing embedded quartets (Zhaxybayeva *et al.*, 2006). One explanation for the conflict between the genome consensus, and the many complex derived characters is rampant gene flow among these genomes, such that the plurality consensus no longer reflects the ecotype, physiology, and organismal history. In support of this hypothesis, cyanophages infecting marine cyanobacteria have been reported to contain genes important for photosynthesis (Mann *et al.*, 2003; Lindell *et al.*, 2004; Millard *et al.*, 2004; Sullivan *et al.*, 2005; Zeidner *et al.*, 2005), and likely mediate transfer and recombination of these genes among the marine cyanobacteria (Zeidner *et al.*, 2005).

Notably, such an observation is not limited to prokaryotes. For example, in incipient species of Darwin's finches frequent introgression can make some individuals characterized as belonging to the same species by morphology and mating behaviour genetically more similar to a sister species (Grant *et al.*, 2004).

9.4 Where is the root of the 'tree of life'?

In phylogenetic analysis using molecular sequences external information is required to identify the root of an inferred tree, i.e., the node from which all other nodes of the tree are descended. Usually, an outgroup is needed to root a tree, i.e., a taxon which is known to diverge earlier than the group of interest. However, this methodology becomes inapplicable to the problem of rooting the tree of life (since all organisms are part of the group of interest). One method of determining the root of the tree of life from molecular data is to use ancient duplication events. If a gene duplication event has occurred before the divergence of all extant organisms, then the phylogenetic tree containing both gene duplicates will be symmetrical, and one set of duplicated genes can be used as an outgroup to the other. In the past, several pairs of anciently duplicated genes have been detected and analysed. The analyses produced a variety of controversial results as summarized in Table 9.1. A large-scale comparison of anciently duplicated genes did not bring any consensus (Zhaxybayeva *et al.*, 2005), as possible locations of the root were observed in various places (Zhaxybayeva *et al.*, 2005). If some groups of organisms evolve slower than others in molecular and physiological characteristics, then their phenotype might be considered as reflecting the phenotype of the most recent common ancestor (MRCA) (Woese, 1987; Xue *et al.*, 2003; Ciccarelli *et al.*, 2006). However, these presumably primitive characteristics do not allow us to place the ancestor in the tree of life, because they do not inform us about the relationship between the slow and fast evolving groups (Cejchan, 2004).

Inspecting phylogenies based on molecular characters one often forgets that the deeper parts of the tree of life are formed by only those lineages that have extant representatives. However, these lineages were not the only ones present in earlier times. Many lineages became extinct, or at least their extant representatives have not been discovered. It is reasonable to assume that at the time of the

Table 9.1. *Different points of view on the location of the root of the tree of life*

Location of the root	Phylogenetic marker used	Citations
on the branch leading to the bacterial domain	V- and F-type ATPases catalytic and regulatory subunits	Gogarten *et al.* (1989)
on the branch leading to the bacterial domain	translation elongation factors EF-tu/1 and EF-G/2	Iwabe *et al.* (1989)
on the branch leading to the bacterial domain	Val/Ile amino-acyl tRNA synthetases	Brown and Doolittle (1995)
on the branch leading to the bacterial domain	Tyr/Trp amino-acyl tRNA synthetases	Brown *et al.* (1997)
on the branch leading to the bacterial domain	internal duplication in carbamoyl phosphate synthetase	Lawson *et al.* (1996)
on the branch leading to the bacterial domain	components of signal recognition particle – the signal recognition particle SRP54 and its signal receptor alpha SRα	Gribaldo and Cammarano (1998)
inconclusive results: Archaea do not appear as a monophyletic group	components of signal recognition particle – the signal recognition particle SRP54 and its signal receptor alpha SRα	Kollman and Doolittle (2000)
on the branch leading to the bacterial domain	aspartate and ornithine transcarbamoylases	Labedan *et al.* (1999)
inconclusive results: no statistical support for the best tree topology	histidine biosynthesis genes hisA/hisF	Charlebois *et al.* (1997)
within Gram-negative bacteria	membrane architecture	Cavalier-Smith (2002)
on the branch leading to the eukaryotic domain	translation elongation factor proteins, EF-1alpha and EF-2	Forterre and Philippe (1999); Lopez *et al.* (1999)
on the branch leading to the bacterial domain, but the authors think it is an artefact due to LBA	elongation factors, ATPases, tRNA synthetases, carbamoyl phosphate synthetases, signal recognition particle proteins	Philippe and Forterre (1999)

Table 9.1. (cont.)

Location of the root	Phylogenetic marker used	Citations
between Archaea and Bacteria	structural features of Gln/Glu tRNA synthetases	Siatecka *et al.* (1998)
within Archaea	16S rRNA	Lake (1988)
within Archaea	transfer RNAs	Xue *et al.* (2003)
under aboriginal trifurcation	various characteristics	Woese *et al.* (1978)
inconclusive results	RNA secondary structure	Caetano-Anolles (2002)
conceptual difficulties	–	Bapteste and Brochier (2004)

organismal MRCA many other organisms were in existence as well (Figure 9.2). If the processes of speciation and extinction can be approximated by a random process, coalescence theory would be an appropriate tool to describe lineages through time plots (Zhaxybayeva and Gogarten, 2004).

Due to the vast time spans needed for the lineages to coalesce to their MRCA, the individual phylogenies from extant lineages usually have two longest branches leading to the MRCA, and therefore these phylogenetic analyses will be subject to potential long branch attraction (LBA) artefacts. Therefore, additional higher order shared derived characters (synapomorphies) are helpful to reinforce the inferred position of the root. Larger insertions/deletions or derived structural features are considered rare and mostly irreversible and may serve as synapomorphies (e.g., see Gupta (2001) for the application of presumably unique insertions/deletions to organismal classification). For example, the domain architecture differences between β and β' RNA polymerase subunits were used to polarize the tree of bacterial RNA polymerases (Iyer *et al.*, 2004). In another study, the position of the root in the Glx-tRNA synthetase family tree was inferred from structural differences in anticodon binding domains between GlnRS and GluRS (Siatecka *et al.*, 1998). Below we examine in detail the use of additional characters in supporting the position of the root in the phylogenetic tree of ATPases.

9.5 Use of higher order characters: example of ATPases

The use of ATPase catalytic and non-catalytic subunits to root the tree of life was originally introduced by Gogarten *et al.* (1989). This pair of anciently duplicated genes places the root on the branch leading to Bacteria with high confidence (see Figure 9.3). Either the catalytic or the non-catalytic subunits can be considered

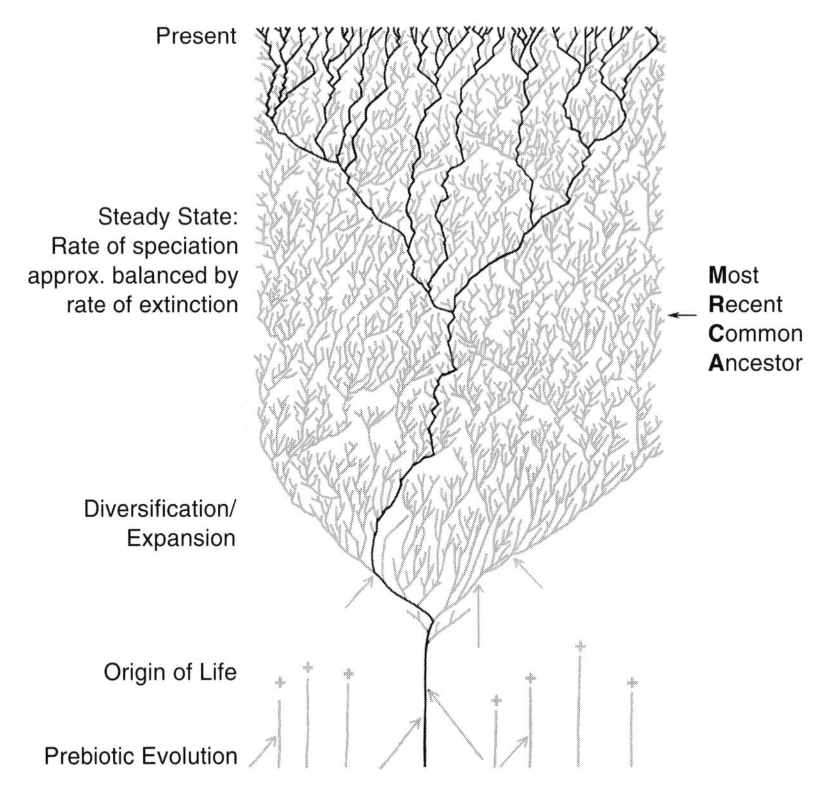

Fig. 9.2. Schematic depiction of a model for the tree/corral of life highlighting the position of the most recent common ancestor. Extinct lineages are shown in grey. Extant lineages at the tip of the tree are traced back to their last common ancestors (in black).

as the ingroup, and the outgroup is provided by the paralogous subunits. The outgroup, a set of sequences rather divergent from the ingroup, joins the ingroup on the longest internal branch. While this placement of the root is recovered using different methods and evolutionary models (Gogarten *et al.*, 1989; Iwabe *et al.*, 1989), it also coincides with the place where the root would be located as the result of the LBA artefact (Philippe and Forterre, 1999; Gribaldo and Philippe, 2002). However, in the case of the ATPases higher order characters exclude placing the outgroup within the archaeal/eukaryotic ATPases subunits (no higher order characters have been recognized for bacterial F-ATPases). The archaeal vacuolar ATPase non-catalytic subunits have lost the canonical Walker motif GxxGxGKT in their ATP binding pocket (Gogarten *et al.*, 1989). This motif is present in the orthologous F-ATPase non-catalytic subunits as well as in all of the ancient paralogues, including the paralogous Rho transcription termination factors (Richardson, 2002) and ATPases involved in assembly of the bacterial flagella

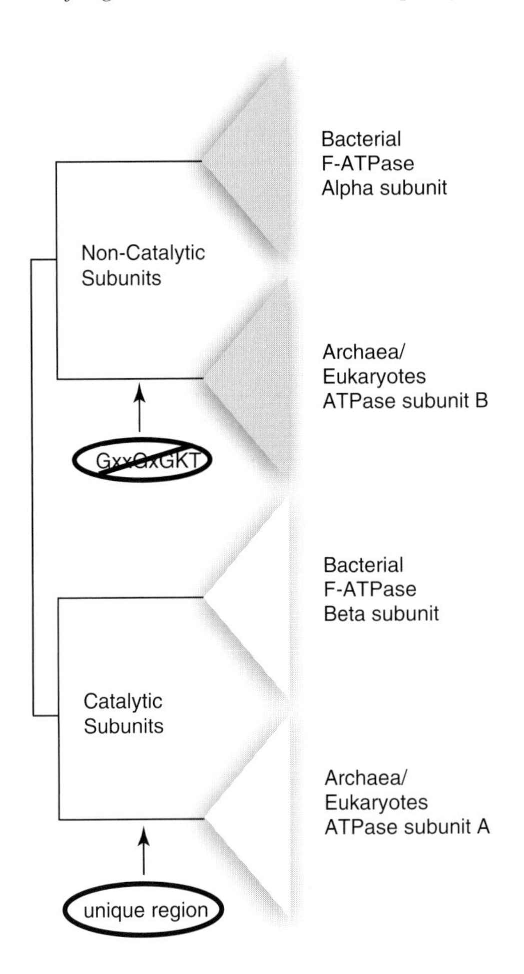

Fig. 9.3. Schematic tree showing the evolution of catalytic and non-catalytic subunits of ATPases (for detailed phylogeny see Fig. 2 in Zhaxybayeva *et al.* (2005)). Higher order characters are mapped to the branch leading to the clade where all the members of the clade possess the character. See text for more details.

(Vogler *et al.*, 1991). Similarly, the catalytic subunits of the archaeal and of the eukaryotic vacuolar type ATPase contain a faster evolving 'non-homologous' region that has no counterpart in the catalytic F-ATPase subunits, nor is this region found in any of the non-catalytic subunits (Zimniak *et al.*, 1988; Gogarten *et al.*, 1989). The absence of the canonical Walker motif in the regulatory subunits and the presence of the non-homologous region in the catalytic subunits thus are shared derived characters of the vacuolar and archaeal ATPases that preclude moving the root of the ATPase phylogeny to a place within the clade constituted by the archaeal and eukaryotic ATPases.

9.6 Conclusions

Consideration of gene transfer makes the analysis of species evolution among prokaryotes similar to population genetics. As is the case for within species analyses, gene trees for diverse groups of prokaryotes do not coalesce to the same ancestor, and the organisms that carried the molecular ancestors often lived at different times. For example, the human Y chromosome of all human males is traced back to an MRCA ('Adam') who lived approximately 50 000 years ago (Thomson *et al.*, 2000; Underhill *et al.*, 2000), while mitochondrial genes trace back to an MRCA ('Eve') who lived about 166 000–249 000 years ago (Cann *et al.*, 1987; Vigilant *et al.*, 1991). Many thousand years separate 'Adam' and 'Eve' from each other. The analogy to gene trees in recombining populations suggests that no single last common ancestor contained all the genes ancestral to the ones shared between the three domains of life. Each contemporary molecule traces back to an individual molecular MRCA, but these molecular ancestors were likely to be present in different organisms at different times. Therefore, even with more accurate phylogenetic reconstruction methods, one should not expect different molecular phylogenies to agree with one another on the placement of the MRCA of all living organisms. Adaptation of population genetics methodology may provide more fruitful results in studying early evolutionary events.

Acknowledgements

Olga Zhaxybayeva is supported through Canadian Institute of Health Research Postdoctoral Fellowship and is an honorary Killam Postdoctoral Fellow at Dalhousie University. This work was supported through the NASA Exobiology, NASA Applied Information Systems Research, and NSF Microbial Genetics (MCB-0237197) grants to J. Peter Gogarten.

References

Bapteste, E., and Brochier, C. (2004). On the conceptual difficulties in rooting the tree of life. *Trends Microbiol.*, **12**, 9–13.

Bapteste, E., Boucher, Y., Leigh, J., and Doolittle, W. F. (2004). Phylogenetic reconstruction and lateral gene transfer. *Trends Microbiol.*, **12**, 406–411.

Beiko, R. J., Harlow, T. J., and Ragan, M. A. (2005). Highways of gene sharing in prokaryotes. *Proc. Natl. Acad. Sci. USA*, **102**, 14332–14337.

Brown, J. R., and Doolittle, W. F. (1995). Root of the universal tree of life based on ancient aminoacyl-tRNA synthetase gene duplications. *Proc. Natl. Acad. Sci. USA*, **92**, 2441–2445.

Brown, J. R., Douady, C. J., Italia, M. J., Marshall, W. E., and Stanhope, M. J. (2001). Universal trees based on large combined protein sequence data sets. *Nat. Genet.*, **28**, 281–285.

Brown, J. R., Robb, F. T., Weiss, R., and Doolittle, W. F. (1997). Evidence for the early divergence of tryptophanyl- and tyrosyl-tRNA synthetases. *J. Mol. Evol.*, **45**, 9–16.

Caetano-Anolles, G. (2002). Evolved RNA secondary structure and the rooting of the universal tree of life. *J. Mol. Evol.*, **54**, 333–345.

Cann, R. L., Stoneking, M., and Wilson, A. C. (1987). Mitochondrial DNA and human evolution. *Nature*, **325**, 31–36.

Cavalier-Smith, T. (2002). The neomuran origin of archaebacteria, the negibacterial root of the universal tree and bacterial megaclassification. *Int. J. Syst. Evol. Microbiol.*, **52**, 7–76.

Cejchan, P. A. (2004). LUCA, or just a conserved Archaeon? Comments on Xue *et al.* *Gene*, **333**, 47–50.

Charlebois, R. L., Sensen, C. W., Doolittle, W. F., and Brown, J. R. (1997). Evolutionary analysis of the hisCGABdFDEHI gene cluster from the archaeon Sulfolobus solfataricus P2. *J. Bacteriol.*, **179**, 4429–4432.

Chisholm, S. W., Olson, R. J., Zettler, E. R., *et al.* (1988). A novel free-living prochlorophyte abundant in the oceanic euphotic zone. *Nature*, **334**, 340–343.

Ciccarelli, F. D., Doerks, T., von Mering C., *et al.* (2006). Toward automatic reconstruction of a highly resolved tree of life. *Science*, **311**, 1283–1287.

Cui, L., Wall, P. K., Leebens-Mack J. H., *et al.* (2006). Widespread genome duplications throughout the history of flowering plants. *Genome. Res.*, **16**, 738–749.

Darwin, C. (1987). *Charles Darwin's Notebooks, 1836–1844*. Cambridge: Cambridge University Press.

Daubin, V., Moran, N. A., and Ochman, H. (2003). Phylogenetics and the cohesion of bacterial genomes. *Science*, **301**, 829–832.

Delsuc, F., Brinkmann, H., and Philippe, H. (2005). Phylogenomics and the reconstruction of the tree of life. *Nat. Rev. Genet.*, **6**, 361–375.

Doolittle, W. F. (1999). Phylogenetic classification and the universal tree. *Science*, **284**, 2124–2129.

Escriva, H., Manzon, L., Youson, J., and Laudet. V. (2002). Analysis of lamprey and hagfish genes reveals a complex history of gene duplications during early vertebrate evolution. *Mol. Biol. Evol.*, **19**, 1440–1450.

Forterre, P., and Philippe, H. (1999). Where is the root of the universal tree of life? *Bioessays*, **21**, 871–879.

Foth, B. J., Goedecke, M. C., and Soldati, D. (2006). New insights into myosin evolution and classification. *Proc. Natl. Acad. Sci. USA*, **103**, 3681–3686.

Ge, F., Wang, L.-S., and Kim, J. (2005). The cobweb of life revealed by genome-scale estimates of horizontal gene transfer. *PLoS Biology*, **3**, e316.

Gogarten, J. P., Kibak, H., Dittrich P., *et al.* (1989). Evolution of the vacuolar H+-ATPase: implications for the origin of eukaryotes. *Proc. Natl. Acad. Sci. USA*, **86**, 6661–6665.

Grant, P. R., Grant, B. R., Markert, J. A., Keller, L. F., and Petren, K. (2004). Convergent evolution of Darwin's finches caused by introgressive hybridization and selection. *Evolution Int. J. Org. Evolution.*, **58**, 1588–1599.

Gribaldo, S., and Cammarano, P. (1998). The root of the universal tree of life inferred from anciently duplicated genes encoding components of the protein-targeting machinery. *J. Mol. Evol.*, **47**, 508–516.

Gribaldo, S., and Philippe, H. (2002). Ancient phylogenetic relationships. *Theor. Popul. Biol.*, **61**, 391–408.

Gupta, R. S. (2001). The branching order and phylogenetic placement of species from completed bacterial genomes. based on conserved indels found in various proteins. *Int. Microbiol.*, **4**, 187–202.

Hilario, E., and Gogarten, J. P. (1993). Horizontal transfer of ATPase genes – the tree of life becomes a net of life. *Biosystems*, **31**, 111–119.

Hughes, A. L., and Friedman, R. (2004). Differential loss of ancestral gene families as a source of genomic divergence in animals. *Proc. Biol. Sci.*, **271** suppl 3, S107–S109.

Iwabe, N., Kuma, K., Hasegawa, M., Osawa, S., and Miyata, T. (1989). Evolutionary relationship of archaebacteria, eubacteria, and eukaryotes inferred from phylogenetic trees of duplicated genes. *Proc. Natl. Acad. Sci. USA*, **86**, 9355–9359.

Iyer, L. M., Koonin, E. V., and Aravind, L. (2004). Evolution of bacterial RNA polymerase: implications for large-scale bacterial phylogeny, domain accretion, and horizontal gene transfer. *Gene*, **335**, 73–88.

Kollman, J. M., and Doolittle, R. F. (2000). Determining the relative rates of change for prokaryotic and eukaryotic proteins with anciently duplicated paralogs. *J. Mol. Evol.*, **51**, 173–81.

Kunin, V., Goldovsky, L., Darzentas, N., and Ouzounis, C. A. (2005). The net of life: reconstructing the microbial phylogenetic network. *Genome Res.*, **15**, 954–959.

Labedan, B., Boyen, A., Baetens, M., *et al.* (1999). The evolutionary history of carbamoyltransferases: A complex set of paralogous genes was already present in the last universal common ancestor. *J. Mol. Evol.*, **49**, 461–73.

Lake, J. A. (1988). Origin of the eukaryotic nucleus determined by rate-invariant analysis of rRNA sequences. *Nature*, **331**, 184–186.

Lawson, F. S., Charlebois, R. L., and Dillon, J. A. (1996). Phylogenetic analysis of carbamoylphosphate synthetase genes: complex evolutionary history includes an internal duplication within a gene which can root the tree of life. *Mol. Biol. Evol.*, **13**, 970–977.

Lerat, E., Daubin, V., and Moran, N. A. (2003). From gene trees to organismal phylogeny in prokaryotes: the case of the gamma-Proteobacteria. *PLoS Biol.*, **1**, 19.

Lindell, D., Sullivan, M. B., Johnson, Z. I., *et al.* (2004). Transfer of photosynthesis genes to and from *Prochlorococcus* viruses. *Proc. Natl. Acad. Sci. USA*, **101**, 11013–11018.

Lopez, P., Forterre, P., and Philippe, H. (1999). The root of the tree of life in the light of the covarion model. *J. Mol. Evol.*, **49**, 496–508.

Mann, N. H., Cook, A., Millard, A., Bailey, S., and Clokie, M. (2003). Marine ecosystems: bacterial photosynthesis genes in a virus. *Nature*, **424**, 741.

Meyer, A., and Van de Peer, Y. (2005). From 2R to 3R: evidence for a fish-specific genome duplication (FSGD). *Bioessays*, **27**, 937–945.

Millard, A., Clokie, M. R., Shub, D. A., and Mann, N. H. (2004). Genetic organization of the psbAD region in phages infecting marine *Synechococcus* strains. *Proc. Natl. Acad. Sci. USA*, **101**, 11007–11012.

Mirkin, B. G., Fenner, T. I., Galperin, M. Y., and Koonin, E. V. (2003). Algorithms for computing parsimonious evolutionary scenarios for genome evolution, the last universal common ancestor and dominance of horizontal gene transfer in the evolution of prokaryotes. *BMC Evol. Biol.*, **3**, 2.

Nam, J., and Nei, M. (2005). Evolutionary change of the numbers of homeobox genes in bilateral animals. *Mol. Biol. Evol.*, **22**, 2386–2394.

Nesbo, C. L., Boucher, Y., and Doolittle, W. F. (2001). Defining the core of nontransferable prokaryotic genes: the euryarchaeal core. *J. Mol. Evol.*, **53**, 340–350.

Partensky, F., Hess, W. R., and Vaulot, D. (1999). *Prochlorococcus*, a marine photosynthetic prokaryote of global significance. *Microbiol. Mol. Biol. Rev.*, **63**, 106–127.

Philippe, H., and Forterre, P. (1999). The rooting of the universal tree of life is not reliable. *J. Mol. Evol.*, **49**, 509–523.

Prentice, R. L., Caan, B., Chlebowski, R. T., *et al.* (2006). Low-fat dietary pattern and risk of invasive breast cancer: the Women's health initiative randomized controlled dietary modification trial. *JAMA*, **295**, 629–642.

Richardson, J. P. (2002). Rho-dependent termination and ATPases in transcript termination. *BBA – Gene Struct. Expr.*, **1577**, 251–260.

Rocap, G., Larimer, F. W., Lamerdin J., *et al.* (2003). Genome divergence in two *Prochlorococcus* ecotypes reflects oceanic niche differentiation. *Nature*, **424**, 1042–1047.

Shimodaira, H. (2002). An approximately unbiased test of phylogenetic tree selection. *Syst. Biol.*, **51**, 492–508.

Shimodaira, H., and Hasegawa, M. (1999). Multiple comparisons of log-likelihoods with applications to phylogenetic inference. *Mol. Biol. Evol.*, **16**, 1114–1116.

Siatecka, M., Rozek, M., Barciszewski, J., and Mirande, M. (1998). Modular evolution of the Glx-tRNA synthetase family-rooting of the evolutionary tree between the bacteria and archaea/eukarya branches. *Eur. J. Biochem.*, **256**, 80–87.

Snel, B., Bork, P., and Huynen, M. A. (2002). Genomes in flux: the evolution of archaeal and proteobacterial gene content. *Genome. Res.*, **12**, 17–25.

Sullivan, M. B., Coleman, M. L., Weigele, P., Rohwer, F., and Chisholm, S. W. (2005). Three *Prochlorococcus* cyanophage genomes: signature features and ecological interpretations. *PLoS Biol.*, **3**, e144.

Thomson, R., Pritchard, J. K., Shen, P., Oefner, P. J., and Feldman, M. W. (2000). Recent common ancestry of human Y chromosomes: evidence from DNA sequence data. *Proc. Natl. Acad. Sci. USA*, **97**, 7360–7365.

Ting, C. S., Rocap, G., King, J., and Chisholm, S. W. (2002). Cyanobacterial photosynthesis in the oceans: the origins and significance of divergent light-harvesting strategies. *Trends Microbiol.*, **10**, 134–142.

Underhill, P. A., Shen, P., Lin, A. A., *et al.* (2000). Y chromosome sequence variation and the history of human populations. *Nat. Genet.*, **26**, 358–361.

Vigilant, L., Stoneking, M., Harpending, H., Hawkes, K., and Wilson, A. C. (1991). African populations and the evolution of human mitochondrial DNA. *Science*, **253**, 1503–1507.

Vogler, A. P., Homma, M., Irikura, V. M., and Macnab, R. M. (1991). Salmonella typhimurium mutants defective in flagellar filament regrowth and sequence similarity of FliI to F0F1, vacuolar, and archaebacterial ATPase subunits. *J. Bacteriol.*, **173**, 3564–3572.

Woese, C. R. (1987). Bacterial evolution. *Microbiol. Rev.*, **51**, 221–271.

Woese, C. R., Magrum, L. J., and Fox, G. E. (1978). Archaebacteria. *J. Mol. Evol.*, **11**, 245–251.

Xue, H., Tong, K. L., Marck, C., Grosjean, H., and Wong, J. T. (2003). Transfer RNA paralogs: evidence for genetic code-amino acid biosynthesis coevolution and an archaeal root of life. *Gene*, **310**, 59–66.

Zeidner, G., Bielawski, J. P., Shmoish, M., *et al.* (2005). Potential photosynthesis gene recombination between *Prochlorococcus* and *Synechococcus* via viral intermediates. *Environ. Microbiol.*, **7**, 1505–1513.

Zhaxybayeva, O. and Gogarten, J. P. (2004). Cladogenesis, coalescence and the evolution of the three domains of life. *Trends Genet.*, **20**, 182–187.

Zhaxybayeva, O., Gogarten, J. P., Charlebois, R. L., *et al.* (2006). Phylogenetic analyses of cyanobacterial genomes: quantification of horizontal gene transfer events. *Genome Res.*, **16**, 1099–1108.

Zhaxybayeva, O., Lapierre, P., and Gogarten, J. P. (2005). Ancient gene duplications and the root(s) of the tree of life. *Protoplasma*, **227**, 53–64.

Zhaxybayeva, O., Lapierre, P., and Gogarten, J. P. (2004). Genome mosaicism and organismal lineages. *Trends Genet.*, **20**, 254–560.

Zimniak, L., Dittrich, P., Gogarten, J. P., Kibak, H., and Taiz, L. (1988). The cDNA sequence of the 69-kDa subunit of the carrot vacuolar H+- ATPase. Homology to the beta-chain of F0F1-ATPases. *J. Biol. Chem.*, **263**, 9102–9112.

10

Evolutionary innovation versus ecological incumbency

Adolf Seilacher

University of Tubingen and Yale University

10.1 The Ediacaran world

The terminal period of the Proterozoic, called Ediacaran after a locality in the Flinders Ranges of South Australia (or Vendian in the Russian terminology), marks the first appearance of undoubted macrofossils. Because they are seemingly complex (and because they have been studied mainly by palaeozoologists), all Ediacaran macrofossils were interpreted originally as early multicellular animals (metazoans). The title of Martin Glaessner's 1984 monumental book, *The Dawn of Animal Life*, expresses this view. The discovery of similar fossils in approximately 30 localities all over the world further contributed to the assumption that the Ediacaran fauna represents simply a prelude to the Cambrian evolutionary explosion of metazoan phyla. Consequently, the Ediacaran period could be considered as the initial stage of the Palaeozoic.

Subsequently, this view has been challenged by the vendobiont hypothesis (Seilacher, 1984, 1992). The challenge started with the observation that most of these organisms represent hydrostatic 'pneu' structures, whose various shapes were maintained by a quilted skin (as in an air mattress) and the internal pressure of the living content. Another feature shared by all vendobionts is the allometric growth of the quilt patterns. No matter whether the addition of new 'segments' continued throughout life (serial mode) or stopped at a certain point, followed by the expansion and secondary subdivision of established quilts (fractal mode), compartments never exceed a certain millimetric diameter. Such allometry is common in oversized *unicellular* organisms, such as certain algae (*Acetabularia*) and larger Foraminifera. It is probably controlled by the structure of the cells and metabolic diffusion distances (a globular cell the size of a walnut physiologically could not exist). Thus, allometric compartmentalization of a multinucleate protoplasm by a chambered architecture may be considered as a strategy of unicellular (but multinucleate) organisms to reach large body sizes (i.e., as an alternative to multicellularity, which

Planetary Systems and the Origins of Life, eds. Ralph E. Pudritz, Paul G. Higgs, and Jonathon R. Stone. Published by Cambridge University Press.

evolved independently in plants and animals). In conclusion, the quilted Ediacaran organisms are considered herein as giant unicells rather than multicellular animals (Seilacher *et al.*, 2003).

Nevertheless, truly *multicellular* organisms have been found in Ediacaran biota. They are documented by: (1) flattened macroalgae in contemporaneous black shales (Knoll, 2003), (2) phosphatized embryos (Donoghue and Dong, 2005), (3) shield-shaped sand sponges (Trilobozoa), (4) tubular shells (e.g., *Cloudina*), (5) worm burrows (Droser *et al.*, 2005; Seilacher *et al.*, 2003), and (6) the ventral death masks of a stem group mollusc (*Kimberella*) together with its radular scratches (*Radulichnus*). Except for the embryos and algae, these fossils are found in the same sandy facies that contain the vendobionts. Compared to them, however, the early metazoans remained rather small, much less numerous, and less diversified.

The title of a paper (Seilacher *et al.*, 2003) expresses the revised scenario: 'The dawn of animal life in the shadow of giant protists.' It also refers tacitly to a similar phenomenon in the evolution of mammals, who spent two-thirds of their geological history in the ecological shadow of the dinosaurs. The present review focuses on this point: why did the evolutionary innovation not lead immediately to ecological success? In other words (inspired by Chris McKay's talk at the conference that was the basis of this book): why did the Cambrian explosion of Metazoa not happen earlier?

Before this question can be addressed, it is necessary to review the observational basis for the new scenario.

10.2 Preservational context

The palaeontological record is largely a matter of preservation. Dissociated bones and shells are the rule; soft-part impressions and articulated skeletons can be expected only in *Konservat-Lagerstaetten* (i.e., where hostile conditions near the bottom, mostly anoxia, fenced off scavengers and delayed decomposition). Rapid burial by event sedimentation also helps. As exemplified by the famous Burgess Shales (see Chapter 11), host sediments usually were muddy, so that the fossils became flattened during compaction.

Ediacaran fossils do not fit this taphonomic scheme. They are found in sandstones, whose sedimentary structures reflect agitation by occasional storms or turbidity currents. Although Ediacarans were soft-bodied and had no stiff or mineralized skeletons, they left three-dimensional impressions, as if cast with plaster of Paris. This paradox has been explained convincingly by Jim Gehling's (1999) 'death mask' model: immediately after death and burial, the carcasses became coated by a bacterial film whose mineralization preserved their relief in three dimensions but only on one side of the body. In the Flinders facies of South Australia and the White

Sea (Narbonne, 2005), masks show the *upper* surface in vendobiont species that were mobile, felled, or attached flatly to the microbially bound bedding surface, while the lower side of the body is depicted in species that lived underneath the biomat, or in immersed anchors. In Newfoundland and England, organisms are also preserved on bedding planes; but, due to the volcanic nature of the covering turbidites, felled as well as attached species preserve the *lower* surfaces of the fronds (but see Narbonne (2005)).

Quite different is the Nama style of preservation in Namibia (Grazhdankin and Seilacher, 2002). There, similarly quilted organisms are three-dimensionally preserved *within* flood-related sands (inundites). Their tops may be deformed by event erosion, but their convex bottoms are never overturned. This suggests that they lived and died within the sediment.

10.3 Vendobionts as giant protozoans

As mentioned above, the view that the largest, most common and most diversified organisms of the Ediacaran biota were giant protozoans (rather than metazoans) was based mainly on their allometric compartmentation. Partly overturned fronds show that quilting patterns were identical on upper and lower surfaces. The penetration of two infaunal *Pteridinium* specimens by another individual and their growth responses to this accident (Figure 10.1) are hard to reconcile with metazoans. So are the resting tracks left behind by *Dickinsonia* and *Yorgia* (Figure 10.2). They destroy the elephant-skin structure of the biomat and are too deep to have been produced simply by the weight of the organism; yet there are no scratches related to mechanical digging. Digestion of the biomat by microscopic pseudopodia best explains the morphology of these resting traces and the lack of a visible trackway connecting them.

Vendobionts, in the present definition, sometimes resemble arthropods and other 'Articulata' by their segmentation. During ontogeny, minute new modules were introduced at the generative pole and grew larger by secondary expansion. Yet, there are fundamental differences:

(i) Some vendobionts (e.g., the spindle-shaped form from Newfoundland; Figure 10.3) had a generative pole at either end. In others, the 'head' segment could become secondarily generative after a traumatic accident (Figure 10.1).

(ii) Secondary growth concerned mainly the long axes of the segments, while their diameters remained fairly constant (sausage-shaped quilts).

(iii) As there is no indication of molting, the flexible skin of these organisms must have been expandable. If the internal septa between quilts consisted of the same semi-rigid material, an arthropod-like ecdysis would have been mechanically impossible, particularly if quilts were further subdivided in a fractal mode.

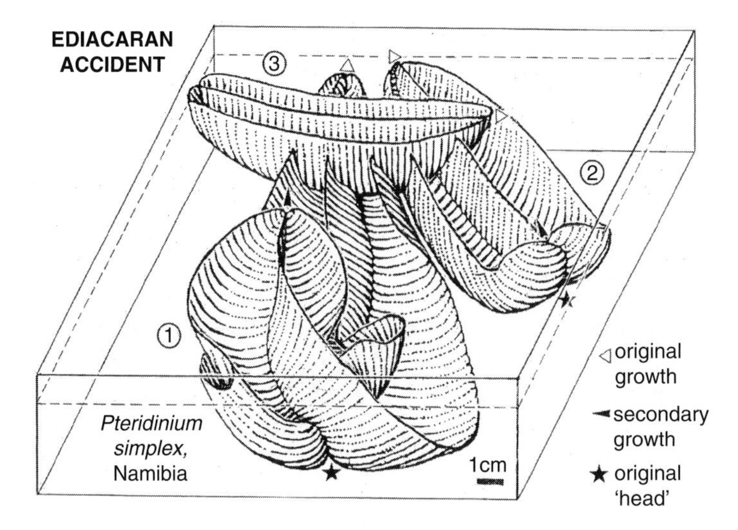

Fig. 10.1. Infaunal vendobionts, such as *Pteridinium*, were immobile. When a newcomer (number 3) grew through resident individuals, the latter responded by growing upwards at the wrong ('head') end. Number 1 first broadened its lateral vanes in the proximal part. As it grew upward, the vanes swapped functions. The left lateral vane retained its relative position, but after having grown a pronounced fold it reversed dorsoventrality; the right vane turned smoothly into the new median vane; and the original median vane became the new right vane, again with a fold. Number 2 had only started to do the same before the whole colony was terminally smothered. As such overfolding by growth is rather common, it may have also been induced by sedimentation alone. (Modified from Grazhdankin and Seilacher (2002).)

(iv) Vendobiont 'segments' never carry a functional earmark, not even the terminal module that would have contained the anus in a metazoan interpretation. Nor was there rigorously determined growth or any fixed countdown programme.
(v) While segmental growth may have been controlled by *Hox* genes, there is no sign of a Para*Hox* cluster (Erwin, 2005) controlling dorso-ventral differentiation and the development of appendages.

There is, however, a major difficulty in considering the vendobionts as unicellular: their quilts are much wider than the chamberlets of large foraminiferans and the cells of metazoans, whose diameters (rather than volumes) were probably restricted by metabolic activities (mainly diffusive) of the contained protoplasm. The solution to this dilemma came from Xenophyophoria that happen to survive on present deep-sea bottoms.

Some members of the unicellular, but multinucleate, xenophyophores (e.g., *Stannophyllum*, Figure 10.4) resemble vendobionts by being foliate and consisting of negatively allometric, sausage-shaped chambers. These chambers, however, have agglutinated walls that cannot secondarily expand, as did the quilts of

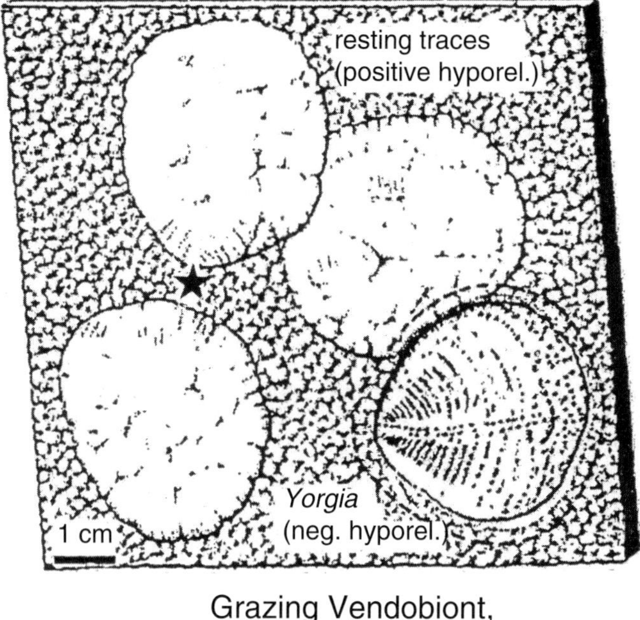

Fig. 10.2. Some prostrate vendobionts were able to creep over the biomat (elephant-skin structure), leaving behind series of resting traces but no connecting trail. Their bumpy surface without scratch marks suggests a digestive mode of burrowing. Occasionally (star), the trace maker impressed its quilts before leaving. Note that the traces and their axes are not aligned. They also have larger diameters (broken line) than the negative hyporelief mask at the last station, suggesting that the organism expanded while feeding. (After Fedonkin (2003).)

vendobionts. Ediacaran xenophyophores (Figure 10.4), which also include strings of globular chambers, lived in shallower waters than do modern ones and were embedded in microbial mats. This is why they are found in situ as positive hyporeliefs on bed soles, reminiscent of trace fossils. These protozoans were also allometrically compartmentalized, but, as in vendobionts, the chambers were too wide for unicellular standards. In xenophyophores, however, the reason for this disproportion is known: rather than being filled by pure protoplasm, their chambers contain a softer *fill skeleton* (stercomare). Consisting of finer sediment particles taken up by the pseudopodia with the food, the function of the stercomare is to further subdivide the protoplasm into strands of permissible diameters (Tendal, 1972). As observed in the Nama style of preservation (Seilacher *et al.*, 2003) and in large *Charniodiscus* (Figure 10.3), the chambers of Vendobionta appear to have been filled to about 50% with loosely bound sand grains that probably were ingested with food and may have served the same purpose as the stercomare.

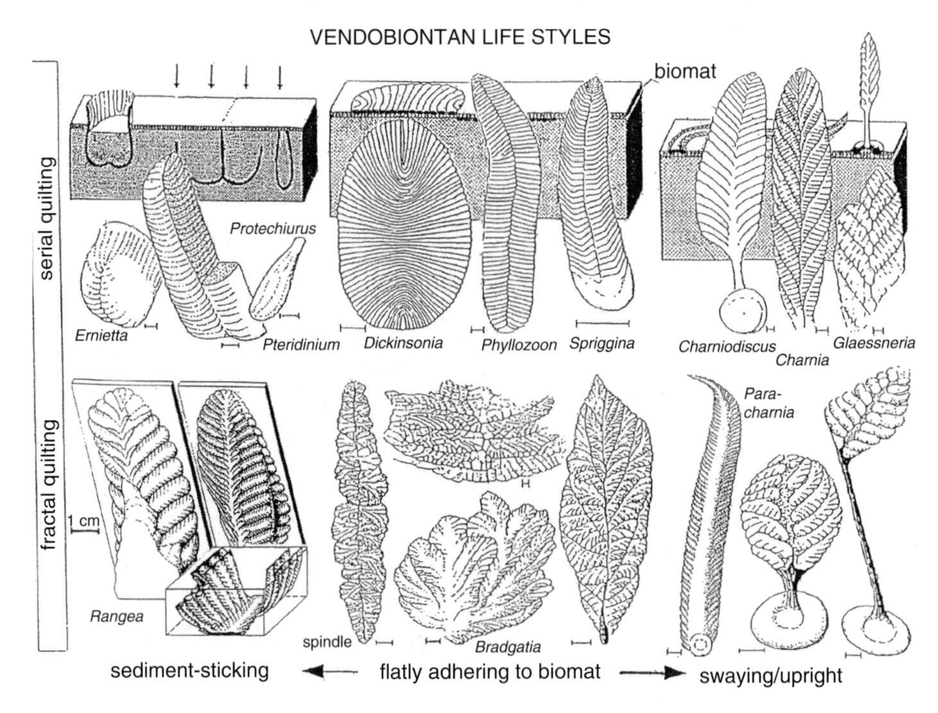

Fig. 10.3. Vendobionts grew unusually large for unicellular organisms. While sharing a quilted foliate construction, they radiated into a variety of shapes and life styles on and below the ubiquitous Precambrian biomats. Note that *Dickinsonia* (as *Yorgia*, Figure 10.2) lived on top of the mat and were mobile, while *Phyllozoon* probably lived below the mat and was immobile (Figure 10.7). Among the anchored forms, *Paracharnia* was built for swaying in a current, while a stiff stem allowed other forms to stand upright in quieter waters. Note also that in *Rangea* the opposite branch of each quilt pair bent alternatingly into one of the two inner vanes and thereby reduced space problems. (Modified from Seilacher (2003).)

In conclusion, it is reasonable to consider the Vendobionta as a coherent group of giant rhizopods that radiated into a variety of morphologies and life styles (Figure 10.3). The epinym 'unicellular dinosaurs' refers to their unusual size, ecological dominance, and morphological disparity, but also to their specialization, which made them more and more vulnerable in the face of global changes of any kind (Seilacher, 1998). The coming of macropredation and the elimination of tough microbial mats by bioturbators were such global events.

10.4 *Kimberella* as a stem-group mollusc

As stated previously, the presence of truly multicellular animals in the Late Proterozoic is documented by various kinds of fossils. While 'worm' burrows and

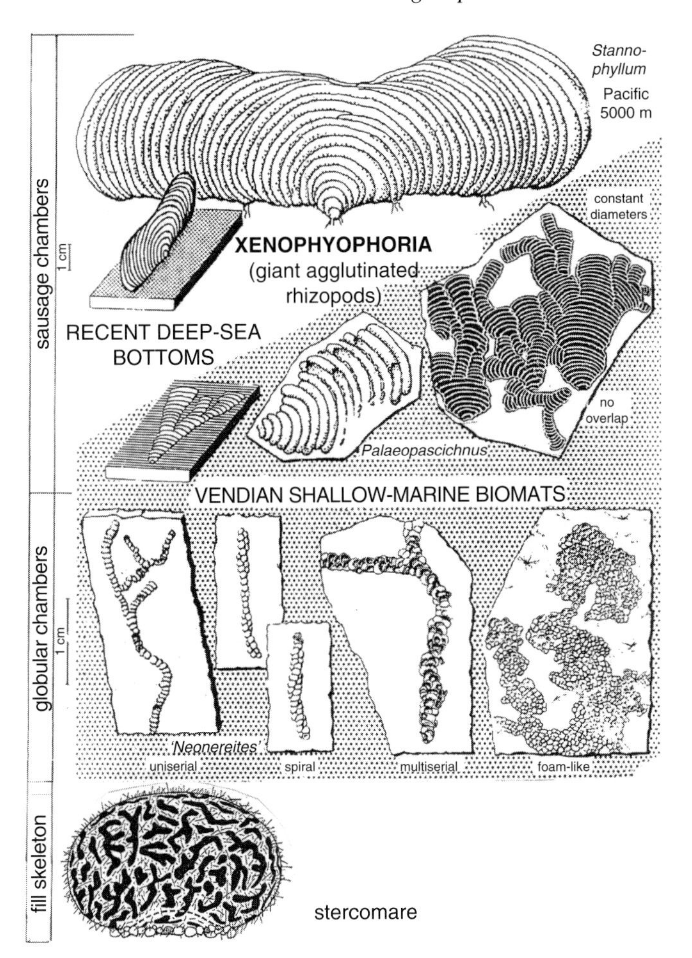

Fig. 10.4. Because they survive in the present deep sea, xenophyophores are known to be giant rhizopods. Their chambers resemble vendobiontan quilts in size, shape, and allometric diameter, but walls are agglutinated and non-expandable. As a silty fill skeleton (stercomare) further subdivides the protoplasm, chambers can be wider than in foraminiferan shells. Shallow-marine Ediacaran representatives, long considered as trace fossils, probably lived embedded in biomats. (Modified from Seilacher *et al.* (2003); enlarged cross section from Tendal (1972).)

embryos are difficult to affiliate, *Kimberella* (originally from South Australia and described as a cubozoan medusa) was clearly a mollusc-like creature. It is more common and better preserved in the White Sea region of Russia (Figure 10.5). Although *Kimberella* certainly lived above the sediment, death masks on sole faces from the White Sea preserve the ventral side. They provide the first earmark of a mollusc: the mucus secreted by the flat foot probably served as a matrix for the bacterial mask makers. Another clue is the morphology of this foot.

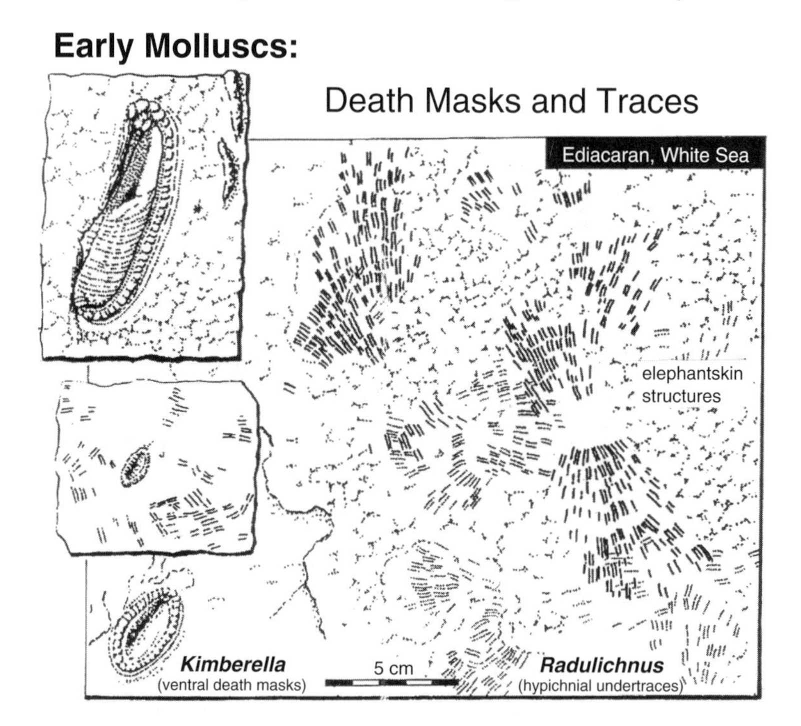

Fig. 10.5. Ventral death masks and associated radula scratches (*Radulichnus*) identify *Kimberella* as a stem-group mollusc. Grazing traces of juveniles are not preserved, because they did not reach the sole of the biomat. *Kimberella* also differed from later chitons and gastropods by grazing with an expandable proboscis in a stationary mode rather than in continuous meanders. (From Seilacher *et al.* (2003).)

It shows fine transverse wrinkles in the central area and a coarser crenulation along the circumference, related either to gills or to longitudinal and circular muscles that contracted upon death. One also recognizes the marginal impression of a limpet-like hood. As can be seen from occasional deformations, this hood was flexible and possibly covered by small sclerodermites, as in Cambrian halkieriids (Conway Morris and Peel, 1990). The ventral mask of *Kimberella* must have formed before the decay of the intestines, which caused an upward collapse in the centre. All these features suggest an animal similar to Cambrian halkieriids or to modern polyplacophorans. However, it had no real shell and grazed algae not on rock surfaces, but on the biomats that were ubiquitous on Precambrian sea bottoms.

To these biomats we also owe a detailed record of feeding habits of *Kimberella*: radular scratch patterns (*Radulichnus*; Figure 10.5). The radula is a feeding apparatus found in all modern mollusc classes except bivalves (where it probably has been lost with the transition to filter-feeding). The chitinous radula teeth scrape

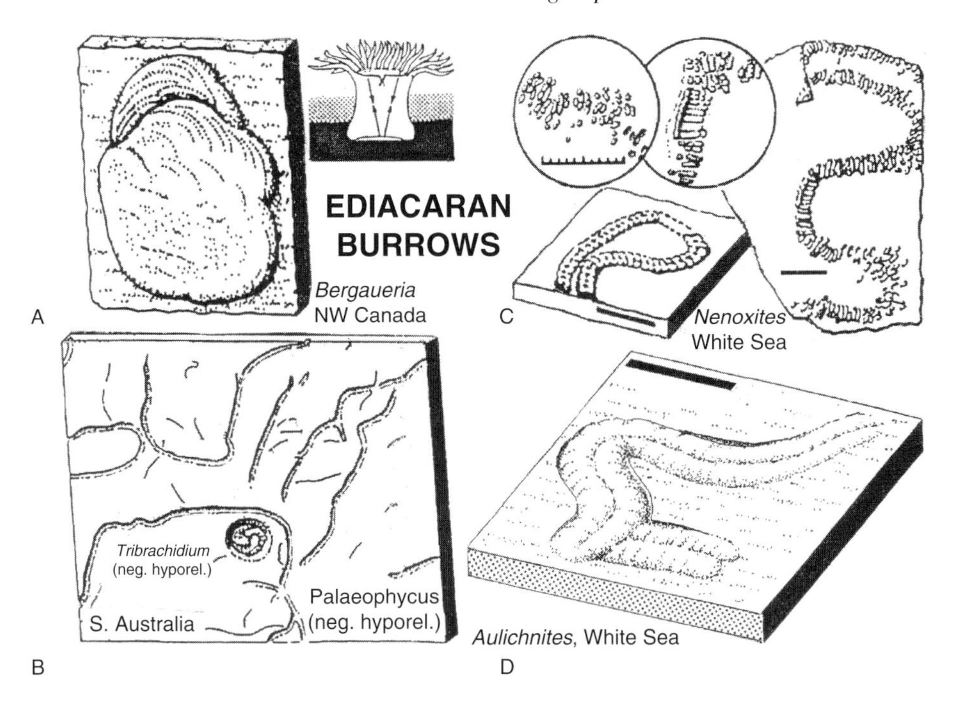

Fig. 10.6. In Ediacaran times, metazoan burrowers did not penetrate more than a few millimetres into the sediment. A. In actinian resting traces (*Bergaueria sucta*), slight lateral movements are expressed by concentric lines. B. *Palaeophycus* sp. records the horizontal movement of a worm-like undermat miner that avoided a possible sand sponge (*Tribrachidium*). C. The 'worm' burrow *Nenoxites* appears to have been lined with fecal pellets. D. *Aulichnites* is probably the backstuffed burrow of a slug-like creature. (A and B from Seilacher *et al.* (2003); C and D after Fedonkin (2003).)

algal films from rock surfaces or an aquarium wall. On soft sediment, however, their characteristic scratch patterns are wiped out when the mollusc's foot crawls over them.

How could such scratches be preserved on Ediacaran soft bottoms? Precambrian sea bottoms were covered by biomats sufficiently tough to carry an animal the size of *Kimberella* without leaving a trail. Radular teeth, on the other hand, penetrated deep enough to produce sharp undertraces at the mat's base, which is commonly marked by distinctive 'elephant-skin' structures (Figure 10.5). Only the grazing activities of juveniles fail to be recorded, because their teeth did not penetrate deep enough to produce an undertrace.

In fact, *Kimberella* would not have wiped out its own raspings anyway, due to a very unusual mode of grazing. Modern molluscs (and cows) move their heads to the left and right while grazing and crawl a step forward after every swing. *Kimberella*, however, stayed in place and foraged with the mouth at the tip of a retractable trunk.

Therefore the bipartite scratches are not arranged in continuous guided meanders but in concentric arcs (Gehling, personal communication, 1995). As the width of the swing increased automatically the further the trunk extended, the scratch field produced during each meal was conical, rather than a continuous band of meanders. Obviously, less energy had to be spent in stationary than in mobile grazing; but an extended trunk was probably too vulnerable after the onset of macropredation in the Cambrian ecologic revolution.

10.5 Worm burrows

After other seemingly complex trace fossils have been identified as either xenophyophores or pseudofossils, there still remains a fair number of distinctive 'worm' burrows (Figure 10.6). For example, *Nenoxites* trails are found on top surfaces with an apparently pelleted wall, while *Palaeophycus* commonly is preserved as a hypichnial furrow. The positive epirelief of *Aulichnites* (Figure 10.6) indicates active backfilling, but the median furrow and angular bending would better fit a short, possibly molluscan trace maker. Specimens of '*Aulozoon*' (Figure 10.7), looking like a highly compressed sand sausage, likely are the backfilled burrows of flatworms (but see Droser *et al.* (2005)).

Although made by infaunal animals, these burrows penetrated no more than a few millimetres below the sediment–water interface. They can be interpreted as the works of undermat miners (Seilacher, 1999), which fed on the decaying lower zone of the biomat. Deeper and more extensive sediment mixing (bioturbation) started only after the Cambrian substrate revolution (Bottjer *et al.*, 2000), when many animals responded to the menace of macropredation by becoming infaunal.

Trace fossils also convey another message. As one knows from later examples, biomats provide an ideal substrate for tracks of arthropods, because their pointed leg tips (mineralized or not) penetrated the mat as easily as did radula teeth. Nevertheless, no such tracks have been found in the Precambrian.

While larger arthropods were evidently absent, other trace fossils (*Bergaueria*, Figure 10.6) document the presence of actinia-like coelenterates. Unlike the resting traces of vendobionts (Figure 10.2), they show concentric scratch patterns, which formed when the bottom parts of the hydraulic polyps actively burrowed for anchoring the body and withdrew it into the sediment when disturbed.

In conclusion, stem groups of various modern animal phyla were certainly present in the benthic communities of Ediacaran times and increasingly so towards the end of the period. Yet they remained rare and small compared to the associated vendobionts and xenophyophores. The earliest animals neither evolved sophisticated search programs, nor ravaged the unprotected giant protists. Overall,

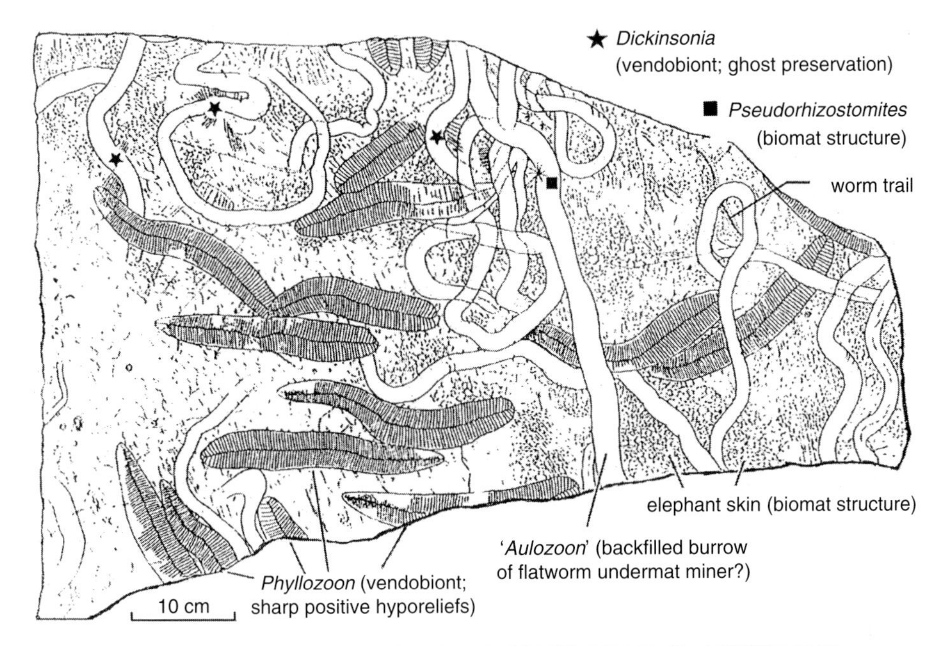

Fig. 10.7. Inverse relief, sharp contours, and the tendency of the trails to hug each other with little overcrossing suggest that the uniformly sized individuals of *Phyllozoon* are preserved in situ (but see Gehling *et al.* (2005), p. 52) below the biomat, while the ghost impressions of *Dickinsonia* are probably pressed through from above the mat. In contrast, the flat sand sausages ('*Aulozoon*') cross over each other but respect *Phyllozoon*. (After Seilacher *et al.* (2003); see Gehling *et al.* (2005a), fig. 5 for a drawing of a complete slab.)

the peaceful 'Garden of Ediacara' (McMenamin, 1986) appears to have been dominated by prokaryote biomats and giant protozoans. Even though the taxonomic composition of the biota may have changed with time (Narbonne, 2005), the general situation remained the same throughout Ediacaran times and in various environments. What made this strange world survive through tens of million years, in spite of the presence of metazoan animals?

10.6 Stability of ecosystems

Even though Darwinian processes call for continuous change, the palaeontological record tells us that macroevolution proceeded in cascades. From the level of local communities (Brett and Baird, 1995) to that of global faunas (Sepkoski and Miller, 1985), one observes long periods of relative stasis interrupted by shorter intervals of rapid change (Gould, 2002). The latter culminate in *mass extinctions*, of which the end of the Ediacaran world is a typical example. Their causation has been the matter of much debate; but it now appears that a variety of events may act as triggers, as long as they have a global effect. Another important factor is probably the readiness of the biosphere. Mass extinctions typically are preceded by greenhouse periods, in which warm climates allowed evolution to reach relatively high levels of niche partitioning and of morphological as well as behavioural specialization. This, in turn, led to an increased vulnerability in the face of environmental changes of any kind (Seilacher, 1998). In a sense, the giant protozoans of Ediacaran times may be compared with Cretaceous ammonites, rudists, and dinosaurs, although their demise and the 'Cambrian Explosion' of metazoans probably was due to radical ecological changes (onset of macropredation; bioturbational substrate revolution) rather than an asteroid impact (see Chapter 11).

The modalities of the Cambrian Explosion, however, do not explain why this turnover happened so late. Why does a superior product not lead to immediate success, as it should in a free market? Obviously the impediment relates to the ecological structure of the biosphere, rather than to the genome. One handy answer could be that metazoan newcomers found the best niches already occupied and had to wait for the demise of the occupants. But, firstly, niches are not independent entities; they can be defined only relative to an organism and its particular requirements. In the present interpretation, the incumbent organisms were so different from later macrofaunas (for instance, they obtained food and oxygen with microscopic pseudopodia spread over the whole body surface, in contrast to metazoan filter feeders) that niches could not be congruent. Secondly, niche diversification and trophic chains were still poorly developed in Ediacaran times, so that the subsequent metazoan radiation could fill an almost empty ecospace.

10.7 The parasite connection

For most of us, endoparasites and pathogens have a negative connotation, and we think that the world would be better without them. In a less anthropocentric view, however, they appear to be essential for the long-term survival of all ecosystems. Like the arms race between predator and prey species at the trophic level, a constant race takes place between hosts and their parasites. But as we know, there is no ultimate winner, because most defence mechanisms (and drugs) work only at the level of the individual host and are eventually superseded by the genetic flexibility of rapidly reproducing parasites.

The positive effect of parasites derives from their vital interest in maintaining the status quo. To this end, all of them dampen fluctuations in population size caused by predator–prey cycles. In addition, heteroxenic parasites (protozoans and metazoans that reproduce asexually in the intermediate host and sexually in a quite different final one) balance the system between the two kinds of hosts by doing less harm to the weaker than the stronger partner. For instance, in the case of the lion–gnu system (Seilacher *et al.*, 2007; Figure 10.8), the herbivores suffer little from their sarcocystid infection, except that the most-loaded individuals become behaviourally conditioned for falling prey. In contrast, only one out of five lion cubs reaches the second year in its natural environment, largely due to the protozoan parasites.

Similarly, trematodes (liver worms) spare the intermediate host, a freshwater snail, by stopping their own reproduction when the host switches to hunger metabolism (Seilacher *et al.* (2007), and literature cited therein).

Another important feature is the *host specificity* of parasites. In the lion–gnu example, another sarcocystid species cohabiting in the same herbivore cannot develop in the lion but needs to reach a hyena sharing the same prey. This specificity, however, is at a high taxonomic level. The lions' parasite may also develop in any other member of the cat family and the one of the hyena in any canid, because all constituent species share a similar histological and physiological outfit. In this way, higher taxonomic categories may become units of selection (for instance in mass extinctions). On the other hand, such flexibility makes parasites tolerant against minor changes of host species due to immigration, speciation, or extinction. Parasitic partnerships also may survive dramatic habitat changes of the host, as long as adequate transfer mechanisms can be established in the new environment. Thus, sarcocystids similar to the ones in the lion–gnu system went to sea with the whales in the Early Tertiary. Now baleens (derived from herbivorous ungulates) serve as their intermediate hosts, and the orcas (probably derived from terrestrial carnivores) as final hosts, while herrings may be the transport hosts back to baleens. The parasites' fidelity may last until a whole clade of hosts becomes eradicated in a mass extinction. On the other hand, their time-constant and well-isolated environment allows endoparasites themselves to radiate in coevolution with their hosts, as shown by the large number of sibling species.

In conclusion, probably no animal (or plant) is without endoparasites and, as the same organism hosts different kinds of them, parasites form networks glueing together established ecosystems.

Although we are far from understanding all these complex relationships (parasites that have no direct or indirect effect on humans remain underexplored), they possibly account for the stasis of ecosystems observed in the fossil record. Cases concerned range from the long-term identity of benthic communities ('coordinated stasis' (Brett and Baird, 1995), to the evolution of 'exclusive clubs' in ancient lakes and semi-restricted basins (snails in the Miocene Steinheim Lake and ceratites in

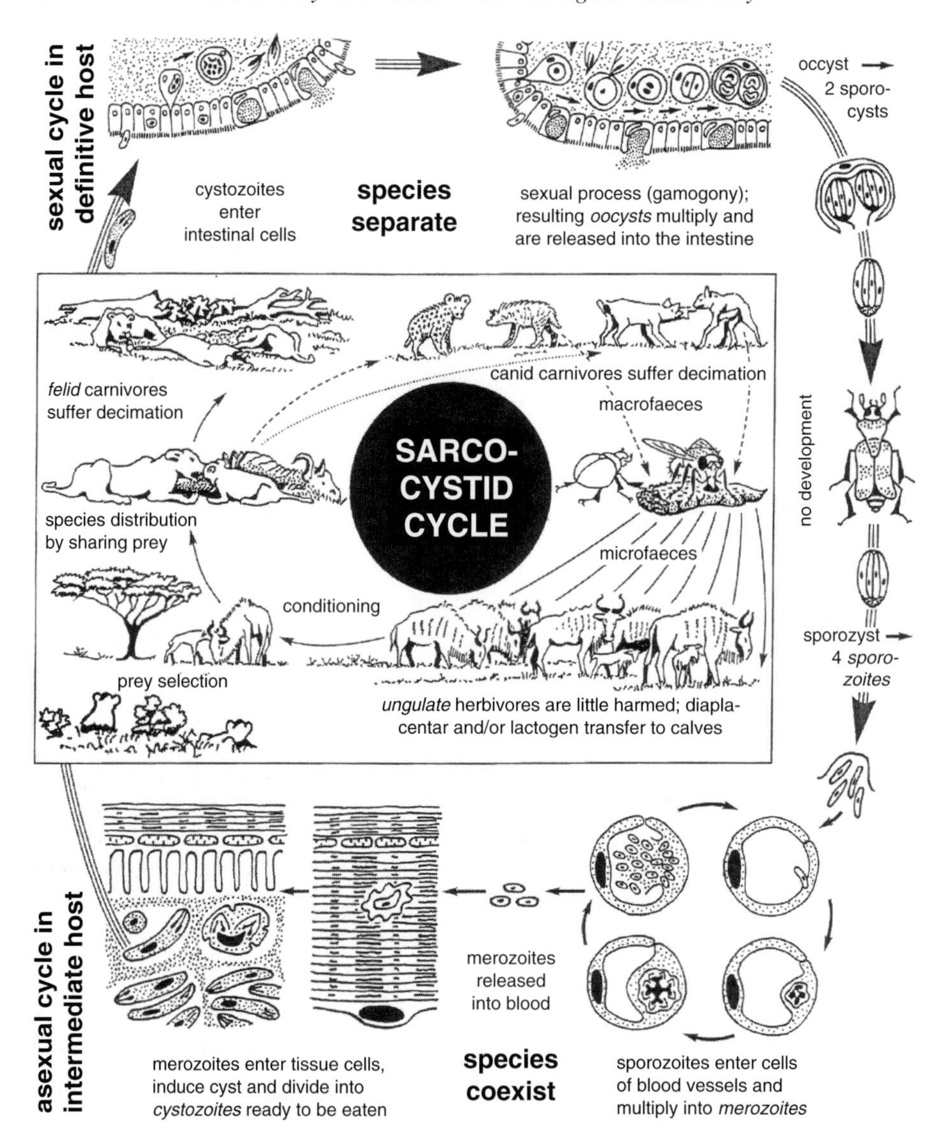

Fig. 10.8. Like other heteroxenic parasites, sarcocystids balance host populations by damaging the stronger carnivores more than the herbivores. (From Seilacher *et al.* (2007). Drawings by P. Wenk.)

the Triassic Muschelkalk Basin (Seilacher *et al.*, 2007)), and to the incumbency of ecological power structures at the largest scale.

Possibly the persistence of the Ediacaran biota has to do with such connections. Although the events in the Precambrian–Cambrian transition were different from the ones that led to the extinction of the dinosaurs, they resulted in the collapse of established ecosystems and allowed multicellular animals to become the rulers.

10.8 Conclusions

The Darwinian foundation of the evolutionary process is well established. The diversity of organisms, their geographic distribution, and molecular data are living testimonies. Yet, this theory fails to explain long-term (macroevolutionary) patterns of stasis observed in the fossil record. Some of them have extrinsic causes. Periods of warmer greenhouse climates kindled evolutionary diversification and made the biosphere more vulnerable. Global catastrophes of various kinds then led to mass extinctions. But they did not happen at random intervals, because ecosystems need time to mature. The stasis observed must have its roots in interactions between species at the level of ecosystems. The processes involved are not recorded by fossils and they are difficult to approach experimentally because of the geologic timescales at which they operate.

Ediacaran biota are not only the earliest (and thereby strangest) ecosystems on Earth that allow a palaeobiological analysis. They also provide a prime example for the incumbency of such systems. The rulers, in terms of numerical dominance and body sizes, were not the members that we would, in retrospect, consider to have had the most derived traits (the metazoans in this case) but highly developed ancient groups, such as the probably unicellular plasmodial vendobionts. The Ediacaran also is unique in lacking macropredation, which modulated trophic chains in later times.

Today, endoparasites of viral, bacterial, protozoan, and metazoan affiliations are present in virtually every organism. From the relatively few cycles that have been studied more extensively because of their epidemological and medical importance, we know that parasites have a major effect on the maintenance of ecological networks. Monoxenic pathogens dampen fluctuations in population size by series of endemic and epidemic states. Heteroxenic parasites, using more than one host, moreover control the balance between the two partners and thereby contribute in their own interest to the persistence of established ecosystems.

In the long-term view, it would be better to give up traditional anthropocentric preconceptions. Pathogens and endoparasites may be detrimental and deadly to their hosts in the short term, but they contribute to the maintenance of the status quo. By the same token, endosymbionts appear to be more advantageous to their hosts; but they also make ecosystems more vulnerable against environmental changes, as can be observed in modern coral reefs.

There is no reason to believe that the rules of macroevolution were basically different in Ediacaran biota. Ecosystems were incumbent despite the absence of macropredation. If life were to start anew or on another planet, the same kind of interactions would probably evolve: competitional change at the visible scale versus conservational players acting behind the ecological scene, just as with mutating selfish genes (Dawkins, 1999) behind the scene of Darwinian evolution. The problem is to scientifically prove or falsify such a scenario.

Acknowledgements

Thanks to Peter Wenk for opening my eyes towards parasites and the organizers of the conference on which this book is based for inviting me and making me think about life beyond our own planet. Leo Hickey and Krister Smith (Yale University) critically read earlier versions and my wife Edith pointed out major inconsistencies. Roger Thomas (Lancaster) noted the analogy with selfish genes.

References

Bottjer, D. J., Hagadorn, J. W., and Dornbos, S. Q. (2000). The Cambrian substrate revolution. *GSA Today*, **10**, 1–9.

Brett, C. E., and Baird, G. C. (1995). Coordinated stasis and evolutionary ecology of Silurian to Middle Devonian faunas in the Appalachian Basin, in *New Approaches to Speciation in the Fossil Record*, eds. H. E. Erwin and R. L. Anstey. New York: Columbia University Press, pp. 285–315.

Conway Morris, S., and Peel, J. S. (1990). Articulated halkieriids from the Lower Cambrian of north Greenland. *Nature*, **345**, 802–805.

Dawkins, R. (1999). *The Selfish Gene*. London: Oxford University Press, p. 352.

Donoghue, P. C. J., and Dong, X. (2005). Embryos and ancestors, in *Evolving Form and Function: Fossils and Development*, ed. D. Briggs. Special Publication, Peabody Museum of Natural History, Yale University, pp. 81–99.

Droser, L. M., Gehling, J. G., and Jensen, S. (2005). *Ediacaran Trace Fossils: True and False*. Special Publication, Peabody Museum of Natural History, Yale University, pp. 125–138.

Erwin, H. D. (2005). The origin of animal body plans, in *Evolving Form and Function: Fossils and Development*, ed. D. Briggs. Special Publication, Peabody Museum of Natural History, Yale University, pp. 67–80.

Gehling, J. G. (1999). Microbial mats in terminal Proterozoic siliciclasics: Ediacaran death masks. *Palaios*, **14**, 40–57.

Gehling, J. G., Droser, M. L., Jensen, S. R., and Runnegar, B. N. (2005). Ediacara Organisms: relating form to function, in *Evolving Form and Function: Fossils and Development*, ed. D. Briggs. Special Publication, Peabody Museum of Natural History, Yale University, pp. 43–66.

Glaessner, M. (1984). *The Dawn of Animal Life: A Biohistoric Study*. Cambridge, UK: Cambridge University Press.

Gould, S. J. (2002). *The Structure of Evolutionary Theory*. London: Belknap Press.

Grazhdankin, D., and Seilacher A. (2002). Underground Vendobionta from Namibia. *Palaeontology*, **45**, 57–78.

Knoll, A. H. (2003). *Life on a Young Planet. The First Three Billion Years of Evolution on Earth.* Princeton, NJ: Princeton University Press.

McMenamin, M. A. S. (1986). The garden of Ediacara. *Palaios*, **1**, 178–182.

Narbonne, G. M., (2005). The Ediacaran biota: Neoproterozoic origin of animals and their ecosystem. *Ann. Rev. Earth Planet. Sci.*, **33**, 13.1–13.22.

Seilacher, A. (1984). Late Precambrian and early Cambrian Metazoa: preservational or real extinctions? In *Patterns of Change in Earth Evolution*, eds. H. D. Holland and A. F. Trendall, Dahlem Konferenzen. Heidelberg: Springer Verlag, pp. 159–168.

Seilacher, A. (1992). Vendobionta and Psammocorallia: lost constructions of Precambrian evolution. *J. Geol. Soc. London*, **149**, 609–613.

Seilacher, A. (1998). Patterns of Macroevolution: how to be prepared for extinction. *Comp. Rends. Sciences de la Terre et des Planetes*, **327**, 431–440.

Seilacher, A. (1999). Biomat-related lifestyles in the Precambrian. *Palaios*, **14**, 86–93.

Seilacher, A., Grazhdankin, D., and Legouta, A. (2003). The dawn of animal life in the shadow of giant protists. *Paleontol. Res.*, **7**, 43–54.

Seilacher, A., Wenk, P., and Reif, W. E. (2007). The parasite connection in Ecosystems and Macroevolution. *Naturwiss.*, in press.

Sepkoski, J. J. Jr., and Miller, A. I. (1985). Evolutionary faunas and the distribution of Paleozoic benthic communities, in *Phanerozoic Diversity Patterns: Profiles in Macroevolution*, ed. J. W. Valentine. Princeton: Princeton University Press, pp. 153–90.

Tendal, O. S. (1972). A monograph of the Xenophyophoria (Rhizopoda, Protozoa). *Galathea Report*, **12**, 7–99.

11

Gradual origin for the metazoans

Alexandra Pontefract and Jonathon Stone
McMaster University

11.1 Introduction

Darwin was obsessed with origins. The book with which he distinguished himself as among the greatest thinkers ever to have walked on Earth, *The Origin of Species by Means of Natural Selection* or, *The Preservation of Favoured Races in the Struggle for Life*, was intended as only an abstract to a 'big species book' (Gould, 2002) about organisms and their environments and how they interact to elicit change over time. In a letter to Hooker (Darwin, 1871), Darwin speculated on how life, itself, might have originated, including the now infamous notion that it started in a 'warm little pond'. As Einstein would do for physics with space and time less than a half-century later (Minkowski, 1952), Darwin, in his *magnum opus* and personal correspondence, accomplished for biology: ecology and evolution were inextricably linked forevermore.

But origins posed problems for Darwin, even concerning particular groups, such as metazoans (i.e., animals): '[t]here is another and allied difficulty, which is much graver. I allude to the manner in which numbers of species of the same group, suddenly appear in the lowest known fossiliferous rocks' (Darwin, 1859; in the section titled 'On the Imperfection of the Geological Record').

In this chapter, we consider the origin of metazoans as a model for the origin of life. Considering only a group within the tree of life offers us three advantages in demonstrating our thesis (which is described in the subsequent paragraph). First, a properly defined group may be hypothesized as constituting a valid evolutionary unit (i.e., an assemblage in which all species are hypothesized to have descended from a common ancestor). Practically, this derives from the methodologies that are involved in conducting phylogenetic systematic analyses: researchers can identify uniquely animal traits that define metazoans, and these traits may be inferred to have originated once. Second, a group must have originated more recently than the more-inclusive phylogenetic tree of which it is a member. Researchers

Planetary Systems and the Origins of Life, eds. Ralph E. Pudritz, Paul G. Higgs, and Jonathon R. Stone.
Published by Cambridge University Press.

estimate that life (defined on the basis of identifiable fossilized cells) originated approximately 3.4 Bya (Knoll and Barghoorn, 1977; Westall *et al.*, 2001) and animals originated approximately 565 Mya (Gould, 1991); consequently, more-accurate estimates for the prevailing ecological conditions are available for the more-recent origin of animals than for the more-distant origins of life. Third, more evidence is available for 'stratigraphilic' than for 'stratigraphobic' groups. Animals have yielded a rich fossil record, documenting changes that transpired from the Ediacaran period (565–544 Mya (Seilacher, 1984, 1992)) through the Cambrian period (543–495 Mya), including biology's notorious 'big bang', the 'Cambrian explosion' (Conway-Morris, 1998; García-Bellido and Collins, 2004; Budd and Jensen, 2000).

We utilize these advantages as 'threads' to weave a 'tapestry' in which biological origins are perceived appropriately only from within an ecological context. Using as specific examples collagen, oxygen, and fossil specimens from the Burgess Shale, we show how radiating patterns like the one that characterizes the Cambrian explosion may be produced by more-gradual underlying processes. We then speculate on whether and how global glaciation events, according to the 'snowball Earth' hypothesis, might have affected a 'gradual origin' of metazoans.

11.2 Collagen as a trait tying together metazoans

The protein collagen is the 'tape' and 'glue' for the animal world. For example it forms the keratose fibres in sponges; it partially constitutes body walls and mesoglea in cnidarians; it supports cuticles in annelids; it anchors adductor muscles, strengthens byssal fibres, and occurs in odontophores in molluscs; it supports lophophores in brachiopoda; it associates with musculature and cartilage in arthropods; it gives rigidity to body walls and sews together calcified elements in echinoderms; and it constitutes connective tissue and organic matrix for bone in humans (Towe, 1970). All animals possess collagen, and collagen may be used to characterize animals (e.g., Brusca and Brusca (1990)).

A single vertebrate collagen molecule comprises three alpha polypeptide chains, each approximately 1000 amino acids long; the polypeptides are arranged in a helical configuration, forming a fibril with a molecular weight of approximately 300 000 u (u = unified atomic mass ($1.660\,538\,86 \times 10^{-27}$ kg)). Invertebrate collagen is similar to vertebrate collagen, although variation can be remarkable (e.g., the nematode genus Ascaris is characterized by three collagens in the cuticle, body wall, and basement membrane with molecular weights of approximately 900 000 u and exhibits a greater range in properties than do collagens among all vertebrates (Willmer, 1990)). The biosynthetic reactions by which collagen is formed, especially the proline hydroxylation step, require oxygen, and synthesizing a complete bona fide collagen molecule would be precluded in an anoxic environment (Towe, 1981).

11.3 The critical oxygen concentration criterion

When life first originated on Earth approximately 3.4 Bya, the atmosphere was oxygen deficient (see Chapter 4). Early organisms (i.e., bacteria and archaeans) gained chemical energy anoxically, excreting ammonia, carbon dioxide, hydrogen sulphide, methane, and sulphur. Approximately 3.5–2.8 Bya, photosynthesis originated (Kardong, 2005), and organisms started intaking carbon dioxide and outputting oxygen. Consequently, carbon dioxide concentrations started to decrease and oxygen concentrations started to increase; oceans changed from reducing to oxidizing; and organisms started using more-efficient and more-energetic aerobic metabolism rather than anaerobic metabolism.

The critical oxygen concentration at which metazoan aerobic metabolic demands were met was approximately 3% of the current atmospheric level (approximately 0.25%). At this level, the ozone layer provided sufficient protection from ultraviolet radiation at the Earth's surface, leading to decreased mutation levels. This allowed an increase in phytoplankton in surface waters and, consequently, an increase in atmospheric oxygen concentration (Towe, 1970). Oxygen concentrations started increasing dramatically approximately 2.4 Bya, predating the first fossilized eukaryotic cells; a second 'great oxidation event' transpired approximately 0.6 Bya, preceding the Ediacaran and Cambrian periods (Kerr, 2005). Researchers have yet to reach consensus about the causes for these great oxidation events (Kerr, 2005), but the events correspond temporally with the two major glacial activity intervals, as theorised by geologists. The most-recent explanation for the first event involves methane-producing bacteria (e.g., Claire *et al.* (2006)); increased methane levels would have facilitated hydrogen release from Earth's atmosphere, enabling photosynthetically produced oxygen to accumulate (Kerr, 2005). The most-recent explanations for the second event involve clays (associated with growth processes in early cells, as mentioned in Chapter 5); increased clay production might have provided surfaces on which organic matter could have been adsorbed and preserved, stifling oxygen consumption in decomposition reactions.

The information in the foregoing paragraphs is suggestive that evolution must be contextualized with reference to ecological setting. In his presentation to the conference on which this book is based, Graham Budd discussed criteria that were necessary for the Cambrian explosion to have occurred and singled-out ecological expansion as paramount. In this respect, Chapter 10 promulgates the same tenet (therein, Seilacher also speculates on an intriguing parasite connection). Herein, we concentrate particularly on oxygen and its importance to metazoan origins (e.g., Nursall (1959)); in the penultimate section, we speculate on whether and how large-scale glaciation events (i.e., ice ages according to snowball Earth hypotheses) influenced metazoan origins.

Early metazoans were forced to utilize oxygen penuriously to carry out physiological processes and were precluded from investing in the relatively inefficient and uneconomical (in biochemical terms) hydroxylation reactions that are required to form collagen (Towe, 1970, 1981; Kaufmann, 1962). But, once oxygen concentrations had surpassed the critical 0.25% level (approximately 0.6 Bya), organisms could begin investing in building collagen-containing constructions, such as cuticles, shells, carapaces, or bones; and then, like a lightning flash, hard parts suddenly appeared for fossilization, and the 'protometazoan exhibition' debuted.

11.4 The Burgess Shale fauna: a radiation on rocky ground

Faced with the menagerie that the fauna that lurked (and continue to dwell) within the Burgess Shale constitute, systematists themselves may be classified (albeit oversimplistically) as adopting one position between two extremes. The strange fossil specimens might represent a bestiary group that perished, leaving post-Cambrian evolution to the incumbent metazoans; or the bizarre fossil specimens may be recognized as members of existing metazoan groups (Figure 11.1). The latter perspective may be appreciated by identifying in the unusual fossil specimens some familiar traits that characterize particular extant groups (stem and crown group designations (Budd and Jensen, 2000)); then, the preternatural fossil specimens may be considered as real-world catalogues for the traits leading to but incompletely characterizing those particular groups.

For instance, velvet worms (i.e., onycophorans) may be considered as the most-closely related extant group to the arthropods (i.e., trilobites, chelicerates, myriopods, crustaceans, and insects). In addition to possessing sclerotized bodies, arthropods are characterized by sharing the derived features: jointed appendages, complex segmentation, lateral lobes, compound eyes, and sclerotized legs. Each among these five traits characterizes and, therefore, might have arisen in, species in the Burgess Shale fossil genera, *Aysheaia*, *Hallucigenia*, *Kerygmachela*, *Opabinia*, and *Anomalocaris,* en route to arthropod origins (Figure 11.1 bottom). The origins of these traits might have been associated with ecological innovations (e.g., jointed appendages for defence; segmentation for increased digestive capacity; and sclerotization for capturing prey), which might have been associated with different feeding modes (e.g., scavenging, macrophagy, and predation).

The Cambrian period was a time interval during which speciation was spectacular, as often is acknowledged. But it also was a time during which extinction was exuberant, and the morphological disparity among phyla might have resulted

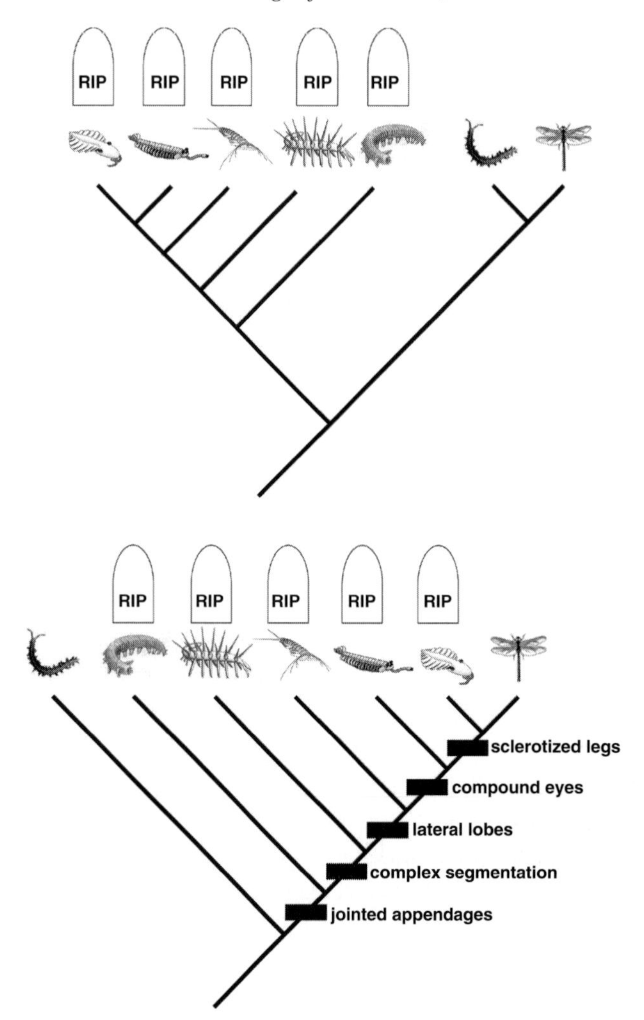

Fig. 11.1. Extreme interpretations for fossil specimens from the Burgess Shale. The phylogenetic tree contains seven taxa. Among these, five are extinct; these are indicated with tombstones. In the graphic at the top, fossil specimens representing species in the genera *Aysheaia*, *Hallucigenia*, *Kerygmachela*, *Opabinia*, and *Anomalocaris* (right to left) compose a valid evolutionary group that shared a most-recent common ancestor with, but is distinct from, another group comprising velvet worms (second from the right) and arthropods (represented by a dragonfly, at the far right). In the graphic at the bottom, fossil specimens representing species in the genera *Aysheaia*, *Hallucigenia*, *Kerygmachela*, *Opabinia*, and *Anomalocaris* (left to right) originated between velvet worms and arthropods; the traits that originated during that period, which are represented as rectangles, were steps en route to arthropod origins. Rectangles represent jointed appendages, complex segmentation, lateral lobes, compound eyes, and sclerotized legs (lower to upper). What might be interpreted as a sudden Arthropod origin and subsequent radiation (top graphic) actually might have comprised a gradual process (bottom graphic).

from excessive (crown group) extinctions (Budd, personal communication). From this perspective, upper-level categories like phyla are old because, in classifying organisms, taxonomists must identify hierarchically nesting, general traits, and this necessitates delving far into the past (Fitch and Sudhaus, 2002); although researchers have yet to reconcile origin dates from fossil-based and molecular-based phylogenies for metazoans (e.g., Wray *et al.*, 1996), consensus about relationships is emerging (see Chapter 8). Researchers must strive to test whether fossil specimens that lack many phylum-level traits might be members of extant phyla, nevertheless.

11.5 Accumulating evidence about snowball Earth

A hot topic in glacial sedimentology, the snowball Earth hypothesis involves the premise that several global ice ages have transpired during Earth's history. Although this theory is contested, it provides an explanation for the observation that diamictites, so-called 'glacial deposits', reside in Palaeoproterozoic and Neoproterozoic rocks across several continents.

Researchers have marshalled evidence for two major glacial activity intervals in Earth's history, the first occurring in the Palaeoproterozoic (approximately 2.4 Bya) and the second occurring during the Neoproterozoic (between 800 Mya and 600 Mya (Kirshvink *et al.*, 2000)). Three main snowball Earth events have been reported from the Neoproterozoic: the Sturtian (approximately 725–710 Mya), the Marinoan (approximately 635–600 Mya), and the Gaskiers (approximately 580 Mya (Narbonne, 2005)). These catastrophic events mark a time when global temperatures plummeted to −50 °C, causing the oceans to freeze, with sea ice thicknesses exceeding 1 km (a theory discussed in a recent paper by Eyles and Januszczack (2004)). These changes are thought to have halted the global hydrological cycle and terminated almost all biological activity (Kirshvink *et al.*, 2002). As evidenced by the geological record, these conditions are thought to have continued for approximately 5 million years, until an equally catastrophic, volcanically generated 'greenhouse effect' caused a total reversal (Hoffman *et al.*, 1998).

The most-cited data supporting the snowball Earth hypothesis are the aforementioned diamictites (Hoffman *et al.*, 1998; Kirschvink *et al.*, 2000; Hoffman and Schrag, 2002; Maconin *et al.*, 2004). According to the dictum that current depositional environments may be used to interpret past environments, the similarity between the Neoproterozoic diamictites and present-day glacial tills is used as hard evidence for a global glaciation between 800 Mya and 600 Mya. Located unconformably atop the diamictites are 'cap carbonates', extensive carbonate rock deposits, termed 'Ccs' (Hoffman and Schrag, 2002; Sankaran, 2003;

Sheilds, 2005). Carbonate sequences typically comprise depositions that represent post-glacial sea-level rises, starting with warm, shallow-water deposits that grade to cool, deep-water shales (Sheilds, 2005). Unique to the Neoproterozoic Ccs, however, are the extreme and abnormal thicknesses that they have attained. In the Otavi group (Namibia), for instance, the carbonate sequence ranges in thickness from 300 to 400 m (Hoffman and Schrag, 2002). These thick sequences are explained as resulting from overturn in an anoxic, deep ocean coupled with accelerated chemical weathering from extraordinary greenhouse conditions that followed global glaciation (Sheilds, 2005).

Additional evidence for several anomalous geological occurrences that can be accounted for by the snowball Earth hypothesis also resides within the diamictite successions, themselves. Analyses involving $\delta^{13}C$ ratios (i.e., ratios quantifying ^{12}C levels relative to ^{13}C levels preserved in strata, which serve as markers for biomass levels) for these successions have revealed an extreme dip in biological activity at several points during the Neoproterozoic (W. Vincent, personal communication). In successions preceding the three main snowball Earth events during the Neoproterozoic, $\delta^{13}C$ levels dip well below normal and start to rise again afterward (Halverson *et al.*, 2002, 2005). Perhaps most interestingly for the hypothesis presented here, the last snowball Earth event ended coincidentally with the Cambrian explosion, which also can be documented by changing $\delta^{13}C$ ratios (Hoffman and Schrag, 2002).

From a biological perspective, snowball Earth events would have imposed strong selection regimes on organisms. As described by Vincent *et al.* (2004), the initial freeze-up phase associated with snowball Earth events would have resulted in a mass extinction and a reduction in genetic diversity; thermophilic (heat-loving) organisms would have been able to survive only near deep-sea hydrothermal vents or geothermal hot springs and volcanoes. Such extreme conditions also would have favoured 'psychrotrophic' (cold-tolerant) organisms and 'psychrophilic' (cold-loving) organisms.

11.6 North of 80°

Research involving psychrotrophic and psychrophilic organisms currently is burgeoning progress and has spawned new-found interest in the Canadian high Arctic. Among the most-ambitious research efforts in this remote region is NASA's Haughton Mars project on Devon Island. Researchers consider this unique ecosystem, situated in the Haughton crater, as an analogue site for the ecological conditions that prevail on the red planet. They seek to identify, collect, and study the organisms living in the frigid environment. Elucidating the mechanisms by which these organisms survive will provide researchers with explanations for how organisms might evolve or might have evolved in similar ecosystems, like those found on Mars and Europa (see Chapter 15).

Fig. 11.2. Ward Hunt Island, Nunavut, Canada. The island is surrounded by the Ward Hunt Ice Shelf to the west (right) and the Markham Ice Shelf to the east (left). (Image captured by A. Pontefract.)

More domestically (from an institutional perspective), the Origins Institute, in a joint effort with Université Laval, has been involved in fieldwork conducted on Ward Hunt Island in Nunavut, Canada, for the past year (A. Pontefract, unpublished data; Figure 11.2). The island is located at 83° N latitude, several kilometres off the coast of Ellesmere Island. Ward Hunt Island was chosen as an ideal analogue site for a snowball Earth environment because extremely low temperatures are experienced by organisms living at that latitude. In addition, the island is surrounded by the Ward Hunt and Markham ice shelves (perennial ice-floats that reach up to 100 m in thickness), and therefore constitutes a unique ecosystem (D. Mueller, personal communication; Figure 11.3). The biota thereon live directly on the ice or snow or in meltwater pools in which sediment is present.

Researchers are interested primarily in determining the temperature tolerances for these organisms, with specific reference to cyanobacteria. Many organisms residing on the ice shelf are psychrotrophic rather than psychrophilic (as hypothesized (A. Pontefract, unpublished data; Vincent *et al.* 2004), Figure 11.4). Typical optimal growth temperatures for cyanobacteria range between 10 °C and 16 °C, with significant growth well past 20 °C (A. Pontefract, unpublished data). Secondary research interests include determining how pigment composition changes with varying temperature regimes and varying ultraviolet b (UVb) radiation levels and exploring microbes' abilities to survive the freeze–thaw process and extended dark-dark photoperiods.

Fig. 11.3. Ward Hunt Ice Shelf viewed aerially north-west from Ward Hunt Island. Crests and troughs are created on the surface through aerial ablation, creating a series of rivers running parallel to the coast-line. (Image captured by A. Pontefract.)

Fig. 11.4. Collecting microbial mat samples on Markham Ice Shelf. (Image captured by Denis Sarrazin.)

Among the more-surprising observations, however, was the discovery that metazoans, such as tardigrades, rotifers, and nematodes, live on the ice shelves! To survive thereon, these animals must withstand high-saline conditions and desiccation in addition to extremely low temperatures (A. Pontefract, unpublished data; Vincent *et al.* 2004). These extreme conditions constitute the basis on which hypotheses that the origin of metazoans was delayed until after the last snowball Earth event, when environmental conditions returned to 'normal', are postulated (Hoffman and Schrag, 2002; Kirshvink, 2002). The Metazoan presence on these ice shelves warrants that this idea should be re-evaluated and raises the possibility that the metazoan lineage underwent several 'revitalizations'.

11.7 Conclusion

Collagen defines metazoans, allowing them to produce traits that fossilize readily. The proline hydroxylation step involved in collagen synthesis could be utilized by early metazoans only after 0.6 Bya, once oxygen concentrations had reached sufficient levels. These organisms may be classified as members in extant phyla and, so, might have constituted the first members of lineages that persist today. These observations reveal that the Cambrian explosion actually might have transpired more gradually than the term would imply (see Chapter 10). If this were the case and snowball Earth events transpired during early metazoan history, therefore subjecting metazoans to extreme selection regimes, then phylogenetic signatures should be discernable, and psychrotrophic and psychrophilic metazoans should appear basally in the groups in which they are members.

Acknowledgements

This contribution was inspired by Graham Budd and 'Modes of evolution in the Cambrian explosion: ecology, not snowballs!', his talk at the conference for which this book constitutes the proceedings. Any valid, Earth-shattering ideas about metazoans contained herein originated with him (and witnessing A. Seilacher share cigars with M. Abou Chakra). We thank W. Vincent and D. Muller for allowing A.P. to traipse with them in the high Arctic.

References

Brusca, R. C., and Brusca, G. J. (1990). *Invertebrates*. Sunderland: Sinauer Associates.

Budd, G. E., and Jensen, S. (2000). A critical reappraisal of the fossil record of the bilaterian phyla. *Biol. Rev.*, **75**, 253–295.

Claire, M. W., Catling, D. C., and Zahnle, K. J. (2006). Biogeochemical modeling of the rise in atmospheric oxygen. *Geobiology*, in press.

Conway-Morris, S. (1998). *Crucible of Creation: The Burgess Shale and the Rise of Animals*. Oxford: Oxford University Press.
Darwin, C. R. (1859). *The Origin of Species by Means of Natural Selection or, The Preservation of Favoured Races in the Struggle for Life*. London: John Murray.
Darwin, C. R. (1871). Letter to J. D. Hooker, [1 February] 1871, in *The Life and Letters of Charles Darwin*, Vol. II, ed. F. Darwin. New York: D. Appleton and Co.
Eyles, N. and Januszczack, N. (2004). 'Zipper-rift': a tectonic model for Neoproterozoic glaciations during the breakup of Rodinia after 750 Ma. *Earth-Sci. Rev.*, **65**, 1–73.
Fitch, D. H. A., and Sudhaus, W. (2002). One small step for worms, one giant leap for 'Bauplane?' *Evol. Dev.*, **4**, 243–246.
Garca-Bellido, D. C., and Collins, D. H. (2004). Moulting arthropod caught in the act. *Nature*, **429**, 40.
Gould, S. J. (1991). *Wonderful Life*. New York: Penguin.
Gould, S. J. (2002). *The Structure of Evolutionary Theory*. Harvard: Bellknap Press.
Halverson, G., Hoffman, P., Schrag, D., and Kauffman, A. (2002). A major perturbation of the carbon cycle before the Ghuab glaciation (Neoproterozoic) in Namibia: Prelude to snowball Earth. *Geochem. Geophys. Geosys.*, **3**, 1–24.
Halverson, G., Hoffman, P., Schrag, D., Maloof, A., and Rice, A. H. N. (2005). Toward a Neoproterozoic composite carbon-isotope record. *Geol. Soc. Am. Bull.*, **117**, 1181–1207.
Hoffman, P. F., and Schrag, D. P. (2002). The snowball Earth hypothesis: testing the limits of global change. *Terra Nova*, **14**, 129–155.
Hoffman, P. F., Kaufman, A. J., Halverson G. P., and Schrag, D. P. (1998). A Neoproterozoic Snowball Earth. *Science*, **281**, 1342–1346.
Kardong, K. V. (2005). *An Introduction to Biological Evolution*. Toronto: McGraw Hill.
Kaufmann, S. (1962). In *Oxygenases*, ed. O. Hayiashi. New York: Academic Press, pp. 129–181.
Kerr, R. A. (2005). The story of O_2. *Science*, **308**, 1730–1732.
Kirshvink, J. L., Gaidos, E. J., Bertaini, L. E., *et al.* (2000). Paleoproterozoic snowball Earth: Extreme climatic and geochemical global change and its biological consequences. *Proc. Natl. Acad. Sci. USA*, **97**, 1400–1405.
Knoll, A. H., and Barghoorn, E. S. (1977). Archaen microfossils showing cell division from the Swaziland system of South Africa. *Science*, **198**, 396–398.
Maconin, M., Besse J., Ader, M., Gilder S., and Yang Z. (2004). Combined paleomagnetic and isotopic data from the Doushantuo carbonates, South China: implications for the Snowball Earth hypothesis. *Earth Plan. Sci. Lett.*, **224**, 387–398.
Minkowski, H. (1952). Space and Time (address delivered at the 80th Assembly of German Natural Scientists and Physicians, Cologne, September 21, 1908), in *The Principle of Relativity* (transl.), eds. W. Perrett and G. B. Jeffery. New York: Dover Publications, p. 75.
Narbonne, G. M. (2005). The Edicara biota: Neoproterozoic origin of animals and their ecosystems. *Annu. Rev. Earth Pl. Sc.*, **33**, 421–442.
Nursall, J. R. (1959). Oxygen as a prerequisite to the Origin of the Metazoa. *Nature*, **183**, 1170–1171.
Sankaran, A. V. (2003). Neoproterozoic 'snowball earth' and the 'cap' carbonate controversy. *Curr. Sci.*, **84**, 871–873.
Seilacher, A. (1984). Late Precambrian and early Cambrian Metazoa: preservational or real extinctions?, in *Patterns of Change in Earth Evolution*, eds. H.D. Holland and A. F. Trendall. Heidelberg: Springer Verlag, pp. 159–168.

Seilacher, A. (1992). Vendobionta and Psammocorallia: lost constructions of Precambrian evolution. *J. Geol. Soc. London*, **149**, 609–613.

Sheilds, G. A. (2005). Neoproterozoic cap carbonates: a critical appraisal of existing models and the plumeworld hypothesis. *Terra Nova*, **17**, 299–310.

Towe, K. M. (1970). Oxygen–collagen priority and the early metazoan fossil record. *Proc. Natl. Acad. Sci. USA*, **65**, 781–788.

Towe, K. M. (1981). Biochemical keys to the emergence of complex life, in *Life in the Universe*, ed. J. Billingham. Cambridge, MA: MIT Press, pp. 297–306.

Vincent, W. F., Mueller, D. M., Van Hove, P., and Howard-Williams, C. (2004). Glacial periods on early Earth and implications for the evolution of life, in *Origins: Genesis, Evolution and Diversity of Life*, ed. J. Seckbach. Dordrecht: Kluwer Academic Publishers, pp. 481–501.

Westall, F., de Wit, M. J., Dann, J., *et al.* (2001). Early Archean fossil bacteria and biofilms in hydrothermally-influenced sediments from the Barberton greenstone belt, South Africa. *Precambrian Res.*, **106**, 93–116.

Willmer, P. (1990). *Invertebrate Relationships: Patterns in Animal Evolution*. Cambridge: Cambridge University Press.

Wray, G. A., Levinton, J. S., and Shapiro, L. (1996). Molecular evidence for deep pre-Cambrian divergences among the metazoan phyla. *Science*, **274**, 568–573.

Part III

Life in the Solar System?

12

The search for life on Mars

Chris P. McKay
NASA Ames Research Center

12.1 Introduction

Mars is the world that has generated the most interest in life beyond the Earth. There are three reasons why Mars is the prime target for a search for signs of life. First, there is direct evidence that Mars had liquid water on its surface in the past, and there is the possibility that there is liquid water in the subsurface at the present time. Second, Mars has an atmosphere, albeit a thin one, that contains CO_2 and N_2. Third, conditions on Mars are cold and dry and thus are favourable for the preservation of evidence of organic remains of life that may have formed under more clement past conditions.

Mars may be cold and dry today but there is compelling evidence that earlier in its history Mars did have liquid water. This evidence comes primarily from the images taken from orbital spacecraft. Figure 12.1 from Malin and Carr (1999) shows an image of a canyon on Mars and represents probably the best evidence for extended and repeated, if not continuous, flows of liquid water on Mars. Water is the common ecological requirement for life on Earth. No organisms are known that can grow or reproduce without liquid water. Thus, the evidence that sometime in its early history Mars had liquid water is the primary motivation for the search for evidence of life (McKay, 1997).

The search for life beyond the Earth is one of the main goals of astrobiology. If life is or was present on Mars it would be important to understand the relationship of Martian life to Earth life. It is possible that Martian life and Earth life are related – part of the same tree of life. This could have resulted either from an exchange of life from one of these worlds to the other via meteorites or by the seeding of both worlds by infalling material carrying life. This later concept, known as panspermia, has been the focus of renewed interest in recent years primarily as an explanation for the origin of life on Earth very soon after the end of the impact bombardment (Davies, 1996).

Planetary Systems and the Origins of Life, eds. Ralph E. Pudritz, Paul G. Higgs, and Jonathon R. Stone.
Published by Cambridge University Press.

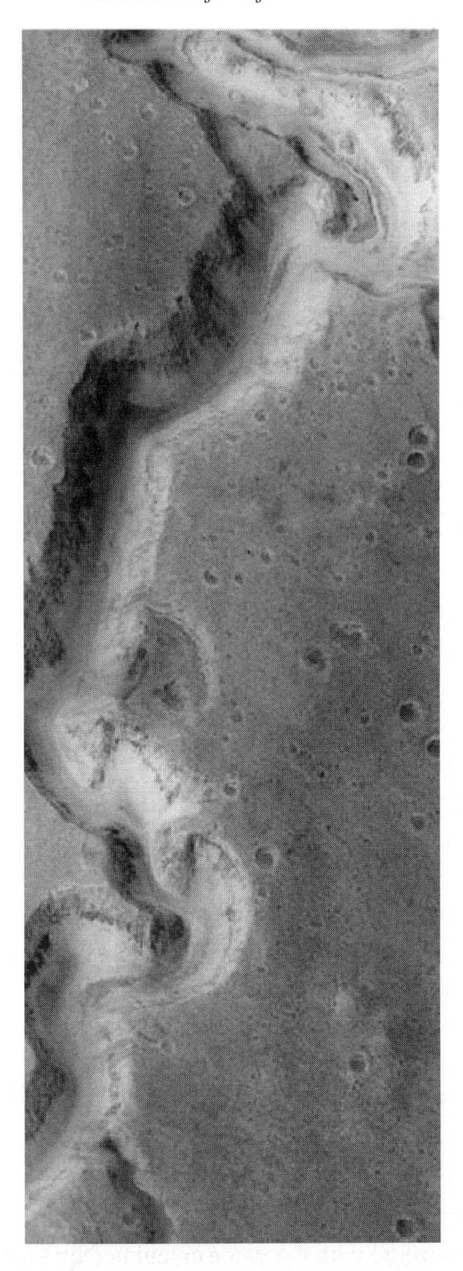

Fig. 12.1. Liquid water on another world. Mars Global Surveyor image showing Nanedi vallis in the Xanthe Terra region of Mars. The image covers an area 9.8 km by 18.5 km; the canyon is about 2.5 km wide. This image is the best evidence we have of liquid water anywhere outside the Earth. Photo from NASA/Malin Space Sciences.

It would be more interesting scientifically and philosophically if Martian life were not related to Earth life but represented a second genesis of life (McKay, 1997). This case is interesting scientifically because we would then have, for the first time, an alternative biochemistry to compare with terrestrial biochemistry. In addition, the fact that life arose independently twice in our Solar System would be persuasive evidence that life is common in the Universe.

Thus, the full astrobiological investigation of Mars goes beyond just a search for signs of life. It is also a study of the nature of that life and its genetic relationship to Earth life. To study the nature of Martian life requires that we access biological material on Mars – organisms either living or dead. In this chapter I discuss the implications of this requirement to access Martian biology and suggest approaches and locations on Mars that may be fruitful in this respect.

12.2 Mars today and the Viking search for life

The surface of Mars today is cold and dry. The surface environmental conditions are summarized in Table 12.1 and compared to the Earth. From the perspective of biology the most important feature of the Martian atmosphere is the low pressure. The time and spatial averaged pressure on Mars is about 0.6 kPa (Haberle *et al.*, 2001). The triple point of water is 0.61 kPa and water does not exist as a liquid when the total pressure is below this level. The low pressure on Mars results in the absence of water as liquid. For pressures slightly above the triple point, the liquid is only marginally stable, since the boiling point and freezing point nearly coincide. At a surface pressure of 1 kPa water will boil at a temperature of 7 °C. Kahn (1985) has shown that if liquid water were present on the surface of Mars it would lose heat rapidly by evaporation, even if it was at 0 °C, since it would be close to its boiling point. This loss of heat would cause the water to freeze since it is so close to the freezing point. The amount of energy input required to maintain the liquid state is larger than the solar constant on Mars for pressures less than about 1 kPa and becomes extremely large as the pressure approaches 0.61 kPa.

Experiments by Sears and Moore (2005) with liquid water exposed to a CO_2 atmosphere at 0.7 kPa and 0 °C in a large environmental chamber imply an evaporation rate for pure water on Mars under these conditions of 0.73 ± 0.14 mm h^{-1} – in agreement with the value calculated by assuming that evaporation depends on diffusion and buoyancy. This corresponds to a latent heat flux of 60 W m^{-2}, which is significant compared to the average solar constant at the top of the Martian atmosphere of 150 W m^{-2}. Brine solutions would have lower vapour pressure than pure water and a corresponding lower evaporation rate.

At the Viking 2 landing site (48° N) frost was observed on the surface of the ground in spring (e.g., Hart and Jakosky (1986)). The considerations above suggest

Table 12.1. *Comparing Mars and Earth*

Parameter	Mars	Earth
surface pressure	0.5–1 kPa	101.3 kPa
average temperature	−60 °C	+15 °C
temperature range	−120 °C to +25 °C	−80 °C to +50 °C
composition	95% CO_2	78% N_2
	2.7% N_2	21% O_2
	1.6% Ar	1% Ar
incident solar radiation	149 W m^{-2}	344 W m^{-2}
surface gravity	3.73 m s^{-2}	9.80 m s^{-2}
solar day	24 h 39 m 35.238 s (1 'sol')	24 h
sidereal year	687 days, 668.6 sols	365.26 days
obliquity of axis	25 deg	23.5 deg
eccentricity	0.0934	0.0167
mean distance to Sun	1.52 AU (2.28×10^8 km)	1 AU (1.49×10^8 km)

that this water would turn to vapour as conditions warmed with little or no transient liquid water phase. However, even a brief transient liquid water film might be important in terms of chemical oxidation reactions even if it is not adequate for biological function.

The only spacecraft to Mars that included a search for life or organic material were the two Viking Landers in 1976. Each lander contained three biology experiments and a device to detect and characterize organic material (a combination gas chromatograph mass spectrometer).

The three Viking biology experiments were: (1) the pyrolytic release experiment (PR), which sought to detect the ability of microorganisms in the soil to consume CO_2 using light (Horowitz and Hobby, 1977); (2) the gas exchange experiment (GEx), which searched for gases released by microorganisms when organic nutrients were added to the soil (Oyama and Berdahl, 1977); and (3) the labelled release experiment (LR), which sought to detect the release of CO_2 from microorganisms when radioactively labelled organic nutrients were added to the sample (Levin and Straat, 1977).

Both the GEx and the LR experiment gave interesting results. When moisture was added to the soil in the GEx experiment, O_2 was released. The release was rapid and occurred when the soil was exposed just to water vapour. The response persisted even if the soil was heated to sterilization levels (Oyama and Berdahl, 1977). The LR experiment indicated the release of CO_2 from the added organics. This response was attenuated with heating and eliminated when the soil temperature was raised to sterilization levels (Levin and Straat, 1977). The release of O_2 in the GEx experiment was not indicative of a biological response. The LR results, however, were precisely what would be expected if microorganisms were present in the Martian soil.

Nonetheless, a biological interpretation of the LR results is inconsistent with the results of the gas chromatograph–mass spectrometer (GCMS). The GCMS did not detect organics in the soil samples at the level of one part per billion (Biemann *et al.*, 1977; Biemann, 1979). One part per billion of organic material would represent more than 10^6 cells existing alone in the soil (Klein, 1978, 1979). However, it is not likely that there are microorganisms in the soil on Mars without associated extracellular organic material. The lack of detection of organics is the main reason for the prevailing view that non-biological factors were the cause of the reactivity of the Martian soil. In the soils of the extreme arid core of the Atacama desert – the most extreme desert location on Earth – even soils with undetectable microbial concentrations have detectable levels of organic material when higher temperatures are used in the pyrolysis than were used by Viking or when the total organic carbon is determined from combustion with oxygen (Navarro-Gonzalez *et al.*, 2003).

It is important to note that all three of the Viking biology experiments were based on providing Martian organism conditions and media for growth. That is, they were incubation experiments, similar to culture experiments. It is now known, but was not known at the time of the Viking missions, that culture experiments fail to detect over 90% of soil microorganisms on Earth (e.g., Kirk *et al.* (2004)). The fact that soil on Earth will grow up in a culture medium is due to the incredible diversity of life in those soils not to the robustness of culturing as a way to detect life. We also now know that there are soils on Earth such as those from the Atacama Desert, as discussed above (Navarro-Gonzalez *et al.*, 2003), where there are low levels of bacteria but nothing grows in any known culture media. Non-culture-dependent methods have been developed since the time of Viking and these have found widespread use in environmental microbiology. However, these methods are keyed to specific features of terrestrial life, e.g., DNA, ATP, which may not be present in Martian life if it represents a second genesis. Nonetheless, future biology experiments on Mars must make use of these culture-independent methods.

12.3 Search for second genesis

As mentioned above, the fundamental question about life on Mars is whether or not it represents a second genesis of life. To determine this we must access intact Martian life. Four possible sources have been considered: (1) dormant life at the surface, (2) subsurface ecosystems, (3) organisms preserved in salt (by analogy with salt and amber on Earth), and (4) organisms preserved in permafrost. We consider each of these. Although it is possible that life on Mars developed beyond the microbial stages (McKay, 1996) a simple comparison with the history of life on Earth would suggest only microbial life on Mars.

12.3.1 Surface life

There are two possible occurrences of surface life on Mars. The first is based on the biological interpretation of the Viking LR experiment. If the LR results are assumed to be due to biology then dormant Martian microbes are widespread in the surface soils of Mars. If the LR results are due to chemical agents, as is widely believed, then no dormant microbes are present.

A more plausible model for surface life on Mars is related to episodic liquid water formed in the north polar regions during favourable orbital conditions. As first pointed out by Murray *et al.* (1973), the conditions in the polar regions of Mars change dramatically in response to changes in the parameters of Mars' orbit. An analysis by Laskar *et al.* (2002) shows that of particular interest over the past 10 My are periodic changes in the obliquity and timing of perihelion. Figure 12.2 shows the obliquity, eccentricity, and summer equinox insulation at the north pole over the past 10 My. It is useful to consider three epochs in this time history; the last 0.5 My, from 0.5 to 6 My ago, and from 6 to 10 My ago.

In the first epoch (the last 0.5 My) the obliquity varied only slightly and changes in polar conditions were dominated by the relative phase of perihelion and equinox. During this epoch, Mars' eccentricity remained high (~0.1) and therefore the solar flux at perihelion was 49% more than at aphelion (see Table 12.2). Today perihelion (longitude of the sun on Mars, Ls = 251) almost coincides with summer solstice (Ls = 270) and the southern summer sun is therefore stronger than the northern summer sun. Procession of the orbit reverses this situation in ~50 000 y. The effect is not symmetrical due to the fact that the north polar regions are at low elevation and higher pressure and therefore, as discussed below, the formation of liquid water by the melting of ground ice is possible. This is not possible in the southern polar regions which are at much higher elevation. Thus ~50 000 y ago conditions in the north polar region of Mars were different than today, the sunlight reaching the polar regions was about 49% stronger.

The second epoch to consider in the orbital history of Mars is the period from about 0.5 My ago to 6 My ago. Laskar *et al.* (2002) showed that during this period the obliquity of Mars varied considerably over the range 15–30° with the average value approximately equal to the value today of 25°. During this epoch there were even larger changes in polar summer sunlight. At the highest obliquity and at high eccentricity, the summer sun could be twice as bright as the value at the present time.

The third epoch to consider began about 6 My ago when the obliquity assumed a larger average value (35°) with excursions as high as 45° (Laskar *et al.*, 2002).

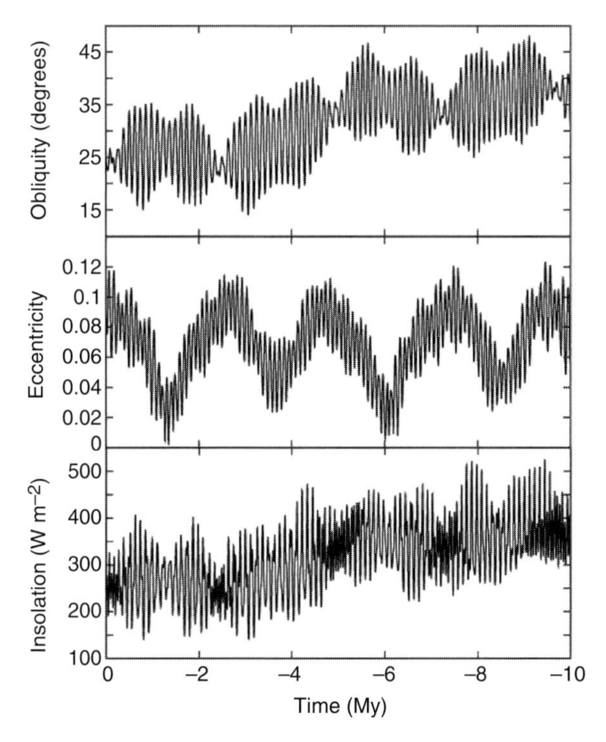

Fig. 12.2. Obliquity cycles and insolation at the north pole of Mars over the past 10 My (adapted from Laskar *et al.* (2002)).

Table 12.2. *Northern polar insolation and depth of ice exchange*

Time period	Summer insolation	Depth of ice exchange
present	200 W m^{-2}	0.1 m
0.5 My	300 W m^{-2}	$<$0.5 m
0.5 to 6 My	400 W m^{-2}	$<$1 m
6 to 10 My	500 W m^{-2}	1–2 m

During this epoch the maximum summer sun in the north polar regions could be 2.5 times brighter than its present value.

Jakosky *et al.* (2003) discussed the potential habitability of Mars' polar regions as a function of obliquity. They concluded that temperatures of ice covered by a dust layer can become high enough (−20 °C) that liquid brine solutions form and microbial activity is possible. Rivkina *et al.* (2000) have shown that microorganisms can function in ice–soil mixtures at temperatures as low as −20 °C.

Costard *et al.* (2002) computed peak temperatures for different obliquities for varying surface properties and slopes. They found that peak temperatures are >0 °C at the highest obliquities, and that temperatures above −20 °C occur for an obliquity as low as 45°. They suggested this as a possible cause of the gullies observed by Malin and Edgett (2000).

Environmental conditions on the surface of Mars today are inauspicious for the survival and growth of even the hardiest terrestrial life forms. The Antarctic cryptoendolithic microbial ecosystems and snow algae found in alpine and polar snowpacks are probably the best candidates for Martian surface life (McKay, 1993). Both ecosystems can grow in environments where the mean air temperatures are below freezing but the temperatures in the substrate (the sandstone rock and the snowpack, respectively) must be at or about melting and liquid water must be present for growth. There are no plausible models for the growth of these systems anywhere on Mars at the present time.

However, conditions at the northern polar regions are the closest to habitability in several respects (Table 12.3). First the low elevation of the northern plains results in atmospheric pressures that are above the triple point of liquid water. Indeed the pressure measured by the Viking 2 lander at 48° N never fell below 0.7 kPa. As Lobitz *et al.* (2001) and Haberle *et al.* (2001) showed, the northern plains of Mars are the main location on the planet where liquid water could be present and stable against boiling due to low pressure. If surface insolation increased in the northern polar regions the surface ice would melt to form liquid. In contrast in the southern polar regions the warmed ice would sublime due to the low pressure. The second factor that favours habitability in the polar regions is the presence of ice near the surface. The third factor arises from the nature of the polar seasons. While orbital average conditions at the polar regions can be quite cold, the summer sun never sets. Indeed, for the range of obliquities considered here the polar regions receive more sunlight per day at solstice than anywhere else on the planet. The polar summer solstice is an energy-rich period and can cause strong seasonally dependent melting. This is observed in the polar regions of Earth. The effect is even stronger as the obliquity increases from its present value of 25° to 45°. Although at the present time liquid water is not expected in the northern latitudes on average, Hecht (2002)

Table 12.3. *Favourable conditions for liquid water at the northern polar region*

1. pressure above triple point (610 Pa)
2. ice near the surface
3. high insolation during summer

has shown that under favourable conditions of solar exposure melting of ice can form liquid. This is consistent with the results of Clow (1987) for the melting of a dusty snowpack. In both cases increased solar heating compared to the present average conditions is required. As the obliquity increases, the average conditions are closer to melting and the variations required to produce liquid water are reduced. Thus liquid water should become more plentiful at the surface, although remaining frozen at depth.

12.3.2 Subsurface life

Although current conditions on Mars suggest there is little chance for life on the surface, there is interest in the possibility of subsurface life on Mars (Boston *et al.*, 1992). Liquid water could be provided by the heat of geothermal or volcanic activity melting permafrost or other subsurface water sources. Gases from volcanic activity deep in the planet could provide reducing power (as CH_4, H_2, or H_2S) percolating up from below and enabling the development of a microbial community based upon chemolithoautotrophy, especially methanogens that use H_2 and CO_2 in the production of CH_4. Stevens and McKinley (1995) and Chapelle *et al.* (2002) have reported on microbial ecosystems deep within basalt rocks on Earth that are based on methanogens and are completely independent of the surface biosphere. In this terrestrial system, H_2 comes from weathering reactions between water and basaltic rocks. With a source of hot water, all the ingredients for this subsurface habitat are present on Mars; CO_2 makes up the bulk of the Martian atmosphere and basaltic rocks are abundant. Lin *et al.* (2005) have reported on a subsurface microbial ecosystem also based on methanogens but with the H_2 produced by radioactive decay.

The possibility of subsurface life on Mars today depends on the existence of hydrothermal systems. While it certainly seems that volcanic activity on Mars has diminished over geological time, crater ages (Hartmann *et al.*, 1999) and the age of the youngest Martian meteorite, 180 My (McSween, 1994), indicates volcanism on Mars as recently as that time. Volcanic activity by itself does not provide a suitable habitat for life – liquid water presumably derived from the melting of ground ice is also required. It is likely that any volcanic source in the equatorial region would have depleted any initial reservoir of ground ice and there would be no mechanism for renewal. Closer to the poles ground ice is stable (Fanale and Cannon, 1974; Squyres and Carr, 1986; Feldman *et al.*, 2002). It is conceivable that a geothermal heat source could result in cycling of water through the cryosphere (Squyres *et al.*, 1987; Clifford, 1993). The heat source would be melting and drawing in water from any underlying reservoir of groundwater or ice that might exist.

The isotopic measurements of water in the SNC meteorites (a class of meteorites thought to have originated on Mars) show an enrichment of D, the heavy isotope of hydrogen, about equal to that in the present Martian atmosphere (Watson *et al.*, 1994). Assuming that this enhancement is due to atmospheric escape then this similarity suggests that there was an exchange between that atmosphere and the rocks from which the SNC meteorites derived. Probably this exchange involved hydrothermal groundwater systems driven by volcanism or impact events (Gulick and Baker, 1989).

Such hypothetical ecosystems are neither supported, nor excluded, by current observations of Mars. Tests for such a subsurface system involve locating active geothermal areas associated with ground ice or detecting trace quantities of reduced atmospheric gases that would leak from such a system. The reports of possible CH_4 in the Martian atmosphere could be an indication of such subsurface hydrothermal activity and possibly biology (Formisano *et al.*, 2004; Krasnopolsky *et al.*, 2004).

12.3.3 Preserved in salt

There are reports of organisms preserved in a viable state over geological time in dehydrating substances such as amber and salt. The oldest well-established preservation of life is found in amber. Cano and Borucki (1995) reported (but this has not yet been independently confirmed) that bacteria can be preserved in amber for 25 My. However, amber is a product of trees and thus is unlikely to occur on Mars. Microorganisms are also found in ancient salt. Vreeland *et al.* (2000) demonstrated retrieval of organisms from salt that is 250 My old. However, it is not clear that the organisms are as old as the salt. The main problem arises because salt, unlike amber, is not impermeable. Small drops of water can migrate through salt in the presence of a temperature gradient leaving the crystal structure of the salt intact with no apparent trace of their movement. Thus, it remains uncertain that the organisms found in the 250 My old salt on Earth are not more recent contaminants.

For Mars the issue of contamination is not so critical. If the desire is to obtain a specimen of Martian life, it does not matter if it is an ancient organism or a geologically recent contamination. Thus the preservation potential of salt deposits should be considered. However, as yet there are no locations on Mars where large salt deposits, similar to salt domes on Earth, are known to exist.

12.3.4 Permafrost

The microbiology of permafrost locations on Earth has been investigated and it has been shown that viable microorganisms can be recovered from Siberian permafrost that is $\sim$3.5 My old (Gilichinsky *et al.*, 1992). New work in Beacon

Valley, Antarctica, indicates the presence of recoverable microorganisms in ice that is thought to be 8 My old.

On Mars there may be extensive permafrost that dates back 3–4 Gy. Data from the Mars Odyssey spacecraft have confirmed the suggestion that the polar regions of Mars are rich in ground ice (Feldmann *et al.*, 2002). The south polar regions, but not the polar cap deposits themselves, are of particular interest because this region contains ancient cratered terrain presumably dating back to the end of the heavy bombardment, 3.8 Gy ago. The actual polar cap deposits are probably much younger. One region of particular interest is centred on 80° S, 180° W (see the map in Smith and McKay (2005)). Here the terrain is heavily cratered, there is ground ice present and, furthermore, there is strong crustal magnetism in the surface materials. The presence of strong crustal magnetism confirms the antiquity of these terrains and suggests that they have been relatively unaltered since their initial deposition. This location may represent the site of the oldest, coldest, undisturbed permafrost on Mars. Martian microorganisms may be trapped and preserved, even if dead, in this permafrost (Smith and McKay, 2005).

12.4 Detecting a second genesis on Mars

There are several ways to search for life. First we can search for life as a collective general phenomenon. However, life might also be a single isolated organism. And that organism might be dead. Finally, signs of life may be fossils, artefacts or other inorganic structures. In the search for life on Mars, any of these would be of interest. Definitions of life typically focus on the nature of the collective phenomenon. In general, such definitions are not useful in an operational search for life on other worlds. The one exception is the proposal by Chao (2000) to modify the Viking LR experiment to allow the detection of organisms that improve their capacity to utilize the provided nutrients. This would, in principle, provide a direct detection of Darwinian evolution and could unambiguously distinguish between biological metabolism and chemical reaction. Chao (2000) argues that Darwinian evolution is the fundamental property of life and other observables associated with life result from evolutionary selection. His method for searching for evolution would be practical if the right medium can be selected to promote the growth of alien microbes. Unfortunately, we now know that only a fraction of microorganisms from an environmental sample grow in culture.

Of course growth experiments of any kind do not detect dead organisms. Yet the remains of dead organisms are potentially important evidence of life on another planet. And so are fossils. However, there is an important distinction between dead organisms and fossils. A fossil is evidence of past life but it does not reveal anything about the biochemical or genetic nature of that life. If we are searching for a second example of life then we need to be able to compare the nature of that life to Earth life. For this, an organism is needed, either dead or alive, but a fossil is not sufficient.

As discussed above, a promising target for the search for biological remains of past life is the subsurface. The deep permafrost on Mars may hold remnants of past life (Smith and McKay, 2005). The organisms in the ground ice are likely to be dead from the accumulated radiation dose but their organic remains could be analysed and compared to the biochemistry of Earth life.

I have argued previously (McKay, 2004) that one way to determine if a collection of organic material is of biological origin is to look for a selective pattern of organic molecules similar to, but not necessarily identical with, the selective pattern of biochemistry in life on Earth. Pace (2001) has argued that life everywhere will be life as we know it. He contends that the biochemical system used by life on Earth is the optimal one and therefore evolutionary pressure will cause life everywhere to adopt this same biochemical system. It is instructive to consider this argument in the context of a conceptual organic phase space. If we imagine all possible organic molecules as the dimensions of a phase space, then any possible arrangement of organic molecules is a point in that phase space. We can define biochemistries as those points in phase space that allow life. The biochemistry of Earth life – life as we know it – represents one point in the organic phase space. We know that this one point represents a viable biochemistry. Pace's (2001) contention that biochemistry is universal is equivalent to stating that in the region of phase space of all possible biochemistries there is only one optimum biochemistry and thus any initial set of biochemical reactions constituting a system of living organisms will move toward that optimum as a result of selective pressure. If Pace (2001) is correct, then the only variation between life forms that we can expect is that associated with chirality. As far as is known, the left and right forms of chiral organic molecules (such as amino acids and sugars) have no differences in their biochemical function. Life is possible that is exactly similar in all biochemical respects to life on Earth except that it has right instead of left amino acids in its proteins and left instead of right sugars in its polysaccharides.

The question of the number of possible biochemistries consistent with life is an empirical one and can only be answered by observations of other life forms on other worlds, or by the construction of other life forms in the laboratory. The observation or construction of even one radically alien life form would suffice to show that biochemistry as we know it is not universal.

The pattern of biochemistry of Earth life follows what I have called the 'Lego principle' (McKay, 2004). This is the unremarkable observation that life on Earth uses a small set of molecules to construct the diverse structures that it needs. This is similar to the children's play blocks known as Lego in which a few different units, repeated over and over again, are used to construct complex structures. The

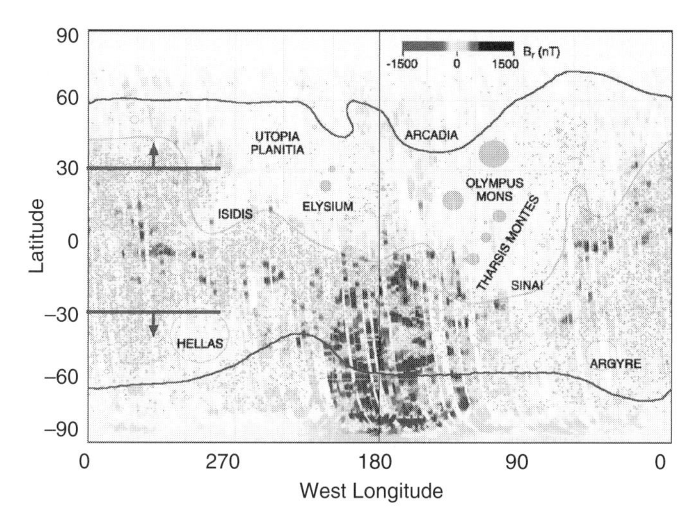

Fig. 12.3. Comparison of biogenic with non-biogenic distributions of organic material. Non-biological processes produce smooth distributions of organic material, illustrated here by the curve. Biology, in contrast, selects and uses only a few distinct molecules, shown here as spikes (e.g., the 20 left-handed amino acids on Earth). Analysis of a sample of organic material from Mars or Europa may indicate a biological origin if it shows such selectivity. Figure from McKay (2004).

biological polymers that construct life on Earth are the proteins, the nucleic acids, and the polysaccharides. These are built from repeated units of the 20 left-handed amino acids, the five nucleotide bases, and the right-handed sugars. The use of only certain basic molecules allows life to be more efficient and selective. Evolutionary selection on life anywhere is likely to result in the same selective use of a restricted set of organic molecules. As discussed above, I believe it is premature to conclude that all life anywhere will use the same set of basic biomolecules. Thus I suggest that life will always use some basic set but it may not be the same basic set used by life on Earth. This characteristic biogenic pattern of organic molecules would persist even after the organism is dead. Given our present state of understanding of biochemistry, we are not able to propose alternative and different biochemical systems that could be the basis for life, but that may reflect a failure of our understanding and imagination rather than a restriction on the possibilities for alien life.

A sample from the deep permafrost in the southern hemisphere of Mars could be analysed for organic material with a fairly simple detection system. If organic material was detected, then it would be of interest to characterize any patterns in that organic material that would indicate a 'Lego principle' pattern. Clearly one such pattern is the identical pattern of all Earth life; 20 amino acids, the five nucleotide bases, A, T, C, G, and U, etc. However, more interesting would be a clear pattern different from the pattern known from Earth life. Figure 12.3 shows a schematic diagram of how a biological pattern would be different from a non-biological pattern.

Implementing this search in practical terms in near term missions will require a sophisticated ability to separate and characterize organic molecules. Currently the instrument best suited for this task is a GC-MS with solvent extraction. However, new methods of fluorescence and Raman spectroscopy could provide similar information and may have a role in future mission applications.

12.5 Conclusions

The search for evidence of past life on Mars is motivated by the direct evidence that Mars once had liquid water. Studies of life on Earth strongly indicate liquid water as the essential requirement for life. Mars presents a challenge and an opportunity. The challenge is to explore a distant planet with a complex history first with robotic probes and eventually with humans. The opportunity is to learn about the nature of life, to search for a possible second type of life in our own Solar System and thereby begin to understand the profound philosophical and scientific issues related to life in the Universe. It is a search worth the best efforts of the planetary science and space engineering community.

References

Acuña, M. H., Connerney, J. E. P., Ness, N. F., *et al.* (1999). Global distribution of crustal magnetism discovered by the Mars Global Surveyor MAG/ER experiment. *Science*, **284**, 790–793.

Barlow, N. (1997). Mars: impact craters, in *Encyclopedia of Planetary Sciences*, eds. J. H. Shirley and R. W. Fairbridge. London: Chapman and Hall.

Biemann, K. (1979). The implications and limitations of the findings of the Viking Organic Analysis Experiment. *J. Mol. Evol.*, **14**, 65–70.

Biemann, K., Oro, J., Toulmin, P., III, *et al.* (1977). The search for organic substances and inorganic volatile compounds in the surface of Mars. *J. Geophs. Res.*, **82**, 4641–4658.

Boston, P. J., Ivanov, M. V., and McKay, C. P. (1992). On the possibility of chemosynthetic ecosystems in subsurface habitats on Mars. *Icarus*, **95**, 300–308.

Cano R. J., and Borucki, M. K. (1995). Revival and identification of bacterial spores in 25- to 40-million-year-old Dominican amber. *Science*, **268**, 1060–1064.

Chao, L. (2000). The meaning of life. *Bioscience*, **50**, 245–250.

Chapelle, F. H., O'Neill, K., Bradley, P. M., *et al.* (2002). A hydrogen-based subsurface microbial community dominated by methanogens. *Nature*, **415**, 312–315.

Clifford, S. M. (1993). A model for the hydrologic and climatic behavior of water on Mars. *J. Geophys. Res.*, **98**, 10 973–11 016.

Clow, G. D. (1987). Generation of liquid water on Mars through the melting of a dusty snowpack. *Icarus*, **72**, 95–127.

Costard, F., Forget, F., Mangold, N., and Peulvast, J. P. (2002). Formation of recent Martian debris flows by melting of near-surface ground ice at high obliquity. *Science*, **295**, 110–113.

Davies, P. C. W. (1996). The transfer of viable micro-organisms between planets, in *CIBA Foundation Symposium 202: Evolution of Hydrothermal Ecosystems on Earth (and Mars?)*, eds. G. R. Bock and J. Goode. Chichester: John Wiley and Sons, pp. 304–317.

Fanale, F. P., and Cannon, W. A. (1974). Exchange of adsorbed H_2O and CO_2 between the regolith and atmosphere of Mars caused by changes in surface insolation. *J. Geophys. Res.*, **79**, 3397–3402.

Feldman, W. C., Boynton, W. V., Tokar, R. L., *et al.* (2002). Global distribution of neutrons from Mars: results from Mars Odyssey. *Science*, **297**, 75–78.

Formisano, V., Atreya, S., Encrenaz, T., Ignatiev, N., and Guiranna, M. (2004). Detection of methane in the atmosphere of Mars. *Science*, **306**, 1758–1761.

Gilichinsky, D. A., Vorobyova, E. A., Erokhina, L. G., Fyordorov-Dayvdov, D. G., and Chaikovskaya, N. R. (1992). Long-term preservation of microbial ecosystems in permafrost. *Adv. Space Res.*, **12(4)**, 255–263.

Gulick, V. C., and Baker, V. R. (1989). Fluvial valleys and martian paleoclimates. *Nature*, **341**, 514–516.

Haberle, R. M., McKay, C. P., Schaeffer, J., *et al.* (2001). On the possibility of liquid water on present-day Mars. *J. Geophys. Res.*, **106**, 23317–23326.

Hart, H., and Jakosky, B. M. (1986). Composition and stability of the condensate observed at the Viking Lander 2 site on Mars. *Icarus*, **66**, 134–142.

Hartmann, W. K., Malin, M., McEwen, A., *et al.* (1999). Evidence for recent volcanism on Mars from crater counts. *Nature*, **397**, 586–589.

Hecht, M. H. (2002). Metastability of liquid water on Mars. *Icarus*, **156**, 373–386.

Horowitz, N. H., and Hobby, G. L. (1977). Viking on Mars: the carbon assimilation experiments. *J. Geophys. Res.*, **82**, 4659–4662.

Jakosky, B. M., Nealson, K. H., Bakermans, C., Ley, R. E., and Mellon, M. T. (2003). Subfreezing activity of microorganisms and the potential habitability of Mars' polar regions. *Astrobiology*, **3**, 343–350.

Kahn, R. (1985). The evolution of CO_2 on Mars. *Icarus*, **62**, 175–190.

Kirk, J. L., Beaudette, L. A., Hart, M., *et al.* (2004). Methods of studying soil microbial diversity. *J. Microbiological Methods*, **58**, 169–188.

Klein, H. P. (1978). The Viking biological experiments on Mars. *Icarus*, **34**, 666–674.

Klein, H. P. (1979). The Viking Mission and the search for life on Mars. *Rev. Geophys. and Space Phys.*, **17**, 1655–1662.

Krasnopolsky, V. A., Maillard, J. P., and Owen, T. C. (2004). Detection of methane in the martian atmosphere: evidence for life? *Icarus*, **172**, 537–547.

Laskar, J., Levrard, B., and Mustard, J. (2002). Orbital forcing of the martian polar deposits. *Nature*, **419**, 375–377.

Levin, G. V., and Straat, P. A. (1977). Recent results from the Viking Labeled Release Experiment on Mars. *J. Geophys. Res.*, **82**, 4663–4667.

Lin, L.-H., Hall, J. A., Lippmann, J., *et al.* (2005). Radiolytic H_2 in the continental crust: Nuclear power for deep subsurface microbial communities. *Geochem. Geophys. Geosys.*, **6:Q07003**. [Online.] doi:10.1029/2004GC000907.

Lobitz, B., Wood, B. L., Averner, M. A., and McKay, C. P. (2001). Use of spacecraft data to derive regions on Mars where liquid water would be stable. *Proc. Nat. Acad. Sci. USA*, **98**, 2132–2137.

Malin, M. C., and Carr, M. H. (1999). Groundwater formation of martian valleys. *Nature*, **397**, 560–561.

Malin, M. C., and Edgett, K. S. (2000). Sedimentary rocks of early Mars. *Science*, **290**, 1927–1937.

McKay, C. P. (1993). Relevance of Antarctic microbial ecosystems to exobiology, in *Antarctic Microbiology*, ed. E. Imre Friedmann. New York: Wiley-Liss, pp. 593–601.

McKay, C. P. (1996). Oxygen and the rapid evolution of life on Mars, in *Chemical Evolution: Physics of the Origin and Evolution of Life*, eds. J. Chela-Flores and F. Raulin. Dordrecht, Holland: Kluwer, pp. 177–184.

McKay, C. P. (1997). The search for life on Mars. *Origins Life Evol. B.*, **27**, 263–289.

McKay, C. P. (2001). The search for a second genesis of life in our Solar System, in *First Steps in the Origin of Life in the Universe*, ed. J. Chela-Flores. Dordrecht, Holland: Kluwer, pp. 269–277.

McKay, C. P. (2004). What is life and how do we search for it on other worlds? *PLoS Biol.*, **2**, 1260–1263.

McSween, H. Y. (1994). What we have learned about Mars from SNC meteorites. *Meteoritics*, **29**, 757–779.

Murray, B. C., Ward, W. R., and Yeung, S. C. (1973). Periodic insolation variations on Mars. *Science*, **180**, 638–640.

Navarro-Gonzalez, R., Rainey, F. A., Molina, P., *et al.* (2003). Mars-like soils in the Atacama Desert, Chile, and the dry limit of microbial life. *Science*, **302**, 1018–1021.

Oyama, V. I., and Berdahl, B. J. (1977). The Viking gas exchange experiment results from Chryse and Utopia surface samples. *J. Geophys. Res.*, **82**, 4669–4676.

Pace, N. (2001). The universal nature of biochemistry. *Proc. Natl. Acad. Sci. USA*, **98**, 805–808.

Rivkina, E. M., Friedmann, E. I., McKay, C. P., and Gilichinsky, D. A. (2000). Metabolic activity of permafrost bacteria below the freezing point. *Appl. Environ. Microbio.*, **66**, 3230–3233.

Sears, D. W. G., and Moore, S. R. (2005). On laboratory simulation and the evaporation rate of water on Mars. *Geophys. Res. Lett.*, **32**, L16202, doi:10.1029/2005GL023443.

Smith, H. D., and McKay, C. P. (2005). Drilling in ancient permafrost on Mars for evidence of a second genesis of life. *Planet. Space Sci.*, **53**, 1302–1308.

Squyres, S. W., and Carr, M. H. (1986). Geomorphic evidence for the distribution of ground ice on Mars. *Science*, **231**, 249–252.

Squyres, S. W., Wilhelms, D. E., and Moosman, A. C. (1987). Large-scale volcano–ground ice interactions on Mars. *Icarus*, **70**, 385–408.

Stevens, T. O., and McKinley, J. P. (1995). Lithoautotrophic microbial ecosystems in deep basalt aquifers. *Science*, **270**, 450–454.

Vreeland, R. H., Rosenzweig, W. D., and Powers, D. W. (2000). Isolation of a 250 million year old bacterium from primary salt crystals. *Nature*, **407**, 897–900.

Watson, L. L., Hutcheon, I. D., Epstein, S., and Stolper, E. M. (1994). Water on Mars: Clues from deuterium/hydrogen and water contents of hydrous phases in SNC meteorites. *Science*, **265**, 86–90.

13

Life in the dark dune spots of Mars: a testable hypothesis

Eörs Szathmáry and Tibor Gánti
Collegium Budapest
Tamás Pócs
Eszterházy Károly College
András Horváth
Konkoly Observatory
Ákos Kereszturi, Szaniszló Bérczi and András Sik
Eötvös University

13.1 Introduction

This chapter presents one of the very rare exobiological hypotheses. The main thesis is that there could be life in the dark dune spots (DDSs) of the southern polar region of Mars, at latitudes between −60° and −80°. The spots have a characteristic annual morphological cycle and it is suspected that liquid water forms in them every year. We propose that a consortium of simple organisms (similar to bacteria) comes to life each year, driven by sunlight absorbed by the photosynthetic members of the consortium. A crucial feature of the proposed habitat is that life processes take place only under the cover of water ice/frost/snow. By the time this frost disappears from the dunes, the putative microbes, named Mars surface organisms (MSOs) must revert to a dormant state. The hypothesis has been worked out in considerable detail, it has not been convincingly refuted so far, and it is certainly testable by available scientific methods. We survey some of the history of, the logic behind, the testable predictions of, and the main challenges to the DDS-MSO hypothesis.

13.2 History

The spots in question were observed on images made by the Mars orbiter camera (MOC) on board the Mars Global Surveyor (MGS) spacecraft between 1998 and 1999 (images are credited to NASA/JPL/Malin Space Science Systems). These features appear in the southern and northern polar regions of the planet in the spring, and range in diameter from a few dozen to a few hundred metres. Malin and Edgett published their first observation on the Internet in 1999, and then in print (Malin and

Planetary Systems and the Origins of Life, eds. Ralph E. Pudritz, Paul G. Higgs, and Jonathon R. Stone. Published by Cambridge University Press.

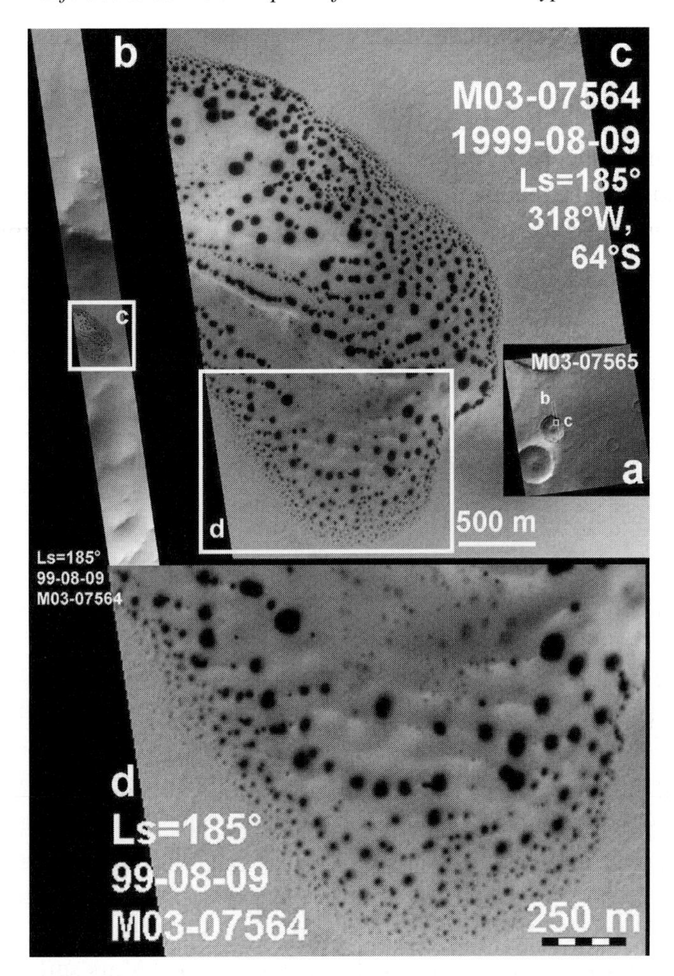

Fig. 13.1. DDSs on a southern dune covered with white frost, photographed by the MOC onboard MGS at L_S = 185 (1999.08.09., image no. M03-07564, M03-07565). Subset images: a. the observed crater, b. the MOC image strip, c. the dark dune d. the magnified image of the spots with high resolution. (Collegium Budapest, Mars Astrobiology Group.)

Edgett, 2000). One of us (A. H.) began to analyse the images in the summer of 2000, based on images from the south that were clear and freely downloadable from the Internet (http://www.msss.com/moc_gallery/). It became clear that one group of these spots was strictly localized to dark dunes, clearly distinguishable from the usual rusty terrain of Mars. Thus, we coined the term dark dune spots (DDSs; Figure 13.1). Nobody can deny that the DDSs are striking or that they call for an explanation. As we shall see, the first explanation (Malin and Edgett, 2000), based on simple frosting and defrosting of the dry ice (carbon dioxide) cover, simply does not work. Noting that the majority of the spots are circular, the suspicion arose that some biological activity may also be involved in spot formation. Indeed, without a scale and the source, the images can be mistaken for pictures of a bacterial culture in a Petri dish. However, one cannot base a hypothesis on an analogy that is so much out of place and scale. Nevertheless, years of work have led to a detailed hypothesis involving biological phenomena that is consistent with all of the observed features. Conversely, we are not aware of any abiotic hypothesis that could explain the full set of observations. This, of course, does not mean that the biological hypothesis is right. Nevertheless, we feel encouraged by the fact that since 2001, when the biological hypothesis was published (Horváth *et al.*, 2001), all new observations and data have made the hypothesis more plausible rather than less so. The main point is that the hypothesis is testable and we are sure to know the answer in the foreseeable future if astrobiological activity continues.

Perhaps the most important intellectual precedent of our hypothesis is the suggestion put forward by Lynn Rothschild (1995), that cryptic photosynthetic microbial mats are a potential basis for life on Mars. These mats can be found on Earth, sometimes under permanent dust cover. She pointed out that such a cover would be very important on Mars as protection from an aggressive environment. Needless to say, some think that proposing extant life on Mars is an extraordinary claim, requiring extraordinary evidence. We shall come back to this issue in the discussion. First, the claim may not be so extraordinary after all, but this is a matter of individual judgement. Second, we do not have extraordinary evidence yet, but we see clearly how such evidence can be obtained.

13.3 Basic facts and considerations about DDSs

Two kinds of frost/snow cover can be distinguished in the polar region: the permanent ice cap and the seasonal cover. Several spot-like features, categorized as spots, fans, blotches, and halos (Christensen *et al.*, 2005), can be distinguished, none of which is fully explained. DDSs can be distinguished by the following features: (1) low albedo, i.e., they are darker than the surrounding frost cover; (2) they occur in a belt in the southern polar region between 60° S and 80° S; (3) they are located predominantly on dark dunes inside craters; (4) they have an internal structure consisting of a dark core and surrounding lighter ring; (5) their diameter is between 5 and 200 m; (6) they show up at the end of local winter, they grow in size and disappear with the frost (seasons on Mars are indicated with solar longitude (L_S)); (7) they reappear annually at the same sites with a probability between 50 and 65% (Figure 13.2).

As mentioned above, the first explanation given for DDS formation was defrosting (sublimation) of dry ice (Malin and Edgett, 2000). As we shall see below, this explanation is far from sufficient, but of course sublimation does play

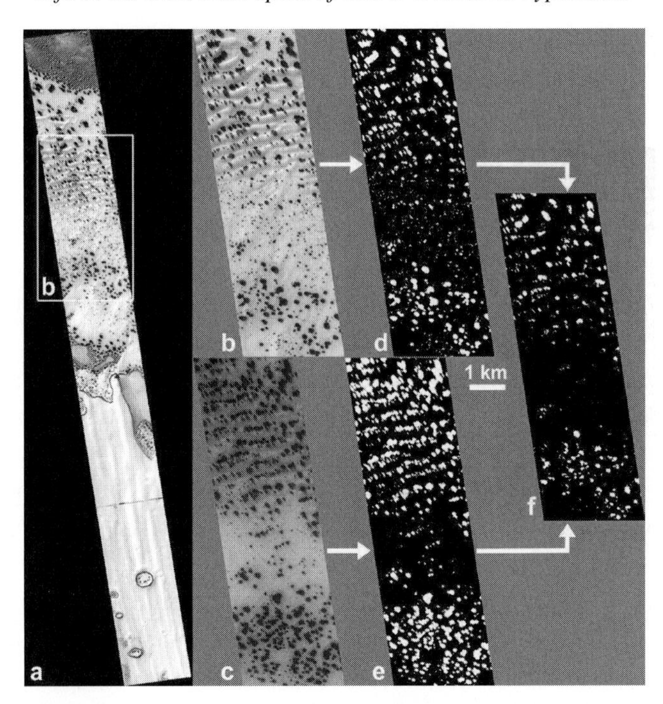

Fig. 13.2. Recurrence of DDSs in 1999 and 2001: a. an MGS MOC narrow-angle image (M07-02775) of the study area; b. part of this area in 1999; c. the same area as on b. one Martian year later in 2001 (E07-00101); d. and e. show only the DDSs with white colour were present in the images b. and c.; while in f. the overlapping area of DDSs that occurred both in 1999 and 2001 is visible. (Collegium Budapest, Mars Astrobiology Group.)

a subsidiary role since the dry ice cover is removed by this process. Kieffer (2003) gave a more involved explanation, according to which carbon dioxide gas is formed between the soil and the dry ice due to the absorption of sunlight, and this then erupts to the surface. As a result the frost diminishes, and the vented gas carries with it fine dust that falls back on the top of the frost. We agree that this is a good candidate as an explanation for the formation of the 'fans', which have only a limited similarity to the DDSs. Fans are fan-shaped spots that predominantly show up between the DDS fields and the permanent ice cap. As we shall see below, this explanation does not seem to be valid for the DDSs. In contrast we suggest that liquid water is an indispensable component of DDS formation.

Over the years (Gánti *et al.*, 2003) we have established the following facts, which strongly indicate that liquid water is involved:

(i) Spots start developing between the ice cover and the soil and they continue to do so until the ice cover disappears in the summer. Sublimation of dry ice occurs only at the surface; hence it cannot explain processes beyond the ice.

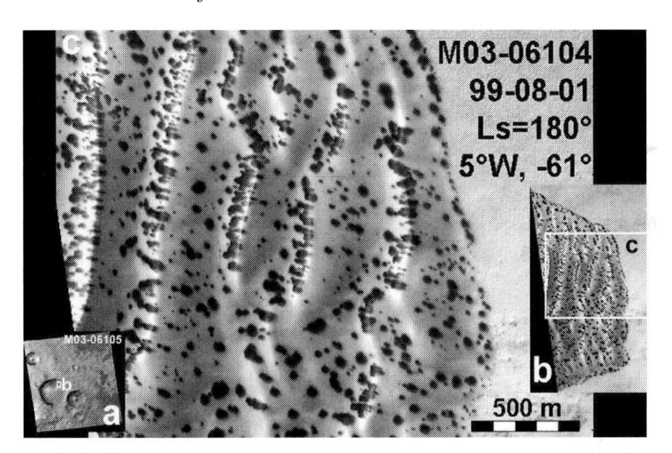

Fig. 13.3. Elongated DDSs: a. the crater, b. the edge of the dune field, c. an enlarged part of the dunefield. The individual dunes follow a nearly vertical stripe pattern. The Sun illuminates from the left and as a result the slopes tilted to the left are bright, while those tilted to the right are in shadow. The slopes facing left are also substantially steeper, and as a result of the higher tilt several spots are elongated there. The darker slopes facing right are longer and less steep. Their tilt is so low that the spots located on them are nearly circular rather than elongated. (M03-06104, M03-06105, 1999.08.01., L_S = 180.3°, 5.35° W 60.77° S.) (Collegium Budapest, Mars Astrobiology Group.)

(ii) Spots appear only on the dark dunes, quite often exactly marking the edge of the dune field. This implies that somehow the dune material affects the formation, development, and maintenance of the spots. Formation of the frost can of course be influenced by surface texture and temperature, but sublimation necessarily proceeds at the surface of the ice. Of course the colour of the soil surface can affect sublimation by heat absorption and conduction, but this effect should be additional to the dark surface of the dunes.

(iii) On horizontal planes the spots are practically circular (Figures 13.1, 13.7), implying some isotropic formative cause. Sublimation depends on the degree of insolation, the wind currents, etc. Hence the location and shape of the spots would be much more erratic if caused by sublimation only. In contrast, the almost perfectly circular shape, the localization, and the growth of the spots imply a mechanism spreading radially from a centre. This observation is consistent with the *in situ* formation of liquid water.

(iv) Importantly, on slopes the spots become elongated and assume an ellipsoid shape, the big axis of which is always parallel with the gradient (Figure 13.3). Though slope winds are supposed to be frequent on Mars (Magalhaes and Gierasch, 1982; Lee *et al.*, 1982), we observed that there are cases where the diffuse fan formed by the wind does not overlap with the darker part of the flow-like structures originating from the DDSs (these seepages will be discussed below). This further suggests that some fluid phase is involved in the formation of the spots: gravitation cannot affect sublimation on this spatial scale. Under the given environmental conditions this could be liquid water (Reiss and Jaumann, 2002).

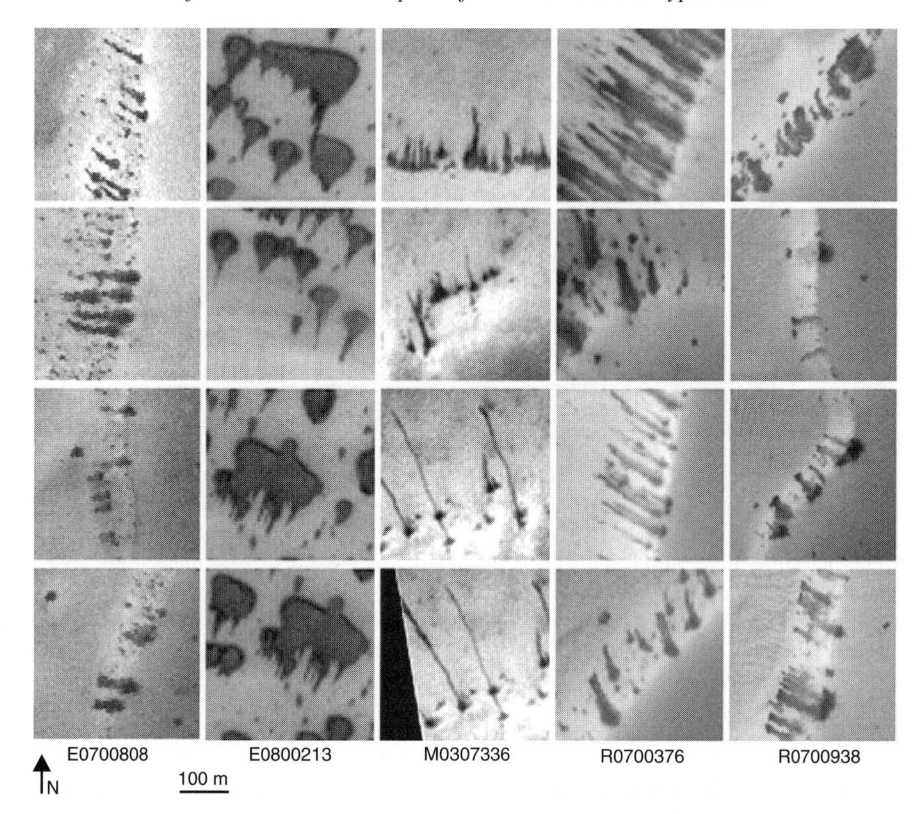

Fig. 13.4. Examples of seepage structures originating from DDSs. Images show an area of 300 m × 300 m width at the top. Parameters of the original images for each column: E07-00808, 69.17° S 150.80° W, 2001-08-13, $L_S = 213.51°$; E08-00213, 70.63° S 16.99° W, 2001-09-03, $L_S = 226.63°$; M03-07336, 69.49°S 17.41° W, 1999-08-07, $L_S = 183.82°$; R07-00376, 69.18° S 150.98° W, 2003-07-06, $L_S = 216.29°$; R07-00938, 69.18° S 150.81° W, 2003-07-13, $L_S = 220.74°$. (Collegium Budapest, Mars Astrobiology Group.)

(v) On steeper slopes 'flows' (called seepages) originate from the elongated spots, and point downwards. This by itself implies that some liquid phase is moving downhill (Figure 13.4).

(vi) The soil of the dunes is covered by white frost during the winter. By early summer this cover has sublimated and the dark material of the dune is exposed. However, some light grey spots can be clearly discerned against this dark surface on the summer images. The distribution of summer grey spots coincides with that of the earlier DDSs (Figure 13.5). This grey remnant on the surface is again evidence against a mere sublimation explanation.

(vii) Finally, the fact that the spots appear in late winter and develop in the spring implies that their formation needs sunlight. In summary, DDS formation below the ice cover is triggered by the sun.

These facts and arguments clearly show that sublimation alone or gas vents are an insufficient explanation. The evidence for liquid phase especially calls for a different mechanism. What may cause the spots then? What allows water below the frost to remain liquid for months?

We think that our DDS-MSO hypothesis offers a good explanation. We assume that in the past Mars was much more hospitable to life than now (Squyres *et al.*, 2004), and a biota did build up on the planet. Drastically changing surface conditions have wiped out most of this ancient living world, but some creatures could have survived provided they adapted to harsh, but still annually recurrent living conditions. Between two favourable periods, these creatures survive by evolved strategies of drying out and freezing. If indeed such organisms survived, there was a strong selection pressure to evolve photosynthetic pigments with very efficient light harvesting. Such organisms could then reactivate themselves each year, provided they are sheltered by a layer of water frost/ice/snow above them. MSOs could thus melt the water ice around them, and provide the liquid water for themselves.

Thus the hypothetical life cycle of the MSOs can be visualized (Figure 13.6) as follows:

(i) In the winter MSOs lay dormant: there are no active life processes. A layer of frost (according to laser MOLA measurements, 0.2–1 m thick (Aharonson *et al.*, 2004)) covers them. Above ground and the dormant MSOs there is a layer of water snow/ice, and above this there is a thicker layer of dry ice.

(ii) In late winter spot formation begins when photosynthetic MSOs absorb light. Cells are reactivated and a liquid water lens is formed about the centre of the colony. This central region appears as a grey spot on the images (Figure 13.6). Sublimation on the surface of the dry ice is facilitated from heat produced at the bottom. It is also possible that geyser-like eruptions in the upper dry ice layer help clear away this upper cover to let the sunshine penetrate more easily into the water ice below. It is important that a thin layer of water vapour is expected to form between the liquid water and the water ice, decreasing the heat conductivity and helping to maintain the liquid phase.

(iii) The centre of a spot loses the water/carbon dioxide cover first. This appears as the black core on the images, surrounded by a still radially extending grey ring. We propose that in the centre the organisms have already reverted to their dormant state, whereas in the grey ring they are still active.

(iv) Finally, by mid-summer all the frost is gone and all MSOs form a dormant surface/subsurface colony that is visible as the summer grey spot.

If the proposed lifecycle is valid, it explains the observed phenomena detailed above. It becomes natural that the spots are initiated below rather than above the frost. The formation of liquid water explains the pattern of growth on flat surfaces and slopes, and it also explains the seepages that originate from DDSs on steep slopes. It then becomes natural, as Clow (1987) suggested, that liquid water will flow each year below the water ice cover. This would be impossible with a bare surface or one covered with dry ice only.

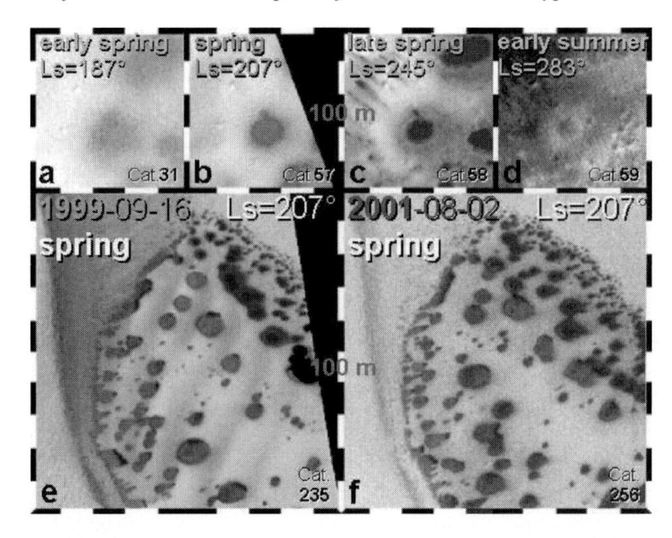

Fig. 13.5. Seasonal changes of DDSs from early spring to early summer according to the progress of seasons: a. $L_s = 187°$ (early spring), M04-00678; b. $L_s = 207°$ (spring), M07-02824; c. $L_s = 245°$ (late spring), M09-03813; d. $L_s = 283°$ (early summer), M11-02076. e. and f. subset images show the annual recurrence of DDSs on the same dune field from two different years: e. $L_s = 207°$, 1999.09.16., M07-02775; f. $L_s = 207°$, 2001.08.02., E07-00101. (Figure courtesy of Malin Space Science Systems.)

A natural outcome of the hypothesis is that MSOs can live only where there is at least a seasonal water frost cover which is thick enough for protection against cold, evaporation, and ultraviolet irradiation, yet is thin enough for photosynthesis. Closer to the pole the temperature will be colder, the frost will be thicker and insolation will be less favourable.

A fact in favour of our hypothesis is that the dunes are the first places to catch frost in the autumn and the last to become defrosted in the summer. This extends the period during which a water ice/snow cover is present. This is, we believe, an explanation of why the DDSs stick to the dunes: the dune surface by virtue of its texture attracts water frost early in the autumn. The fact that surfaces catch frost with varying efficiency is well known here on Earth: vegetation, for example, is a particularly efficient frost attractor.

A further consideration favouring dunes over the red regolith may have to do with their different chemical composition. The red colour comes from oxidized iron, predominantly haematite (Fe_2O_3). After the Viking missions it has become clear that ultraviolet radiation and haematite together lead to the formation of aggressive oxidants (e.g., Möhlmann (2004)), which destroy organic matter efficiently. The colour of the dunes is dark blue/dark violet, and they are assumed to consist of

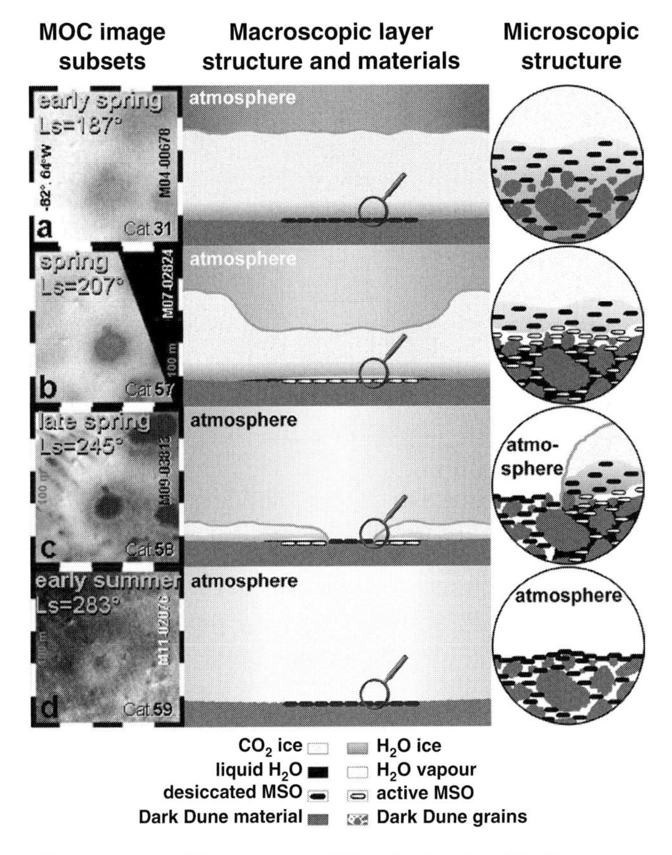

Fig. 13.6. Hypothetical life cycle of MSOs in DDSs, MOC photographs (left, M04-00678, M07-02824, M09-03813, M11-02076), hypothetical cross-section (middle, right) during a. early spring, b. spring, c. late spring, d. summer. (Figure courtesy of Malin Space Science Systems.)

basaltic sand (Herkenhoff and Vasavada, 1999). If so, then the dune surface will be a milder environment for life than the usual red surface.

Finally, the DDS hypothesis can explain the fine structure of the spots that is observed in some of the best images (Figure 13.7). There is a very thin whitish ring between the black centre and the grey ring. If the model in Figure 13.6 holds, then there is liquid water under the grey ring, which will leak towards the black centre. But there is no longer a protecting ice shield, so water will quickly evaporate, and it will also cool quickly. Thus some refrosting of liquid water is expected where the black interior and the grey ring meet.

This hypothesis may be considered attractive but for the moment it looks very qualitative. The devil is hiding in the details. Therefore, in the next section we consider the challenges to the hypothesis.

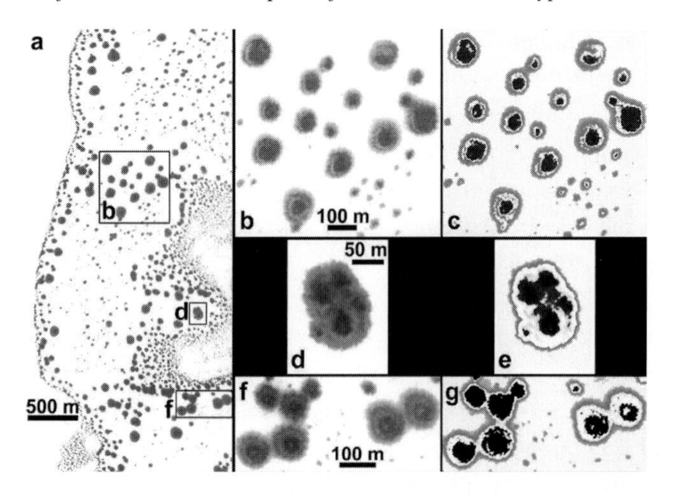

Fig. 13.7. Size and inner structure of Dark Dune Spots on a Martian intracrater dune field at 65° S and 15.55° W with enhanced contrast. The left hand image is an MGS MOC narrow-angle photo (M07-00853) acquired on 1999-09-05, L_S = 200.6°. Subset images show the structural units of DDSs. (Collegium Budapest, Mars Astrobiology Group.)

13.4 Challenges and answers

There are three major challenges to the hypothesis. (1) If there is photosynthesis, what is the hydrogen donor? If it is oxygen, where is it? (2) How is it possible to resist the harsh ultraviolet irradiation on the surface? (3) It may just be too cold even under the frost. We consider these problems in turn.

13.4.1 The nature of photosynthesis and the microbial consortium

Central to our hypothesis is the assumption that the primary productivity of this ecosystem rests on photosynthesis. Since water plays a central role, it is natural to assume that the source of reducing power is the photolysis of water. Such a process happens in cyanobacteria, for example, and oxygen is produced as a by-product. Where is the oxygen then? This objection could be very serious, since, according to Lovelock (1979), the best astronomical sign for life is an atmosphere out of equilibrium. On this basis some would say our hypothesis cannot hold. But this is not a valid objection, for a number of reasons. (i) We are dealing not with what people call 'abundant life'; rather, we are postulating life in 'pockets' (refugia), meaning that the present biomass is tiny. (ii) The Martian atmosphere is out of equilibrium. There is some oxygen (Krasnopolsky *et al.*, 1996.) and also some methane (Krasnopolsky *et al.*, 2004; Formisano *et al.*, 2004). The source of the latter is unknown, and the former is constantly being produced by the ultraviolet lysis of water. This fact leads to the conclusion that there must be a great oxygen sink in the Martian surface (Kolb *et al.*, 2002). Thus the small amount of oxygen produced by such a small biota could remain unnoticed. (iii) We are postulating not only photosynthetic, but also heterotrophic microbes. Thus oxygen could easily be recycled by respiration. The reason oxygen was allowed to accumulate in the terrestrial atmosphere is that a large amount of organic material was buried by tectonic processes. Without such a process oxygen would not have accumulated here either. Since there is no plate tectonics on Mars (Carr, 1981), and there is the aforementioned large oxygen sink, it becomes understandable why the concentration of oxygen in the air is so low.

It may just be that the presence of methane can also be explained at least in part by the action of some MSOs. It is known that in some hydrothermal systems on Earth photosynthetic organisms and methanogens occur close to each other, where methane is the endproduct of decomposition of organic matter from the main algal–bacterial mat (Ward, 1978). Dark dunes are not a hydrothermal system, however. The finding of Tung *et al.* (2005) offers another solution: they found methanogenic archaebacteria at a depth of 3 km below the Greenland ice sheet that operate at a survival metabolism level of −9 °C. The same authors calculated that on Mars a habitat of around 0 °C could host enough methanogens to account for the amount of methane if a 10-m-thick layer existed with a density of around 1 cell per millilitre. An alternative could be production in the spots. The rate of loss from the Martian atmosphere is 270 tons per year (Krasnopolsky *et al.*, 2004), which must be offset by production. If there are 10 million active spots, then each must produce 27 g methane per year, and with an average area of 2500 m^2 per spot, this amounts to 10.8 mgm^{-2} per year of production, a modest value. Kral *et al.* (2004) showed that methanogens can grow on a Mars soil simulant when supplied with carbon dioxide, molecular hydrogen, and varying amounts of water. They say that ‘Currently, the surface of Mars is probably too cold, too dry, and too oxidizing for life, as we know it, to exist’ (p. 615). If the DDS-MSO hypothesis is correct, this does not hold. Our new methanogenic MSO hypothesis rests on the assumption that the structure of the mat allows the activity of photosynthesizers and methanogens as well, and the source of hydrogen would also be organic material from primary production.

13.4.2 The menace of ultraviolet irradiation

As mentioned above, the surface of the dunes may be a more hospitable environment because of the reduced level of aggressive oxidizing chemical species. Yet the ultraviolet radiation at wavelengths shorter than 290 nm is a serious challenge. This radiation without a special protective mechanism is sterilizing for terrestrial organisms. Various strategies exist for microbes to escape from this threat: (1) protection by the ice cover; (2) protection by screening compounds and by dead cells; (3) protection by a surface layer of sand/dust; and (4) escape into the soil when appropriate. We consider these options in turn.

(i) Protection by ice layers. This case has been thoroughly analysed by Córdoba-Jabonero *et al.* (2005); in this discussion we closely follow their treatment. Fortunately, the high latitude by itself reduces surface radiation levels due to lower solar zenith angles. Too much light is bad because of ultraviolet radiation and too little light is bad for photosynthesis. Based on terrestrial organisms, Córdoba-Jabonero *et al.* (2005) have defined a radiative habitable zone (RHZ) and considered its applicability to the Martian polar environments. Although ionizing radiation also reaches the surface, tolerance of this hazard can be taken for granted. It was concluded that the CO_2 ice cover alone is an insufficient shield against ultraviolet radiation, since its thickness does not exceed 1 m. (Their model is valid only for CO_2 ice and not for CO_2 'snow'. Unfortunately the microphysical structure of the condensed CO_2 is unknown, just like the probable difference between the ultraviolet-shielding capacity of CO_2 snow, relative to CO_2 ice.) In contrast, a few centimetres of water snow would also be sufficient to provide protection to the MSOs at a level that would also be tolerable for terrestrial microbes. The snowy condition appears crucial: if we replace it by water ice, then an ice sheet about 1 m thick is required. Sadly, we do not know anything about the nature or thickness of water frost precipitating on the dunes in autumn: altimetry provides the thickness of all layers together. Whereas a water frost sheet of several centimetres to at most 1 or 2 dm is not excluded, other means of reducing the ultraviolet radiation hazard must be considered.

(ii) Ultraviolet radiation protection by pigments is an obvious solution. On the Earth cyanobacteria have special pigments (like scytonemin and gloeocapsin) in their mucilaginous sheath that shield the cells from ultraviolet radiation (Garcia-Pichel *et al.* 1992; Garcia-Pichel and Castenholz, 1994; Garcia-Pichel and Bebout, 1996). In addition, dead cells can leave this sheath behind, and it remains there as a passive protection for the other cells. The ratio of light-harvesting and light-screening pigments is regulated as a function of photosynthetically active radiation (PAR), ultraviolet and temperature in the mat-forming cyanobacterium Phormidium murrayi (Roos and Vincent, 1998). It would strongly support our hypothesis if the French operated OMEGA reflectance spectrometer could detect ultraviolet screening pigments in the area of DDSs freshly opened up from the snow/ice cover.

(iii) Also, it is conceivable that if we look at the case more closely we are dealing with subsurface organisms, albeit at the millimetre scale. Lynn Rothschild (1995) analysed an interesting ecosystem called a 'cryptic' microbial mat. Photosynthetic organisms live there under the protection of sand and gravel. The colour of the algae becomes visible if one scrapes away the surface sand. As Rothschild emphasized, this lifestyle is ideal for protection against ultraviolet radiation and any surface oxidants, thus it offers a good model for a possible Martian lifestyle. Experiments on dried monolayers of the desiccation-tolerant cyanobacterium *Chroococcidiopsis* sp. 029 showed that even

1 mm of Martian-like soil provides efficient protection against a simulated ultraviolet flux (Cockell *et al.*, 2005).

(iv) Terrestrial analogues are considered in the next section; here it suffices to say that one type of the so-called cryptobiotic crust consists of cyanobacteria that take shelter under the surface under harsh conditions and return to it when conditions improve. This lifestyle is a combination of a surface-bound classical microbial mat lifestyle and the near-surface cryptic mat lifestyle.

We believe that these mechanisms taken together are more than enough to yield the sufficient protection mechanism against ultraviolet radiation. Now we turn to the challenge of temperature.

13.4.3 The plausibility of liquid water and the challenge of low temperature

Our hypothesis is qualitative at present, and a physical model of heat balance will be mandatory in the future. The main question is whether liquid water can indeed form in the DDSs, as suggested by the morphology of their development; and whether temperatures permit a *sustainable population dynamics* of the mat. The latter is by no means an automatic consequence of the former.

Surface temperatures above the frost depend on the season. In summer they can rise above 0 °C (Supulver *et al.*, 2001; Reiss and Jaumann, 2002) which is obviously permissive, but there is a problem in late winter and early spring, when they can be around −60 to −70 °C. Is it credible that liquid water can form below the frost, where the insulating layer is at most 1 m of water and dry ice? The dune surface and, if we are correct, the MSOs absorb the sunlight. Temperatures are bound to rise and the solid state greenhouse effect (Matson and Brown, 1989) helps the process. The temperature at which liquid water appears depends strongly on the salinity of the soil. If the dunes undergo yearly wetting and drying, as we suggest, then the surface must be salty. Melting temperatures of salty solutions can decrease to below −50 °C depending on their composition (Mellon and Phillips, 2001; Kargel and Marion, 2004). Of course, we know nothing about the degree of salinity on the dunes, and in any case we consider it unlikely that the effect could be so strong as to keep a macroscopic body of fluid water liquid at −50 °C. In contrast, an effect down to −30 °C would be more realistic and could help the proposed life cycle. Intense microbial activity was discovered in the Discovery deep hypersaline basin of the Mediterranean Sea where the water is almost saturated (5 M) with $MgCl_2$ (van der Wielen *et al.*, 2005).

It is important to note that for the initiation of the life cycle a macroscopic body of liquid water is not necessary. We know that Antarctic lichens start photosynthesizing at −18 °C below the snow because enough molecules of water leave the surface of the snow cavity and enter the cells at this low temperature (Kappen *et al.*, 1996). Also, liquid water at the micro- and nanoscale is markedly different from macroscopic water. Thin films of liquid water exist well below subzero temperatures and can be in contact with individual soil grains and cells. Adsorption water can exist on Mars in liquid form down to at least −40 °C (Möhlmann, 2004, 2005). Price (2000) discusses how organisms continue to have access to water and nutrients even well below the freezing point of pure water, by utilizing aqueous veins at triple junctions of ice grains. So, MSOs may never be in a completely desiccated medium.

It is not unlikely that photosynthetic organisms can actively contribute to heating by their light-harvesting apparatus. The light phase of photosynthesis is not dependent on temperature. Thus at low temperatures more light is absorbed than can be consumed by the dark photosynthetic reactions (including carbon fixation). Photons may thus be converted to heat by photosystem II in cyanobacteria (Morgan-Kiss *et al.*, 2006). We postulate that in MSOs analogous processes contribute to heating in late winter.

We conclude that the appearance of a sufficient quantity of liquid water to kick-start metabolism in the late winter in the MSOs is not unrealistic, but we also acknowledge that the issue is very quantitative and proper modelling is needed to settle it in the future. For the sake of the argument, in what follows we assume that liquid water is available.

Price and Sowers (2004) presented a thoughtful analysis of temperature dependence of microbial metabolism. They differentiate between growth (through reproduction), maintenance (active metabolism and maintenance of ion levels, etc. with no net growth), and mere survival (when macromolecular damage is repaired only). For the same organism decreasing temperature pushes the cell further down this list. Low temperature has, of course, several adverse effects. A general consequence is the slow-down of metabolism due to Arrhenius' law: other things being equal, rates of chemical reactions decrease exponentially with decreasing temperature (Vincent, 1998). There is an advantage from this for psychrophilic organisms that is commonly not realized: the rates of adverse reactions also go down with temperature. Therefore the most serious type of adverse effect is one that does not decrease, or even increases with decreasing temperature. Membrane properties, formation of ice crystals, and radiation damage belong to this category. Generally, recurrent freezing–thawing cycles are expected to be very adverse. They apply to MSOs in the first part of their growth period.

Regarding metabolism, Price and Sowers (2004) conclude, based on experimental evidence and inference, that there is 'no evidence of a threshold or cutoff metabolic rate at temperatures down to −40 °C' (p. 4636). This is very important because (if they exist) MSOs must be at least as low-temperature tolerant as terrestrial organisms. The boundary temperatures between survival, maintenance, and growth are a matter of species-specific adaptations.

Liquid water and energy input should allow metabolism but low temperatures are not favourable for the net growth of the mat, as analysed by Friedmann *et al.* (1993). Long-term productivity of cryptoendolithic microbial communities in Antarctica is very low (net ecosystem productivity is as low as 3 mg Cm^{-2} per year). However, one part of this may be the stress of frequent dehydration–rehydration in a desert that would not apply to MSOs.

This is clearly an issue where detailed physical modelling and adequate Mars chamber experiments must be carried out in the future.

13.5 Partial analogues on Earth

It is good if an exobiological hypothesis can be supported by evidence from analogous organisms and life cycles from Earth. Yet any such analogy is bound to be partial, considering the differences in past and present environmental conditions of the different planets and the differences in the paths that evolution may have taken on the two planets. Clearly, a dark dune type of environment is found nowhere on Earth. Extremophiles from various extreme habitats serve as a worst-case approach to Martian hypothetical organisms like the MSOs. They present a worst case in the sense that if organisms have evolved and are still alive on Mars, they are expected to have better adapted to Martian conditions than terrestrial organisms.

A possible objection to this argument is that terrestrial extremophiles may have discovered the specific boundaries of life using nucleic acids and proteins as informational molecules. If the two living worlds are related (which is not unlikely), then terrestrial extremophiles almost exactly delineate what would be possible on Mars in terms of occupying extreme habitats. This counter-argument misses one important point about evolution, however. In the case of Earth these extreme habitats are rare and the population numbers of the occupying organisms are low. In other words, only a small fraction of the total living population has been exposed to this extreme form of directional selection. In contrast, on Mars environmental deterioration of living conditions has been widespread. If we accept that life did exist on Mars more than 3 billion years ago, then it could have attained population numbers similar to those on the ancient Earth. It follows that when conditions deteriorated, a large population was exposed to directional selection. Favourable mutations are more common in a large than in a small population (see Maynard Smith (1998) for background reading on evolutionary genetics). Consequently, extremophiles can be expected to be more 'extreme' in their tolerance on Mars than on our home planet, at least in certain ecological dimensions. Even in the case of the Earth, would anybody have guessed that an organism like *Deinococcus*

(Englander *et al.*, 2004; Cox and Battista, 2005) could exist? We should expect many more such striking cases on Mars if there is life there at all.

Another factor is time. Nadeau *et al.* (2001) determined the phylogenetic relations of psychrophilic oscillatorians (cyanobacteria). They diverged from their relatives about 20 million years ago when Antarctica cooled down. Most other psychrophiles are much younger than this (Morgan-Kiss *et al.*, 2006). In contrast, Martian organisms must have been under directional selection for billions of years.

This general argument does not tell us anything about the real tolerance of Martian organisms, however. All the calculations on ultraviolet radiation, salinity, and temperature tolerance in the previous sections have been based on terrestrial organisms since we do not have anything better at hand. Bearing this in mind we now turn to some partial terrestrial analogues to the hypothetical MSOs.

Salt tolerance is important because of the supposed salinity and temperatures of the dark dunes. There are some deep hypersaline basins of the eastern Mediterranean, where a rich biota is found in almost saturated salt solution (van der Wielen *et al.*, 2005). Note that life was previously thought impossible under such conditions. Recently, in the Bannock basin a very rich prokaryotic community was discovered in the oxic–anoxic transition zone, where a chemocline with varying conditions is found (Daffonchio *et al.*, 2006). A much steeper chemocline could exist in the upper few centimetres of the dark dunes.

Perhaps it is the cyanobacteria that come closest to the MSOs in their properties. They are everywhere in extreme habitats, including hot and cold deserts. Apparently they can survive for tens of thousands of years in the permafrost (Vishnivetskaya *et al.*, 2001). In northeastern Siberia the oldest viable cells date back 2–3 million years (Gilichinsky *et al.*, 1995). Several species are multiple extremophiles, e.g., desiccation and ultraviolet tolerant at the same time. The expression of the protein WspA in Nostoc is modulated by ultraviolet irradiation and desiccation, and it modulates the three-dimensional extracellular matrix by binding to the ultraviolet-absorbing pigment complexes mycosporine and scytonemin (Wright *et al.*, 2005).

A special case of microbial mats is the so-called cryptobiotic crust or biological soil crust (Belnap and Lange, 2003). They are found in arid environments on the soil or rock surface. Remarkably, their assimilation intensity rivals that of higher plants in the same region. Some cyanobacteria are extreme halophiles as well (e.g., Garcia-Pichel *et al.* (1998)). Large masses of them can occur in soda lakes of saturated sodium carbonate solution, such as African Magadi Lake in the Rift Valley or in the shotts of the Sahara desert. This is important because, as explained above, we can expect considerable salt deposition on the dark dunes, which (according to our hypothesis) undergo a regular wetting–drying cycle. Efficient Na^+/H^+ antiporters in the membrane help the life of the halotolerant (up to 3.0 M NaCl) cyanobacterium *Aphanothece halophytica* (Wutipraditkul *et al.*, 2005).

Chroococcidiopis is not only desiccation and ultraviolet tolerant, but somewhat similar to *Deinococcus*, it is also able to repair extensive DNA damage following ionizing radiation, which ability is linked to its desiccation tolerance (Billi *et al.*, 2000). In contrast, the genome of the *Nostoc* commune is protected against oxidation and damage by a different strategy after decades of desiccation, aided by the non-reducing disaccharide trehalose (Shirkey *et al.*, 2003).

We find remarkable survival strategies in various forms of the cryptobiotic crust (Pócs *et al.*, 2004). We find pigment-rich species in the upper layer of deserts, which shield the less ultraviolet-tolerant species below them. An even more exciting case is that of the *Microcoleus* species, which during the dry period leave their mucilaginous sheath and glide down into the soil a few millimetres. There they develop a new sheath and with the advent of the rainy season they glide back to the surface again.

The main objection to these examples is that ice and a prolonged low temperature do not play a role in them. For such a comparison we must look for different analogues. Rivkina *et al.* (2000) analysed the metabolism of permafrost bacteria below the freezing point. They showed that these (non-photosynthetic) bacteria are active down to -18 °C, where their minimum doubling time is ca. 160 days. Moreover, metabolic activity is directly proportional to the thickness of the liquid, adsorbed water layer.

A remarkable case of a permanently ice-enclosed consortium of partly photosynthetic bacteria was described by Priscu *et al.* (1998). Organisms become active under sunlight in water inclusions around aeolian-derived sediments; photosynthesis, nitrogen fixation, and decomposition occur in parallel. The main driving process is photosynthesis at a depth of a few metres. MSOs would be better off because they would be in direct contact with underlying soil. Under the ice cover (about 5 m thick) there is liquid water where a psychrophilic phytoplankton can be found, living at less than 1% of normal PAR (Morgan-Kiss *et al.*, 2006).

13.6 Discussion and outlook

According to the late Carl Sagan, an astronomer, 'Extraordinary claims require extraordinary evidence'. The DDS-MSO hypothesis is sometimes refuted on the grounds that it is an extraordinary claim without extraordinary evidence. This objection may not hold, for a number of reasons.

First, 'extraordinariness' is in the eye of the beholder. In the particular case we often find that in the end opponents think that life itself is unlikely, and they do not really believe in extraterrestrial life, past or present. Since we lack an accepted scenario for the origin of life, this stance is legitimate but it is good to be aware of it. Second, we do not take it as fact that there is life on the Martian dunes, but we present a testable hypothesis for a phenomenon. We agree that in a way confirmatory evidence would be extraordinary. What could such evidence be?

Of course it would be best to go there and have a look. This presents substantial financial and engineering challenges, but ultimately – should other evidence prompt it – it must be done. The second best option is to obtain a weird spectroscopic signal from the DDSs. Spectrometry should operate in a broad spectrum and with good spatial resolution. This has not yet been achieved but some good ideas have been suggested: (1) using auto fluorescence (Schoen and Dickinsheelz, 2000) with spectroscopic instruments (Wynn-Williams *et al.*, 2002) to identify acytonemin, chlorophyll, and phycocyanin; (2) using a combination of *in situ* Fourier transform microwave, near-infrared and visible spectrometers (Hand *et al.*, 2005), and (3) using *in situ* analysis of hopanoids as long-lived bacterial cell wall products and photosynthetic pigments (Ellery and Wynn-Williams, 2003).

The most recent American mission, the Mars Reconnaisance Orbiter, could provide us with very suggestive optical imagery at a resolution of up to 25 cm with HiRSE and 6.55 nm per channel spectral resolution between 370 and 3920 nm with CRISM (Bergstrom *et al.*, 2004). Although in biology morphological evidence is often misleading, new images could help a lot in guiding future research.

There is the possibility of building appropriate chambers and testing various elements of the DDS-MSO hypothesis in them. Despite ambitious attempts, we have to say that the chambers known to us are inappropriate for this purpose: something (the ultraviolet radiation, or the temperature, or the frost, or the gas) is always missing. We encourage well-equipped laboratories to test our hypothesis.

Finally, a shrewd objection to the DDS story is the following. If the hypothesis assumes that we are looking at the refugial remnants of a flourishing past Martian biota, then it is very surprising that they remained alive just long enough to allow us to find them. This is indeed somewhat striking, but we know other similar examples. Geographical, linguistic, and genetic distances still show very good correlations on the Earth and they have been the subject of exciting studies (e.g., Cavalli-Sforza, 1997). It has also been pointed out that they have been analysed in the 'last hour' at many sites, because increased mobility will soon undermine the pattern. 'Chance favors the prepared mind' (Louis Pasteur).

Acknowledgements

This work was supported by the ESA ECS-project No. 98004.

References

Aharonson, O., Zuber, M. T., Smith, D. E., *et al.* (2004). Depth, distribution, and density of CO_2 deposition on Mars. *J. Geophys. Res.*, **109**, E5, CiteID E05004.

Belnap, J., and Lange, O. L. (2003). *Biological Soil Crusts: Structure, Function, and Management.* Revised second printing. Berlin: Springer, p. 503.

Bergstrom, J. W., Delamere, W. A., and McEwen, A. (2004). MRO high resolution imaging science experiment (HiRISE): instrument test, calibration and operating constraints, *55th International Astronautic Federation Congress*, IAC-04-Q.3.b.02 unpublished.

Billi, D., Friedmann, E. I., Hofer, K. G., Caiola, M. G., and Ocampo-Friedmann, R. (2000). Ionizing-radiation resistance in the desiccation-tolerant cyanobacterium *Chroococcidiopsis*. *Appl. Env. Microbiol.*, **66**, 1489–1492.

Carr, M. H. (1981). *The Surface of Mars.* New Haven: Yale University Press.

Cavalli-Sforza, L. L. (1997). Genes, peoples, and languages. *Proc. Natl Acad. Sci. USA*, **94**, 7719–7724.

Christensen, P. R., Kieffer, H. H., and Titus, T. N. (2005). Infrared and visible observations of south polar spots and fans. *American Geophysical Union*, Fall Meeting 2005, #P23C-04.

Clow, G. D. (1987). Generation of liquid water on Mars through the melting of a dusty snowpack. *Icarus*, **72**, 95–127.

Cockell, C. S., Schuerger, C., Billi, D., Friedmann, E. I., and Panitz, C. (2005). Effects of a simulated martian UV flux on the cyanobacterium *Chroococcidiopsis* sp. 029. *Astrobiology*, **5**, 127–140.

Córdoba-Jabonero, C., Zorzano, M.-P., Selsis, F., Patel, M. R., and Cockell, C. S. (2005). Radiative habitable zones in martian polar environments. *Icarus*, **175**, 360–371.

Cox, M. M., and Battista, J. R. (2005). *Deinococcus radiodurans* – the consummate survivor. *Nat. Rev. Microbiol.*, **3**, 882–892.

Daffonchio, D., Borin, S., Brusa, T., *et al.* (2006). Stratified prokaryote network in the oxic-anoxic transition of a deep-sea halocline. *Nature*, **440**, 203–207.

Ellery, A., and Wynn-Williams, D. (2003). Why Raman spectroscopy on Mars? A case of the right tool for the right job. *Astrobiology*, **3**, 565–579.

Englander, J., Klein, E., Brumfeld, J., *et al.* (2004). DNA toroids: framework for DNA repair in Deinococcus radiodurans and in germinating bacterial spores. *J. Bacteriol.*, **186**, 5973–5977.

Formisano, V., Atreya, S., Encrenaz, T., Ignatiev, N., and Giuranna, M. (2004). Detection of methane in the atmosphere of Mars. *Science*, **306**, 1758–1761.

Friedmann, E. I., Kappen, L., Meyer, M. A., and Nienow, J. A. (1993). Long-term productivity in the cryptoendolithic microbial community of the Ross Desert, Antarctica. *Microbial Ecol.*, **25**, 51–69.

Gánti, T., Horváth, A., Bérczi, Sz., Gesztesi, A., and Szathmáry, E. (2003). Dark Dune Spots: Possible biomarkers on Mars? *Origins Life Evol. B.*, **33**, 515–557.

Garcia-Pichel, F., and Bebout, B. M. (1996). The penetration of UV radiation into shallow water sediments: high exposure for photosynthetic communities. *Mar. Ecol. Progress Ser.*, **131**, 257–261.

Garcia-Pichel, F., and Castenholz, R. W. (1994). On the significance of solar ultraviolet radiation for the ecology of microbial mats, in *Microbial Mats. Structure, Development and Environmental Significance*, eds. L. J. Stal and P. Camuette. Springer, Heidelberg, pp. 77–84.

Garcia-Pichel, F., Nubel, U., and Muyzer, G. (1998). The phylogeny of unicellular, extremely halotolerant cyanobacteria. *Arch. Microbiol.*, **169**, 469–482.

Garcia-Pichel, F., Sherry, N. D., and Castenholz, R. W. (1992). Evidence for a UV sunscreen role of the extracellular pigment scytonemin in the terrestrial cyanobacterium *Chlorogloeopsis* spp. *Photochem. Photobiol.*, **56**, 17–23.

Gilichinsky, D. A., Wagener, S., and Vishnivetskaya, T. A. (1995). Permafrost microbiology. *Permafrost Periglac.*, **6**, 281–291.

Hand, K. P., Carlson, R., Sun, H., *et al.* (2005). Waves of the future (for Mars): in-situ mid-infrared, near-infrared, and visible spectroscopic analysis of antarctic cryptoendolithic communities. *American Geophysical Union*, Fall Meeting 2005, #P51D-0959.

Herkenhoff, K. E., and Vasavada, A. R. (1999). Dark material in the polar layered deposits and dunes on Mars. *J. Geophys. Res.*, **104**, E7, 16487–16500.

Horváth, A., Gánti, T., Gesztesi, A., Bérczi, Sz., and Szathmáry, E. (2001). Probable evidences of recent biological activity on Mars: appearance and growing of dark dune spots in the South Polar Region. *37th Lunar and Planetary Science Conference*, #1543.

Kappen, L., Schroeter, B., Scheidegger, C., Sommerkorn, M., and Hestmark, G. (1996). Cold resistance and metabolic activity of lichens below 0 °C. *Adv. Space Res.*, **18**, 119–128.

Kargel, J. S., and Marion, G. M. (2004). Mars as a salt-, acid-, and gas-hydrate world. *35th Lunar and Planetary Science Conference*, #1965.

Kieffer, H. H. (2003). Behaviour of CO_2 on Mars: real zoo. *Sixth International Conference on Mars*, #3158.

Kolb, C., Lammer, H., Abart, R., *et al.* (2002). The Martian oxygen surface sink and its implications for the oxidant extinction depth, in *Proceedings of the First European Workshop on Exo-Astrobiology, Graz, Austria*, ed. Huguette Lacoste. ESA SP-518. Noordwijk, Netherlands: ESA Publications Division, pp. 181–184.

Krasnopolsky, V. A., Maillard, J. P., and Owen, T. C. (2004). Detection of methane in the martian atmosphere: Evidence for life? *Icarus*, **172**, 537–547.

Krasnopolsky, V. A., Mumma, M. J., Bjoraker, G. L., and Jennings, D. E. (1996). Oxygen and carbon isotope ratios in Martian carbon dioxide: measurements and implications for atmospheric evolution. *Icarus*, **124**, 553–568.

Kral, T. A., Bekkum, C. R., and McKay, C. P. (2004). Growth of methanogens on a Mars soil simulant. *Origins Life Evol. B.*, **34**, 615–626.

Lee, S. W., Thomas, P. C., and Veverka, J. (1982). Wind streaks in Tharsis and Elysium – Implications for sediment transport by slope winds. *J. Geophys. Res.*, **87**, 10025–10041.

Lovelock, J. (1979). *Gaia, A New Look at Life on Planet Earth*. Oxford: Oxford University Press.

Magalhaes, J., and Gierasch, P. (1982). A model of Martian slope winds – implications for eolian transport. *J. Geophys. Res.*, **87**, 9975–9984.

Malin Space Science Systems, Mars Orbiter Camera Image Gallery, Images at http://www.msss.com/moc_gallery/

Malin, M. C., and Edgett, K. S. (2000). Frosting and defrosting of Martian polar dunes. *31st Annual Lunar and Planetary Science Conference*, #1056.

Matson, D. L., and Brown, R. H. (1989). Solid-state greenhouses and their implication for icy satellites. *Icarus*, **77**, 67–81.

Maynard Smith, J. (1998). *Evolutionary Genetics*. Oxford: Oxford University Press.

Mellon, M. T., and Phillips, R. J. (2001). Recent gullies on Mars and the source of liquid water. *J. Geophys. Res.*, **106**, 1–15.

Möhlmann, D. (2004). Water in the upper martian surface at mid- and low-latitudes: presence, state, and consequences. *Icarus*, **168**, 318–323.

Möhlmann, D. (2005). Adsorption water-related potential chemical and biological processes in the upper Martian surface. *Astrobiology*, **5**, 770–777.

Morgan-Kiss, R. M., Priscu, J. C., Pocock, T., Gudynaite-Savitch, L., and Huner, N. P. (2006). Adaptation and acclimation of photosynthetic microorganisms to permanently cold environments. *Microbiol. Mol. Biol. Rev.*, **70**, 222–252.

Nadeau, T. L., Milbrandt, E. C., and Castenholz, R. W. (2001). Evolutionary relationships of cultivated Antarctic oscillatorians (cyanobacteria). *J. Phycol.*, **37**, 650–654.

Pócs, T., Horváth, A., Gánti, T., Bérczi, Sz., and Szathmáry, E. (2004). Possible cripto-biotic crust on Mars? *ESA*, **SP-545**, 265–266.

Price, P. B. (2000). A habitat for psychrophiles in deep Antarctic ice. *Proc. Natl. Acad. Sci. USA*, **97**, 1247–1251.

Price, P. B., and Sowers, T. (2004). Temperature dependence of metabolic rates for microbial growth, maintenance, and survival. *Proc. Natl. Acad. Sci. USA*, **101**, 4631–4636.

Priscu, J. C., Fritsen, C. H., Adams, E. E., *et al.* (1998). Perennial Antarctic lake ice: an oasis for life in a polar desert. *Science*, **280**, 2095–2098.

Reiss, D., and Jaumann, R. (2002). Spring defrosting in the Russel crater dune field - recent surface runoff within the last martian year. *33rd Lunar and Planetary Science Conference*, #2013

Rivkina, E. M., Friedmann, E. I., McKay, C. P., and Gilichinsky, D. A. (2000). Metabolic activity of permafrost bacteria below the freezing point. *Appl. Env. Microbiol.*, **66**, 3230–3233.

Roos, J. C., and Vincent, W. F. (1998). Temperature dependence on UV radiation effects on Antarctic cyanobacteria, *J. Phycol.*, **34**, 118–125.

Rothschild, L. J. (1995). A 'cryptic' microbial mat: a new model ecosystem for extant life on Mars. *Adv. Space Res.*, **15**, 223–228.

Schoen, C. H., and Dickensheets, D. L. (2000). Tools for robotic *in situ* optical microscopy and Raman spectroscopy on Mars, in *Concepts and Approaches for Mars Exploration*, Houston: Lunar and Planetary Institute, p. 275.

Sheehan, W. (1996). *The Planet Mars: A History of Observation and Discovery.* Tuscon, AZ: The University of Arizona Press.

Shirkey, B., McMaster, N. J., Smith, S. C., *et al.* (2003). Genomic DNA of Nostoc commune (Cyanobacteria) becomes covalently modified during long-term (decades) desiccation but is protected from oxidative damage and degradation. *Nucleic. Acids Res.*. 31, 2995–3005.

Squyres, S. W., Grotzinger, J. P., Arvidson, R. E., *et al.* (2004). *In situ* evidence for an ancient aqueous environment at Meridiani Planum, Mars. *Science*, **306**, 1709–1714.

Supulver, K. D., Edgett, K. S., and Malin, M. C. (2001). Seasonal changes in frost cover in the Martian south polar region: Mars Global Surveyor MOC and TES monitoring the Richardson crater dune field. *32nd Lunar and Planetary Science Conference*, #1966.

Tung, H. C., Bramall, N. E., and Price, P. B. (2005). Microbial origin of excess methane in glacial ice and implications for life on Mars. *Proc. Natl. Acad. Sci. USA*, **102**, 18292–18296.

van der Wielen, P. W., Bolhuis, H., Borin, S., *et al.* (2005). The enigma of prokaryotic life in deep hypersaline anoxic basins. *Science*, **307**, 121–123.

Vincent, W. F. (1988). *Microbial Ecosystems of Antarctica*. Cambridge: Cambridge University Press.

Vishnivetskaya, T. A., Rokhina, L. G., Spirina, E. V., *et al.* (2001). Ancient viable phototrops within the permafrost. *Nova Hedwigia*, Beiheft **123**, 427–441.

Ward, D. M. (1978). Thermophilic methanogenesis in a hot-spring algal-bacterial mat (71 to 30°C). *Appl. Env. Microbiol.*, **35**, 1019–1026.

Wynn-Williams, D. D., Edwards, H. G. M., Newton, E. M., and Holder, J. M. (2002). Pigmentation as a survival strategy for ancient and modern photosynthetic microbes under high ultraviolet stress on planetary surfaces. *Int. J. Astrobiol.*, **1**, 39–49.

Wright, D. J., Smith, S. C., Joardar, V., *et al.* (2005). UV irradiation and desiccation modulate the three-dimensional extracellular matrix of *Nostoc commune* (Cyanobacteria). *J. Biol. Chem.*, **280**, 40271–4081.

Wutipraditkul, N., Waditee, R., Incharoensakdi, A., *et al.* (2005). Halotolerant cyanobacterium Aphanothece halophytica contains NapA-Type Na+/H+ antiporters with novel ion specificity that are involved in salt tolerance at alkaline pH. *Appl. Env. Microbiol.*, **71**, 4176–4184.

14

Titan: a new astrobiological vision from the Cassini–Huygens data

François Raulin

LISA, CNRS, and Universités Paris 7 et Paris 12

14.1 Introduction

The Earth is certainly, so far, the most interesting planetary body for astrobiology since it is still the only one where we are sure that life is present. However, there are many other bodies of astrobiological interest in the Solar System. There are planetary bodies where extraterrestrial life (extinct or extant) may be present, and which thus would offer the possibility of discovering a second genesis, the nature and properties of these extraterrestrial living systems, and the environmental conditions which allowed its development and persistence. Mars and Europa seem to be the best places for such a quest. On the other hand, there are planetary bodies where a complex organic chemistry is going on. The study of such chemistry can help us to better understand the general chemical evolution in the Universe and more precisely the prebiotic chemical evolution on the primitive Earth. Comets are probably the best example, especially considering that their organic content may have also been involved in the prebiotic chemistry on the primitive Earth.

Titan, which is the largest satellite of Saturn, may cover these two complementary aspects and is thus an interesting body for astrobiological research. Moreover, with an environment very rich in organics, it is one of the best targets on which to look for prebiotic chemistry at a full planetary scale. This is particularly important when considering that Titan's environment presents many analogies with the Earth. Studying Titan today may give us information on the conditions and processes which occurred on the Earth 4 billion years ago. In addition, models of the internal structure of Titan strongly suggest the presence of a large permanent subsurface water ocean, and the potential for extant life.

Since the Voyager fly-bys of Titan in the early 1980s our knowledge of this exotic place, the only satellite of the Solar System having a dense atmosphere, has indeed drastically improved. The vertical atmospheric structure has been determined,

Planetary Systems and the Origins of Life, eds. Ralph E. Pudritz, Paul G. Higgs, and Jonathon R. Stone.
Published by Cambridge University Press.

and the primary chemical composition, trace compounds, and especially organic constituents described. Several other atmospheric species have also been identified by ground-based observation and by ESA's Infrared Space Observatory (ISO). Other ground-based and Hubble observations have also allowed a first mapping of the surface, showing a heterogeneous milieu. However, at the turn of the twentieth century, many questions still remained concerning Titan and its astrobiological aspects. What is the origin of its dense atmosphere? What is the source of methane? How complex is the organic chemistry? What is the chemical composition of the aerosols which are clearly present in the atmosphere (and even mask the surface in the visible wavelengths)? What is the chemical composition of Titan's surface? What is the nature of the various potential couplings between the gas phase, the aerosol phase, and the surface and their role in the chemical evolution of the satellite and its organic chemistry? How close are the analogies between Titan and the primitive Earth? Is there life on Titan?

The NASA-ESA Cassini–Huygens mission was designed to explore the Saturn system in great detail, with a particular focus on Titan, and to find answers to these questions. Indeed, since the successful Saturn orbital insertion of Cassini on 1 July 2004 and the release of the Huygens probe in Titan's atmosphere on 14 January 2005 (Lebreton *et al.*, 2005), many new data have already been obtained which are essential for our vision and understanding of Titan's astrobiological characteristics.

This chapter reviews three main aspects of Titan with astrobiological importance, on the basis of these new data provided by Cassini–Huygens (Table 14.1), and complemented by theoretical modelling and laboratory experimental studies.

14.2 Analogies between Titan and the Earth

With a diameter of more than 5100 km, Titan is the largest moon of Saturn and the second largest moon of the Solar System. It is also the only one to have a dense atmosphere. This atmosphere, as is clearly evidenced by the presence of haze layers (Figures 14.1, 14.2) extends up to approximately 1500 km (Fulchignoni *et al.*, 2005). Like Earth, Titan's atmosphere is mainly composed of dinitrogen (= molecular nitrogen), N_2. The other main constituents are methane, CH_4, with a mole fraction of about 0.016–0.02 in the stratosphere, as measured by the composite infrared spectrometer (CIRS) instrument on Cassini (Flasar *et al.*, 2005) and the gas chromatograph-mass spectrometer (GC-MS) on Huygens (Niemann *et al.*, 2005) and dihydrogen (= molecular hydrogen,) H_2, with a mole faction of the order of 0.001. With a surface temperature of approximately 94 K, and a surface pressure of 1.5 bar, Titan's atmosphere is nearly five times denser than the Earth's. Despite

Table 14.1. *Cassini–Huygens science instruments and interdisciplinary scientists (IDS) and the potential astrobiological return of their investigation*

Cassini instruments and interdisciplinary programs	PI, team leader, or IDS	Country	Astrobiological return
Optical remote sensing instruments			
composite infrared spectrometer (CIRS)	V. Kunde / M. Flasar	USA	+++
imaging science subsystem	C. Porco	USA	+++
ultraviolet imaging spectrograph (UVIS)	L. Esposito	USA	++
visual and IR mapping spectrometer	R. Brown	USA	++
Fields particles and waves instruments			
Cassini plasma spectrometer	D. Young	USA	+
cosmic dust analysis	E. Grün	Germany	+
ion and neutral mass spectrometer	H. Waite	USA	+++
magnetometer	D. Southwood/ M. Dougherty	UK	
magnetospheric imaging instrument	S. Krimigis	USA	
radio and plasma wave spectrometer	D. Gurnett	USA	
Microwave remote sensing			
Cassini radar	C. Elachi	USA	+++
radio science subsystem	A. Kliore	USA	++
Interdisciplinary science			
magnetosphere and plasma	M. Blanc	France	+
rings and dust	J.N. Cuzzi	USA	+
magnetosphere and plasma	T.I. Gombosi	USA	+
atmospheres	T. Owen	USA	+++
satellites and asteroids	L. A. Soderblom	USA	+
aeronomy and solar wind interaction	D. F. Strobel	USA	++
Huygens instruments and interdisciplinary programs	**PI or IDS**	**Country**	**Exobiological return**
gas chromatograph-mass spectrometer	H. Niemann	USA	+++
aerosol collector and pyrolyser	G. Israël	France	+++
Huygens atmospheric structure instrument	M. Fulchignoni	Italy	++
descent imager/spectral radiometer	M. Tomasko	USA	+++
Doppler wind experiment	M. Bird	Germany	+

Table 14.1. (cont.)

surface science package	J. Zarnecki	UK	+ + +
Interdisciplinary science			
aeronomy	D. Gautier	France	++
atmosphere/surface interactions	J. I. Lunine	USA	++
chemistry and exobiology	F. Raulin	France	+ + +

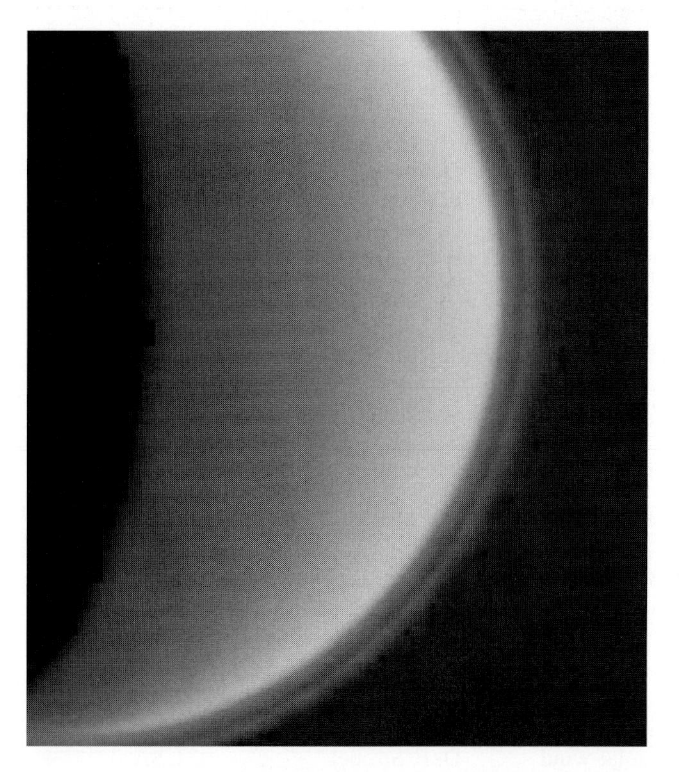

Fig. 14.1. This image taken with the Cassini spacecraft narrow-angle camera shows the complex structure of Titan's atmospheric haze layers. It was taken on 3 July 2004, at a distance of about 789 000 kilometres from Titan. Image courtesy of NASA/JPL/Space Science Institute.

these differences between Titan and the Earth, there are several analogies that can be drawn between the two planetary bodies.

The first resemblance concerns the vertical structure of its atmosphere (see Table 14.2). Although Titan is much colder, with a troposphere (~94–70 K), a tropopause (70.4 K), and a stratosphere (~70–175 K), its atmosphere presents a similar complex structure to that of the Earth and also includes, as evidenced by

Fig. 14.2. This ultraviolet image of Titan's night side limb, also taken by the narrow-angle camera, shows many fine haze layers extending several hundred kilometres above the surface. Image courtesy of NASA/JPL/Space Science Institute.

Table 14.2. *Main characteristics of Titan (including the HASI-Huygens data)*

surface radius	2.575 km
surface gravity	1.35 m s^{-2} (0.14 Earth's value)
mean volumic mass	1.88 kg dm^{-3} (0.34 Earth's value)
distance from Saturn	20 Saturn radii ($\sim 1.2 \times 10^6$ km)
orbit period around Saturn	~16 days
orbit period around Sun	~30 years

Atmospheric data

	Altitude (km)	Temperature (K)	Pressure (mbar)
surface	0	93.7	1470
tropopause	42	70.4	135
stratopause	~250	~187	$\sim 1.5 \times 10^{-1}$
mesopause	~490	~152	$\sim 2 \times 10^{-3}$

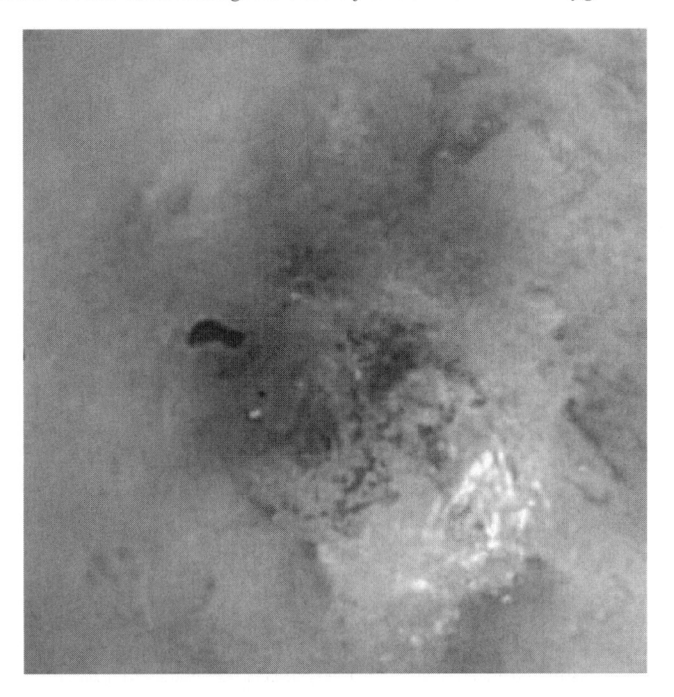

Fig. 14.3. A picture of Titan taken near the south pole by the ISS narrow-angle camera on Cassini shows, in addition to bright clouds, the presence of a dark feature on the upper left which could be a liquid hydrocarbon lake. The pole is in the middle of the picture. Image courtesy of NASA/JPL/Space Science Institute.

Cassini–Huygens, a mesosphere and a thermosphere. Because of a much higher density in the case of Titan, the mesosphere extends to altitudes higher than 400 km (instead of only 100 km for the Earth), but the shape looks very much the same.

These analogies are linked to the presence in both atmospheres of greenhouse gases and antigreenhouse elements. Methane has strong absorption bands in the medium and far infrared regions corresponding to the maximum of the infrared emission spectrum of Titan and is transparent in the near ultraviolet and visible spectral regions. It can thus be a very efficient greenhouse gas in Titan's atmosphere. H_2, which also absorbs in the far infrared (through bimolecular interaction – its dimers) plays a similar role. In the pressure–temperature conditions of Titan's atmosphere, CH_4 can condense but H_2 cannot. Thus, on Titan, CH_4 and H_2 are equivalent respectively to terrestrial condensable H_2O and non-condensable CO_2. In addition, the haze particles and clouds in Titan's atmosphere play an antigreenhouse effect similar to that of the terrestrial atmospheric aerosols and clouds (McKay *et al.*, 1991).

Indeed, CH_4 on Titan seems to play the role of water on the Earth, with a complex cycle which still has to be understood. Although the possibility that Titan is covered with hydrocarbon oceans (Lunine, 1993) is now ruled out (West *et al.*, 2005), Cassini data show that Titan's surface includes lakes of methane and ethane, as detected by the radar instrument in the north polar regions of Titan (Stofan *et al.*, 2007; Sotin, 2007). In addition, the imaging science subsystem (ISS) camera on Cassini has detected dark surface features near the south pole (Figure 14.3) which could be such liquid bodies. Moreover, the descent imager/spectral radiometer (DISR) instrument on Huygens has provided pictures of Titan's surface which clearly show dentritic structures (Figure 14.4) which look like a fluvial net in a relatively young terrain (fresh crater impacts), strongly suggesting recent liquid flow on the surface of Titan (Tomasko *et al.*, 2005). In addition, GC-MS data show that the methane mole fraction increases in the low troposphere (up to 0.05) and reaches the saturation level at approximately 8 km altitude, allowing the possible formation of clouds and rain (Niemann *et al.*, 2005). Furthermore, GC-MS analyses recorded an $\sim$50% increase in the methane mole fraction at Titan's surface, suggesting the presence of condensed methane on the surface near the lander.

Other observations from the Cassini instruments clearly show the presence of various surface features of different origins indicative of volcanic, tectonic, sedimentological, and meteorological, processes, as we find on Earth (Figure 14.5).

Analogies between Titan and the Earth have even been pushed further by comparing Titan's winter polar atmosphere and the terrestrial Antarctic ozone hole (Flasar *et al.*, 2005), although these implied different chemistry.

Another important comparison concerns the noble gas composition and the origin of the atmosphere. The ion neutral mass spectrometer (INMS) on Cassini and GC-MS on Huygens have detected argon in the atmosphere. Similarly to the Earth's atmosphere, the most abundant argon isotope is ^{40}Ar, which comes from the radioactive decay of ^{40}K. Its stratospheric mole fraction is about 4×10^{-5}, as measured by GC-MS (Niemann *et al.*, 2005). The abundance of primordial argon (^{36}Ar) is about 200 times smaller. Moreover, the other primordial noble gases have a mixing ratio smaller than 10 ppb. This strongly suggests that Titan's atmosphere, like that of the Earth, is a secondary atmosphere produced by the degassing of trapped gases. Since N_2 cannot be efficiently trapped in the icy planetesimals which accreted and formed Titan, unlike NH_3, this also indicates that its primordial atmosphere was initially made of NH_3. Ammonia was then transformed into N_2 by photolysis and/or impact driven chemical processes (Owen, 2000; Gautier and Owen, 2002). The $^{14}N/^{15}N$ ratio measured in the atmosphere by INMS and GC-MS (183 in the stratosphere) is less than the primordial N and indicates that several times the present mass of the atmosphere has probably been lost during the history of the satellite (Niemann *et al.*, 2005). Since such evolution may also imply methane transformation into organics, this may also be an indication of large deposits of organics on Titan's surface.

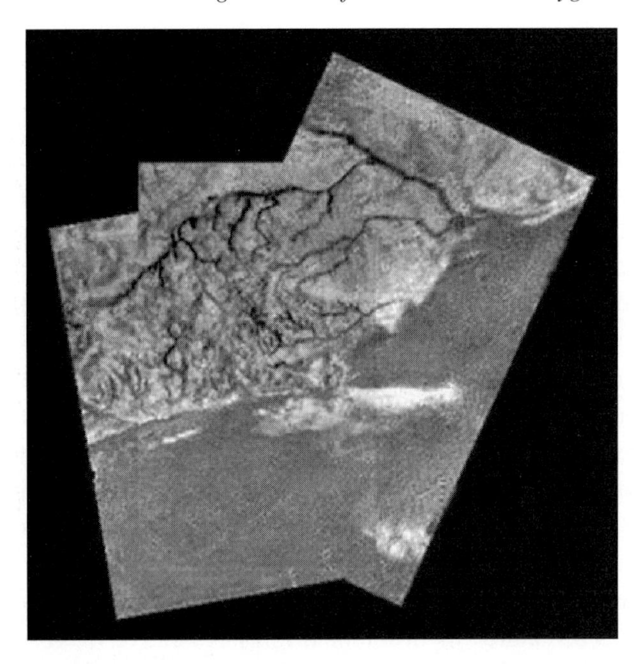

Fig. 14.4. Channel networks, highlands, and dark–bright interface seen by the DISR instrument on Huygens at 6.5 km altitude. Image courtesy of ESA/NASA/JPL/University of Arizona.

Fig. 14.5. Titan, seen by the Cassini spacecraft narrow-angle camera, shows a very diverse surface, with bright (like the so-called 'Xanadu' region in the middle of the picture) and darker areas. Image courtesy of NASA/JPL/Space Science Institute.

Analogies can also be made between the organic chemistry which is now very active on Titan and the prebiotic chemistry which was active on the primitive Earth. In spite of the absence of permanent bodies of liquid water on Titan's surface, both chemistries are similar. Several of the organic processes which are occurring today on Titan imply the presence of the organic compounds which are considered as key molecules in the terrestrial prebiotic chemistry, such as hydrogen cyanide (HCN), cyanoacetylene (HC_3N) and cyanogen (C_2N_2). In fact, the atmosphere of Titan (dinitrogen with a small percentage of methane) is one of the most favourable atmospheres for prebiotic synthesis, as shown by Miller's experiments. Until recently, such atmosphere composition was supposed to be far from that of the primitive Earth. However, a new model of the hydrogen escape in the primitive atmosphere of the Earth suggests that it may have been much richer in hydrogen and methane than previously thought (Tian *et al.*, 2005). This suggests that Titan may be even more similar to the primitive Earth than we thought.

Indeed, a complex organic chemistry seems to be present in the three components of what one can call, always by analogy with our planet, the 'geofluids' of Titan: air (gas atmosphere), aerosols (solid atmosphere) and surface (lakes).

14.3 A complex prebiotic-like chemistry

In the atmosphere of Titan, CH_4 chemistry is coupled with N_2 chemistry resulting in the formation of many organics – hydrocarbons and N-containing organic compounds – in the gas and particulate phases. These compounds are hydrocarbons, nitriles, and complex refractory organics. Several photochemical models describing the chemical and physical pathways involved in the chemical evolution of the atmosphere of Titan and the resulting vertical concentration profiles of the different molecules involved have been published during the last 20 years. For a review, see the most recent publications and the included references (Lebonnois *et al.*, 2001; Wilson and Atreya, 2004; Hebrard *et al.*, 2005).

The whole chemistry starts with the dissociation of N_2 and CH_4 through electron and photon impacts. The primary processes allow the formation of C_2H_2 and HCN in the high atmosphere. These molecules play a key role in the general chemical scheme: once they are formed, they diffuse down to the lower levels where they allow the formation of higher hydrocarbons and nitriles. Additional CH_4 dissociation probably also occurs in the low stratosphere through photocatalytic processes involving C_2H_2 and polyynes.

Another approach to the study of organic chemistry on Titan that is complementary to photochemical modelling is to develop simulation experiments in the laboratory. These experiments seem to mimick the real processes well, since experiments, carried out in particular at the Laboratoire Interuniversitaire des Systèmes Atmosphériques (LISA), produced all the gas phase organic species already detected in Titan's atmosphere, within the right orders of magnitude of relative concentration for most of them. Such observations demonstrate the validity of these experimental simulations. The experiments also produce many other organics which can be assumed also to be present in Titan's atmosphere. Thus, simulation experiments appear to be a very useful guide for further searches (by both remote sensing and *in situ* observations). The gas phase as well as the aerosol phases are affected by such an extrapolation.

In the gas phase, more than 150 different organic molecules have been detected in the simulation experiments (Coll *et al.*, 1998, 1999a). These global simulations of Titan's atmospheric chemistry use an open reactor and a low pressure N_2–CH_4 gas mixture. The energy source is a cold plasma discharge producing mid-energy electrons (around 1–10 eV). The gas phase endproducts (molecules) are analysed by IRFTS (infrared Fourier transform spectroscopy) and GC-MS techniques; the transient species (radicals and ions) are determined by on-line ultraviolet–visible spectroscopy. The evolution of the system is also theoretically described using coupled physical and chemical (ions and neutrals) models. The identified organic products are mainly hydrocarbons and nitriles. The absence at a detectable level of molecules carrying amino groups, like amines, with the exception of ammonia, must be highlighted. These experiments have allowed the detection of all gaseous organic species observed on Titan, including C_4N_2 (Coll *et al.*, 1999b). Among the other organics formed in these experiments and not yet detected in Titan's atmosphere, one should note the presence of polyynes (C_4H_2, C_6H_2, C_8H_2) and probably cyanopolyyne HC_4CN. These compounds are also included in photochemical models of Titan's atmosphere, where they could play a key role in the chemical schemes allowing the transition from gas phase products to aerosols. Also of astrobiological interest is the formation of organic compounds with asymmetric carbon such as

$$CH_3C^*H(C_2H_5)CH = CH_2 \quad \text{and} \quad CH_3C^*H(CN)CH = CH_2$$

Experiments on N_2–CH_4 mixtures including CO at the 100 ppm level (Bernard *et al.*, 2003; Coll *et al.*, 2003) show the incorporation of oxygen atoms in the produced organics, with an increasing diversity of the products (more than 200 were identified). The main oxygen-containing organic compound is neither formaldehyde nor methanol, as expected from theoretical models (both thermodynamic and kinetic), but oxirane (also called ethylene oxide), $(CH_2)_2O$. Oxirane thus appears to be a good candidate to search for in Titan's atmosphere. These studies also show the formation of ammonia at noticeable concentrations, opening new avenues in the chemical schemes of Titan's atmosphere.

Simulation experiments also produce solid organics, as mentioned above, usually called tholins (Sagan and Khare, 1979). These 'Titan tholins' are supposed to be laboratory analogues of Titan's aerosols, those tiny solid particles which are present in Titan's atmosphere and mask the surface of the satellite in the visible. They have been extensively studied since the first work by Sagan and Khare more than 20 years ago (Khare *et al.*, 1984; 1986 and included references). These laboratory analogues show very different properties depending on the experimental conditions (Cruikshank *et al.*, 2005). For instance, the average C/N ratio of the product varies from less than 1 to more than 11, in the published reports. More recently, dedicated experimental protocols allowing a simulation closer to the real conditions have been developed at LISA using low pressure and low temperature (Coll *et al.*, 1998, 1999a) and recovering the laboratory tholins without oxygen contamination (from the air of the laboratory) in a glove box purged with pure N_2. Representative laboratory analogues of Titan's aerosols have thus been obtained and their complex refractive indices have been determined (Ramirez *et al.*, 2002), with – for the first time – error bars. These data can be seen as a new point of reference for modellers who compute the properties of Titan's aerosols. Systematic studies have been carried out on the influence of the pressure of the starting gas mixture on the elemental composition of the tholins. They show that two different chemical–physical regimes are involved in the processes, depending on the pressure, with a transition pressure around 1 mbar (Bernard *et al.*, 2002; Imanaka *et al.*, 2004).

The molecular composition of the Titan tholins is still poorly known. Several possibilities have been considered such as HCN polymers or oligomers, HCN–C_2H_2 cooligomers, HC_3N polymers, HC_3N–HCN cooligomers (Tran *et al.*, 2003 and included references). However, it is well established that they are made of macromolecules of largely irregular structure. Gel filtration chromatography of the water soluble fraction of Titan tholins shows an average molecular mass of about 500–1000 Da (McDonald *et al.*, 1994). Information on the chemical groups included in their structure has been obtained from their infrared and ultraviolet spectra and from analysis by pyrolysis-GC-MS techniques (Ehrenfreund *et al.*, 1995; Coll *et al.*, 1998; Imanaka *et al.*, 2004; and included references). The data show the presence of aliphatic and benzenic hydrocarbon groups, of CN, NH_2, and C$=$NH groups. Direct analysis by chemical derivatization techniques before and after hydrolysis allowed the identification of amino acids or their precursors (Khare *et al.*, 1986). Their optical properties have been determined (Khare *et al.*, 1984; McKay, 1996; Ramirez *et al.*, 2002; Tran *et al.*, 2003; Imanaka *et al.*, 2004), because of their importance for retrieving observational data related to Titan. Finally, it is obviously of astrobiological interest to mention that Stoker *et al.* (1990) demonstrated the nutritious properties of Titan tholins for microorganisms.

Nevertheless, there is still a need for improved experimental laboratory simulations to better mimic the chemical evolution of the atmosphere, including the dissociation of dinitrogen by electron impact with energies close to the case of Titan's atmosphere, and the dissociation of methane through photolysis processes. Such an experiment is currently under development at LISA, with the SETUP (Simulation Exprimentale et Théorique Utile à la Planétologie) programme which, in a dedicated low-temperature flow reactor, couples N_2 dissociation by electron and CH_4 photodissociation by two-photon (248 nm) laser irradiation, and theoretical studies, in order to improve the chemical schemes. The preliminary results demonstrate the dissociation of methane through the two-photon process (Romanzin *et al.*, 2005).

Several organic compounds have already been detected in Titan's stratosphere (Table 14.3). The list includes hydrocarbons (with both saturated and unsaturated chains) and nitrogen-containing organic compounds, exclusively nitriles, as expected from laboratory simulation experiments. Most of these detections were performed by Voyager observations, with the exception of the C_2 hydrocarbons which were observed in spectra in the 1970s (see Strobel (1974)), acetonitrile, which was detected by ground observation in the millimetre wavelength, and water and benzene, which were tentatively detected by ISO. Since the arrival of Cassini in the Saturn system, the presence of water and benzene has been unambiguously confirmed by CIRS. In addition, the direct analysis of the ionosphere by INMS during the low-altitude Cassini fly-bys of Titan shows the presence of many organic species at detectable levels (Figure 14.6), in spite of the very high altitude (1100–1300 km) (Waite *et al.*, 2007).

Surprisingly, GC-MS on board Huygens has not detected a large variety of volatile organic compounds in the low atmosphere. The mass spectra collected during the descent show that the medium and low stratosphere and the troposphere are poor in volatile organic species, with the exception of methane. Condensation of these species on the aerosol particles is a probable explanation for these atmospheric characteristics (Niemann *et al.*, 2005). These particles, for which no direct data on the chemical composition were available before, have been analysed by the aerosol collector and pyrolyser (ACP) instrument. ACP was designed to collect the aerosols during the descent of the Huygens probe on a filter in two different regions of the atmosphere. Then the filter was heated in a closed oven at different temperatures and the produced gases were analysed by the GC-MS instrument. The results show that the aerosol particles are made of refractory organics: non-volatile compounds which are only vapourized after degradation at high temperature. The ACP data also show that these organics release HCN and NH_3 during pyrolysis (Israel *et al.*, 2005). This strongly supports the tholin hypothesis: from these new and first *in situ* measurement data it seems very likely that the aerosol particles are

Table 14.3. *Main composition of Titan's stratosphere, trace components already detected and comparison with the products of laboratory simulation experiments (maj. = major product; ++: abundance smaller by one order of magnitude; +: abundance smaller by two orders of magnitude.*

Compounds	Stratosphere mixing ratio E = equator; N = north pole		Production in simulation experiments
Main constituents			
Nitrogen N_2	0.98		
Methane CH_4	0.02		
Hydrogen H_2	0.0006–0.00014		
Hydrocarbons			
Ethane C_2H_6	1.3×10^{-5}	E	maj.
Acetylene C_2H_2	2.2×10^{-6}	E	maj.
Propane C_3H_8	7.0×10^{-7}	E	++
Ethylene C_2H_4	9.0×10^{-8}	E	++
Propyne C_3H_4	1.7×10^{-8}	N	+
Diacetylene C_4H_2	2.2×10^{-8}	N	+
Benzene C_6H_6	few 10^{-9}		+
N-organics			
Hydrogen cyanide HCN	6.0×10^{-7}	N	maj.
Cyanoacetylene HC_3N	7.0×10^{-8}	N	++
Cyanogen C_2N_2	4.5×10^{-9}	N	+
Acetonitrile CH_3CN	few 10^{-9}		++
Dicyanoacetylene C_4N_2	Solid phase	N	+
O-compounds			
Carbon monoxide CO	2.0×10^{-5}		
Carbon dioxide CO_2	1.4×10^{-8}	E	
Water H_2O	few 10^{-9}		

made of a refractory organic nucleus, covered with condensed volatile compounds (Figure 14.7). The nature of the pyrolysates provides information on the molecular structure of the refractory complex organics: it indicates the potential presence of nitrile groups (–CN), amino groups (–NH_2, –NH– and –N<) and/or imino groups (–C=N–).

Furthermore, comparison of the data obtained for the first (mainly stratospheric particles) and second (midtroposphere) samplings indicate that the aerosol composition is homogeneous (Israel *et al.*, 2005). This also fits with some of the data obtained by the descent imager and radial spectrometer, DISR, relative to the

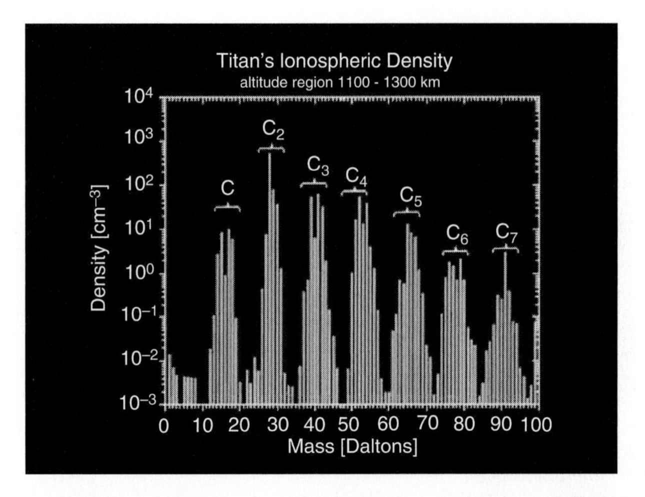

Fig. 14.6. Mass spectrum of Titan's ionosphere at an altitude of about 1200 km. The spectrum shows the signature of organic compounds including up to seven carbon atoms. (Image courtesy of NASA/JPL/University of Michigan.)

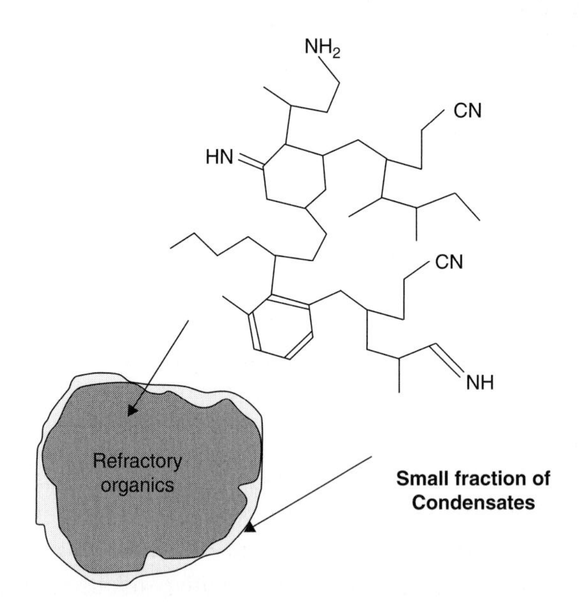

Fig. 14.7. Model of the chemical composition of Titan's aerosol from the Huygens–ACP data.

aerosol particle which indicates a fairly constant size distribution of the particles with altitude (with a mean dimension of the order of 1 μm).

These particles sediment down to the surface where they likely form a deposit of complex refractory organic and frozen volatile compounds. DISR collected the infrared reflectance spectra of the surface with the help of a lamp, illuminating the

Fig. 14.8. The surface of Titan as seen by the Huygens DISR camera. Image courtesy of ESA/NASA/JPL/University of Arizona.

surface before the Huygens probe touched down. These data show the presence of water ice, but no clear evidence – so far – of tholins. The presence of water ice is also suggested by the data of the SSP instrument (Zarnecki *et al.*, 2005). Its accelerometer measurements can be interpreted as due to the presence of small water ice pebbles on the surface where Huygens has landed, in agreement with the DISR surface pictures (Figure 14.8). On the other hand, GC-MS was able to analyse the atmosphere near the surface for more than one hour after the touch down. The corresponding mass spectra show the clear signature of many organics, including cyanogen, C_3 and C_4 hydrocarbons, and benzene, indicating that the surface is much richer in volatile organics than the low stratosphere and the troposphere (Niemann *et al.*, 2005). These observations are in agreement with the hypothesis that in the low atmosphere of Titan, most of the organic compounds are in the condensed phase.

Thus, altogether, these new data show the diversity of the locations at which organic chemistry is taking place on Titan. Surprisingly the high atmosphere looks very active, with neutral and ion organic processes; the high stratosphere, where many organic compounds have been detected also shows an active organic chemistry in the gas phase. In the lower atmosphere this chemistry seems mainly concentrated in the condensed phase. Titan's surface is probably covered with frozen volatile organics together with refractory, tholin-like organic materials.

Irradiating effects of cosmic rays reaching Titan's surface may induce additional organic syntheses, particularly if these materials are partly dissolved in some small liquid bodies made of low molecular weight hydrocarbons (mainly methane and ethane). This could indeed allow the additional formation of reactive compounds such as azides as well as the polymerization of HCN (Raulin *et al.*, 1995). Moreover, the interface between the liquid phase and the solid deposits at the surface may include sites of catalytic activity favourable to these additional chemical reactions.

In spite of the surface temperatures, even the presence of liquid water is not excluded. Cometary impacts on Titan may melt surface water ice, offering possible episodes as long as about 1000 years of liquid water (Artemieva and Lunine, 2003). This provides conditions for short terrestrial-like prebiotic syntheses at relatively low temperatures. Low temperatures reduce the rate constants of prebiotic chemical reactions, but may increase the concentration of reacting organics by the eutectic effect which increases the rate of the reaction. In addition, the possible presence of a water–ammonia ocean in the depths of Titan, as expected from models of its internal structure (Tobie *et al.*, 2005, and included references), may also provide an efficient way to convert simple organics into complex molecules, and to reprocess chondritic organic matter into prebiotic compounds. These processes may have occurred very efficiently at the beginning of Titan's history (with even the possibility of the water–ammonia ocean exposed to the surface) allowing a CHNO prebiotic chemistry that resulted in the evolution of compounds of terrestrial biological interest.

Even if these liquid water scenarios are false, the possibility of a pseudo biochemistry evolving in the absence of a noticeable number of O atoms cannot be ruled out, with an N-chemistry, based on 'ammono analogues' replacing the O-chemistry (Raulin and Owen, 2002). Such alternatives of terrestrial biochemistry where, in particular, the water solvent could be replaced by ammonia or other N-compounds have been re-examined by Benner (2002) and by Schulze-Makuch and Irwin (2004).

14.4 Life on Titan?

There are therefore several ways that Titan's environment could have evolved to support a prebiotic chemistry and perhaps even biotic systems. But if life emerged on Titan, are Titan's conditions compatible with sustaining life? The surface is too cold and not energetic enough to provide the right conditions. However, the (hypothetical) subsurface oceans may be suitable for life. Fortes (2000) has shown that there are no insurmountable obstacles. With a possible temperature of this ocean as high as about 260 K and the possible occurrence of cryovolcanic hotspots (where volcanic activity increases the temperature) allowing 300 K, the temperature conditions in Titan's subsurface oceans could allow the development of living systems. Even at a depth of 200 km, the expected pressure of about 5 kbar is not incompatible with life, as shown by terrestrial examples. The expected pH of an aqueous medium made of 15% by weight of NH_3 is equivalent to a pH of 11.5. Some bacteria can grow on Earth at pH 12. Even the limited energy resources do not exclude the sustaining of life. Interestingly, the situation seems similar to that of Europa. However, it must be pointed out that Europa's ocean may be much closer to the surface (see Chapter 15). Moreover, Europa's ocean is more likely to be in direct contact with a silicate floor than Titan's, and more likely to include hydrothermal vents that Titan's.

Taking into account only the potential radiogenic heat flow ($\sim 5 \times 10^{11}$ W) and assuming that 1% of that is used for volcanic activity and 10% of the latter is available for living system metabolism, Fortes (2000) estimates an energy flux available in the subsurface oceans of about 5×10^8 W. Such a flux corresponds to the production of about 4×10^{11} mol of ATP per year and about 2×10^{13} g of biomass per year. If we assume an average turnover for the living systems in the order of one year, the biomass density would be 1 g m^{-2}. This is very small compared to the lower limit of the value of the biomass for the Earth (about 1000–10 000 g m^{-2}). Nevertheless, this indicates the possible presence of a limited but not negligible bioactivity on the satellite. The biota on Titan, if any, assuming that the living systems are similar to the ones we know on Earth, and based on the chemistry of carbon and the use of liquid water as solvent, would thus be localized in the subsurface deep ocean. Several possible metabolic processes such as nitrate/nitrite reduction or nitrate/dinitrogen reduction, sulphate reduction, and methanogenesis have been postulated (Simakov, 2001) as well as the catalytic hydrogenation of acetylene (Abbas and Schulze-Makuch, 2002). As expected, no sign of macroscopic life has been detected by Huygens as it approached the surface or after it landed. This can be concluded in particular from the many pictures taken by DISR of the same location on Titan during more than one hour after landing. But this does not exclude the possibility of the presence of a microscopic life. The metabolic activity of the corresponding biota, even if it is localized far from the surface, in the deep internal structure of Titan, may produce chemical species which diffuse through the ice mantle covering the hypothetical internal ocean and feed the atmosphere. It has even been speculated in several publications that the methane we see in the atmosphere today is the product of biological activity (Simakov, 2001). If this was the case, the atmospheric methane would be notably enriched in light carbon. Indeed, on Earth, biological processes induce an isotopic fragmentation producing an enrichment in ^{12}C: $^{12}C/^{13}C$ increases from 89 (the reference value, in the Belemnite of the Pee Dee Formation) to about 91–94 depending on the biosynthesis processes. The $^{12}C/^{13}C$ ratio in atmospheric methane on Titan, as determined by the GC-MS instrument on Huygens, is 82 (Niemann *et al.*, 2005). Although we do not have a reference for $^{12}C/^{13}C$ on Titan, this low value suggests that the origin of methane is probably abiotic.

14.5 Conclusions

Although exotic life, such as methanogenic life in liquid methane, cannot be fully ruled out (McKay and Smith, 2005), the presence of extant or extinct life on Titan seems very unlikely. Nevertheless, with the new observational data provided by the Cassini–Huygens mission, the largest satellite of Saturn looks more than ever like a very interesting object for astrobiology. The several analogies of this exotic and cold planetary body with the Earth and the complex organic chemical processes which are going on now on Titan provide a fantastic means to better understand the prebiotic processes which are no longer reachable on the Earth, at the scale and within the whole complexity of a planetary environment.

The origin and cycle of methane on Titan illustrates the whole complexity of Titan's system. Methane may be stored in large amounts in the interior of the satellite, in the form of clathrates (methane hydrates) trapped during the formation of the satellite from the Saturnian subnebula where it was formed by Fisher–Tropsch processes (Sekine *et al.*, 2005). It may also be produced through high-pressure processes, like serpentinization, allowing the formation of H_2 by reaction of H_2O with ultramafic rocks, or by cometary impact (Kress and McKay, 2004). Interestingly, those processes have rarely been considered in the case of the primitive Earth, although they may have contributed to a possible reducing character of the primordial atmosphere of our planet, as mentioned earlier in this chapter. This is an example of how Titan's study is indeed providing new insights into terrestrial chemical evolution.

In Titan's atmosphere, methane is photolysed by solar ultraviolet radiation, producing mainly ethane and tholin-like organic matter. The resulting lifetime of methane in Titan's atmosphere is relatively short (about 10–30 My). Thus methane stored in Titan's interior may be continuously replenishing the atmosphere, through degassing induced by cryovolcanism. This volcanic activity at low temperature, where terrestrial lava is replaced by water, has been clearly evidenced from the first images of Titan's surface provided by the visual and infrared mapping spectrometer (VIMS), ISS, and radar instruments on Cassini (Sotin *et al.*, 2005). It may also be released episodically to the atmosphere, as suggested by Tobie *et al.* (2006). In any case, the methane cycle should result in the accumulation of large amounts or complex organics on the surface and large amounts of ethane, which mixed with the dissolved atmospheric methane should form liquid bodies on the surface or

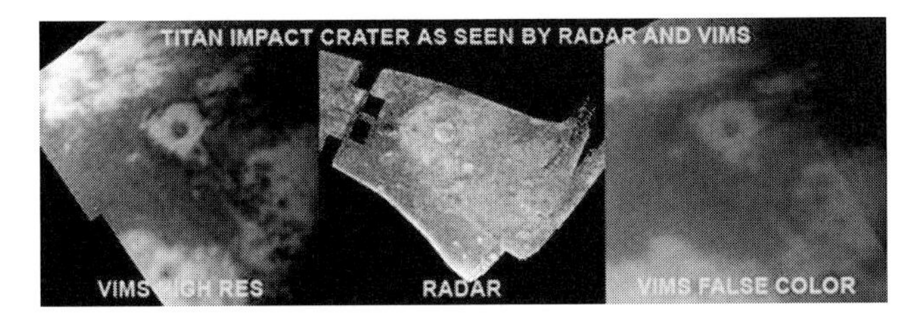

Fig. 14.9. One of the largest impact craters (about 80 km diameter) observed on Titan's surface by two Cassini instruments: VIMS infrared (left), radar image (centre) and false-colour image (right). The faint halo, slightly darker than the surrounding parts, is probably somewhat different in composition. Since it is made of material excavated when the crater was formed, this indicates that the composition of Titan's upper crust varies with depth, Image courtesy of NASA/JPL/University of Arizona.

in the near subsurface of the satellite. It is possible that the dark feature seen in Figure 14.3 is one of these expected liquid bodies.

The Cassini–Huygens mission is far from complete. It will continue its systematic exploration of the Saturnian system up to 2008, and probably to 2011 if the extended mission is accepted. Numerous data of paramount importance for astrobiology are still expected from several of its instruments (Table 14.1). The CIRS spectrometer should be able to detect new organic species in the atmosphere during the future limb observation of Titan, especially at the pole. ISS and VIMS should provide a detailed picture of Titan's surface revealing the complexity but also the physical and chemical nature of this surface and its diversity, as evidenced by the discovery of a 5-μm bright spot (Barnes *et al.*, 2005). Radar observation will also continue the systematic coverage of Titan's surface which shows contrasted regions of smooth and rough areas, suggesting a possible shoreline. The coupled observation of the same regions by these instruments, which has already started (Figure 14.9), will surely bring essential new information to our understanding of this new, exotic and astonishing world.

Acknowledgements

The author wishes to thanks the Cassini–Huygens teams, and particularly the Huygens project scientist, Jean-Pierre Lebreton, his colleague Olivier Witasse, and the principal investigators of the Huygens instruments, especially Guy Israel and Hasso Niemann, for making available several of the data used in this chapter. The author also wishes to thank his colleagues of the LISA-GPCOS team, especially Patrice Coll, Marie-Claire Gazeau, Eric Hebrard, and Mai-Julie Nguyen, for useful discussions during the preparation of the manuscript. Final thanks go to ESA and the French Space Agency, CNES, for financial support.

References

Abbas, O., and Schulze-Makuch, D. (2002). Acetylene-based pathways for prebiotic evolution on Titan. *ESA SP*, **518**, 345–348.

Artemieva, N., and Lunine, J. (2003). Cratering on Titan: impact melt, ejecta, and the fate of surface organics. *Icarus*, **164**, 471–480.

Barnes, J. W., Brown, R. H., Turtle, E. P., *et al.* (2005). A 5-micron-bright spot on Titan: evidence for surface diversity. *Science*, **310**, 92–95.

Benner, S. (2002). Weird life: chances versus necessity, Communication at Weird Life. Planning session for NRC Committee on the Origins and Evolution of Life, National Academies of Sciences, USA. http://www7.nationalacademies.org/ssb/weirdlife.html

Bernard, J.-M., Coll, P., Coustenis, A., and Raulin, F. (2003). Experimental simulation of Titan's atmosphere detection of ammonia and ethylene oxide. *Planet. Space Sci.*, **51**, 1003–1011.

Bernard, J.-M., Coll, P., and Raulin, F. (2002). Variation of C/N and C/H ratios of Titan's aerosols analogues. *Proc. Second European Workshop on Exo-/Astro-Biology*. Noordwijk: ESA publications division, pp. 623–625.

Coll, P., Bernard, J.-M., Navarro-González, R., and Raulin, F. (2003). Oxirane: An exotic oxygenated organic compound in Titan? *Astrophys. J.*, **598**, 700–703.

Coll, P., Coscia, D., Gazeau, M.-C., and Raulin, F. (1998). Review and latest results of laboratory investigation of Titan's aerosols. *Origins Life Evol. B.*, **28**, 195–213.

Coll, P., Coscia, D., Smith, N., *et al.* (1999a). Experimental laboratory simulation of Titan's atmosphere: aerosols and gas phase. *Planet. Space Sci.*, **47**, 1331–1340.

Coll, P., Guillemin, J. C., Gazeau, M. C., and Raulin, F. (1999b). Report and implications of the first observation of C4N2 in laboratory simulations of Titan's atmosphere. *Planet. Space Sci.*, **47**, 1433–1440.

Cruikshank, D. P., Imanaka, H., and Dalle Ore, C. M. (2005). Tholins as coloring agents on outer Solar System bodies. *Adv. Space Res.*, **36**, 178–183.

Ehrenfreund, P., Boon, J. J., Commandear, J., *et al.* (1995). Analytical pyrolysis experiments of Titan aerosol analogues in preparation for the Cassini–Huygens mission. *Adv. Space Res.*, **15**, 335–342.

Flasar, F. M., Achterberg, R. K., Conrath, B. J., *et al.* (2005). Titan's atmospheric temperatures, winds, and composition. *Science*, **308**, 975–978.

Fortes, A. D. (2000). Exobiological implications of a possible ammonia-water ocean inside Titan. *Icarus*, **146**, 444–452.

Fulchignoni, M., Ferri, F., Angrilli, F., *et al.* (2005). Titan's physical characteristics measured by the Huygens Atmospheric Structure Instrument (HASI). *Nature*, **438**, 785–791.

Gautier, D., and Owen, T. (2002). Touring the Saturnian system: the atmosphere of Titan and Saturn. *Space Sci. Rev.*, **104**, 347–376.

Hebrard, E., Benilan, Y., and Raulin, F. (2005). Sensitivity effects of photochemical parameters uncertainties on hydrocarbon production in the atmosphere of Titan. *Adv. Space Res.*, **36**, 268–273.

Imanaka, H., Khare, B. N., Elsila, J. E., *et al.* (2004). Laboratory experiments of Titan tholin formed in cold plasma at various pressures: implications for nitrogen-containing polycyclic aromatic compounds in Titan haze. *Icarus*, **168**, 344–366.

Israel, G., Szopa, C., Raulin, F., *et al.* (2005). Evidence for the presence of complex organic matter in Titan's aerosols by in situ analysis. *Nature*, **438**, 796–799.

Khare, B. N., Sagan, C., Arakawa, E. T., *et al.* (1984). Optical constants of organic tholins produced in a simulated Titanian atmosphere: from soft X-rays to microwave frequencies. *Icarus*, **60**, 127–137.
Khare, B. N., Sagan, C., Ogino, H., *et al.* (1986). Amino acids derived from Titan tholins. *Icarus*, **68**, 176–184.
Kress, M. E., and McKay, C. P. (2004). Formation of methane in comet impacts: implications for Earth, Mars, and Titan. *Icarus*, **168**, 475–483.
Lebonnois, S., Toublanc, D., Hourdin, F., and Rannou, P. (2001). Seasonal variations of Titan's atmospheric composition. *Icarus*, **152**, 384–406.
Lebreton, J.-P., Witasse, O., Sollazzo, C., *et al.* (2005). Huygens descent and landing on Titan: mission overview and science highlights. *Nature*, **438**, 758–764.
Lunine, J. I. (1993). Does Titan have an ocean? A review of current understanding of Titan's surface. *Rev. Geophys.*, **31**, 133–149.
McDonald, G. D., Thompson, W. R., Heinrich, M., Khare, B. N., and Sagan, C. (1994). Chemical investigation of Titan and Triton tholins. *Icarus*, **108**, 137–145.
McKay, C. P. (1996). Elemental composition, solubility, and optical properties of Titan's organic haze. *Planet. Space Sci.*, **44**, 741–747.
McKay, C. P., Pollack, J. B., and Courtin, R. (1991). The greenhouse and antigreenhouse effects on Titan. *Science*, **253**, 1118–1121.
McKay, C. P., and Smith, H. D. (2005). Possibilities for methanogenic life in liquid methane on the surface of Titan. *Icarus*, **178**, 274–276.
Niemann, H. B., Atreya, S. K., Bauer, S. J., *et al.* (2005). Implications for the origin and evolution of the Titan atmosphere from measurements by Huygens GCMS. *Nature*, **438**, 779–784.
Owen, T. (2000). On the origin of Titan's atmosphere. *Planet. Space Sci.*, **48**, 747–752.
Ramirez, S. I., Coll, P., Da Silva, A., Navarro-Gonzalez, R., Lafait, J., and Raulin, F. (2002). Complex refractive index of Titan's aerosol analogues in the 200-900 nm domain. *Icarus*, **156**, 515–530.
Raulin, F., and Owen, T. (2002). Organic chemistry and exobiology on Titan. *Space Sci., Rev.*, **104**, 379–395.
Raulin, F., Bruston, P., Paillous, P., and Sternberg, R. (1995). The low temperature organic chemistry of Titan's geofluid. *Adv. Space Res.*, **15**, 321–333.
Romanzin, C., Gazeau, M.-C., Bénilan, Y., *et al.* (2005). Laboratory investigations on CH4 photochemistry in the frame of a new experimental program of Titan's atmosphere simulation. *Adv. Space Res.*, **36**, 258–267.
Sagan, C., and Khare, B. N. (1979). Organic chemistry of interstellar grains and gas. *Nature*, **277**, 102–107.
Schulze-Makuch, D., and Irwin, L. N. (2004). *Life in the Universe, Expectations and constraints*. New York: Springer.
Sekine, Y., Sugita, S., Shido, T., *et al.* (2005). The role of Fischer–Tropsch catalysis in the origin of methane-rich Titan. *Icarus*, **178**, 154–164.
Simakov, M. B. (2001). The possible sites for exobiological activities on Titan. *ESA SP*, **496**, 211–214.
Sotin, C. (2007). Titan's lost seas found. *Nature*, **445**, 29–30.
Sotin, C., Jaumann, R., Buratti, B. J., *et al.* (2005). Release of volatiles from a possible volcano from near-infrared imaging of Titan. *Nature*, **435**, 786–789.
Stofan, E. R., Elachi, C., Lunine, J. I., *et al.* (2007). The lakes of Titan. *Nature*, **445**, 61–64.
Stoker, C. R., Boston, P. J., Mancinelli, R. L., *et al.* (1990). Microbial metabolism of tholins. *Icarus*, **85**, 241–256.

Strobel, D. A. (1974). The photochemistry of hydrocarbons in the atmosphere of Titan. *Icarus*, **21**, 466–470.

Tian, F., Toon, O. B., Pavlov, A. A., and De Sterck, H. (2005). A hydrogen-rich early earth atmosphere. *Science*, **308**, 1014–1017.

Tobie, G., Grasset, O., Lunine, J. I., Mocquet, A., and Sotin, C. (2005). Titan's internal structure inferred from a coupled thermal-orbital model. *Icarus*, **175**, 496–502.

Tobie, G., Lunine, J. I., and Sotin, C. (2006). Titan's internal structure inferred from a coupled thermal-orbital model. *Nature*, **440**, 61–64.

Tomasko, M. G., Archinal, B., Becker, T., *et al.* (2005). Rain, winds and haze during the Huygens probe's descent to Titan's surface. *Nature*, **438**, 765–778.

Tran, B. N., Joseph, J. C., Ferris, J. P., Persans, P. D., and Chera, J. J. (2003). Simulation of Titan haze formation using a photochemical flow reactor: The optical constants of the polymer. *Icarus*, **165**, 379–390.

Waite Jr., J. H., Young, D. T., Cravens, T. E., *et al.* (2007). The process of tholin formation in Titan's upper atmosphere. *Science* **316**, 870–875.

West, R. A., Brown, M. E., Salinas, S. V., Bouchez, A. H., and Roe, H. G., (2005). No oceans on Titan from the absence of a near-infrared specular reflection. *Nature*, **436**, 670–672.

Wilson, E. H., and Atreya, S. K. (2004). Current state of modeling the photochemistry of Titan's mutually dependent atmosphere and ionosphere. *J. Geophys. Res.*, **109**, (E6), E06002.

Zarnecki, J. C., Lesse, M. R., Hathi, B., *et al.* (2005). A soft solid surface on Titan at the Huygens landing site as measured by the Surface Science Package (SSP). *Nature*, **438**, 792–795.

15

Europa, the ocean moon: tides, permeable ice, and life

Richard Greenberg
University of Arizona

15.1 Introduction: life beyond the habitable zone

As the horizons of the field of astrobiology have extended to the rapidly growing population of known extrasolar planetary systems, the concept of a habitable zone around each star has been of particular interest. With each discovery, the question is raised of whether the new planet is in the so-called *habitable zone*, or whether hypothetical orbits within that zone would be stable. In this context, the *habitable zone* is commonly defined as a range of distances from the star where temperatures would be in a range for life to be viable and for water to exist near the surface in a phase able to support living organisms (Kasting *et al.*, 1993; Menou and Tabachnik, 2003; Raymond and Barnes, 2005). This restrictive definition can contribute to a relatively pessimistic view of the potential for extra-terrestrial life (e.g., Ward and Brownlee (2000)). It has even been used to advocate planetary exploration objectives, purportedly motivated by the search for life, that are restricted to the terrestrial planets in our Solar System (e.g., Hubbard (2005)).

Unless corrected, this widespread semantic usage of the term *habitable zone* would leave out one of the most likely places for exploration to reveal extraterrestrial life: Jupiter's moon Europa. The analyses cited above neglect tidal friction, which is known in the case of Jupiter's Galilean satellites to provide significant heating, comparable in fact to the solar heating in the habitable (and inhabited) terrestrial-planet zone. (Ward and Brownlee (2000) implicitly acknowledge this broader range of habitability in using an image of Conamara Chaos on Europa as the frontispiece of their book, while still tending to define the habitable zone in the more restricted sense.) Of course, heat alone may not provide energy accessible to organisms, but the heat can promote a physical setting that facilitates derivation of metabolic energy from chemical or solar (or stellar) sources.

While the observational astrophysical community is not yet ready to worry about satellites, and some political interests might prefer to focus on the Moon and Mars

Planetary Systems and the Origins of Life, eds. Ralph E. Pudritz, Paul G. Higgs, and Jonathon R. Stone.
Published by Cambridge University Press.

in our own Solar System, a true description of the habitable zone of any planetary system should include any appropriate satellite systems whose heat source may be relatively independent of a nearby star.

This chapter reviews what is known about the physical character of Europa's potential biosphere, specifically the 100-km-deep liquid water ocean, and the thin layer of ice that lies above it. The global ocean just below the ice probably has received substances released from the deep silicate interior, while the surface ice has been bombarded by cometary and asteroidal material, as well as suffering impacts from the swarm of circum-Jovian particles. The dynamic ice crust controls the interface of these endogenic and exogenic substances, maintaining disequilibrium conditions but allowing interaction over various spatial and temporal scales, as shown in this chapter.

The permeability of the ice layer is thus critical to providing a setting that might support life, both in the ocean and within the crust itself. For a complete discussion of the physical character and likely processes operating on Europa, and their implications for life, see Greenberg (2005). Another review from a biological perspective is Lipps and Rieboldt (2005).

An aspect of Europa's physical character is that the environment at a given location in the ice evidently changes on various timescales. Short-term change provides mixing and disequilibrium conditions that could supply any individual organisms with daily needs; stability over thousands of years could help an ecosystem to flourish; and changes over longer timescales (still very rapid compared with usual geological change) would force adaptation, perhaps serving as a driver for biological evolution.

In order to help visualize how the physical character of Europa might support life, a hypothetical ecosystem within a crack in the ice is described in Section 15.4. But first we review the appearance of Europa and interpret what is seen in the context of the effects of tides.

15.2 The surface of Europa

15.2.1 Global scale

Europa's surface is composed predominantly of water ice according to reflectance spectra (Pilcher *et al.*, 1972), and thus the moon would appear to the human eye as a nearly uniform white sphere, 1565 km in radius. The gravitational figure measured by spacecraft indicates that an outer layer as thick as ~150 km has the density of H_2O (Anderson *et al.*, 1998). Much of that thick layer is probably liquid water according to estimates of tidal heating (Section 15.3.3), while the surface is frozen due to radiative cooling.

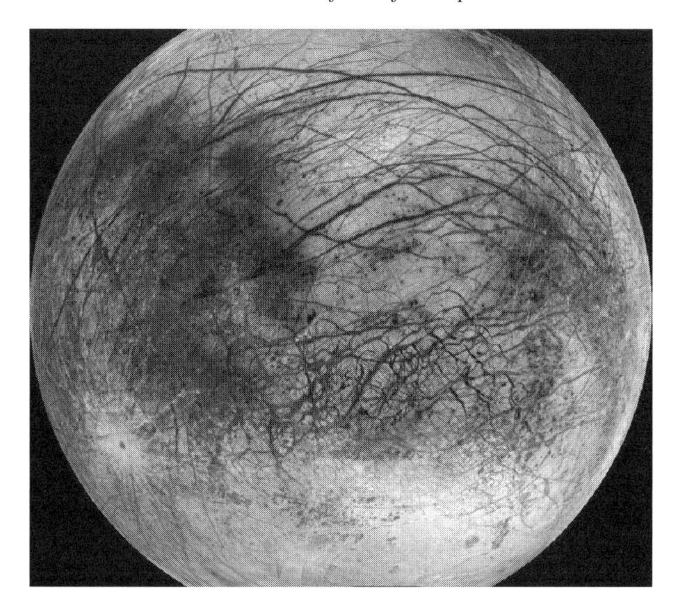

Fig. 15.1. A full-disk view of Europa shows large-scale lineaments indicative of tectonic cracks, dark splotches marking thermally created chaotic terrain. Craters are rare, although a prominent rayed crater, Pwyll, is evident to the left. The equator runs across the centre here. The dark wedge-shaped and tightly curved features just south of the equator are mostly dilational bands.

The surface does contain more than pure water ice. Even global-scale pictures (with resolution >10 km per pixel) taken with Voyager and Galileo spacecraft cameras show (when the contrast is enhanced as in Figure 15.1) orange-brown markings of still-unidentified substances, which likely include hydrated salts (McCord *et al.*, 1998a) and sulphur compounds (Carlson *et al.*, 1999). The patterns include splotches ranging from tens to hundreds of kilometres across and a network of narrow lines. When the splotches and lines are observed at high resolution (as discussed below), they prove to represent the two major resurfacing processes on Europa: formation of chaotic terrain and tectonics, respectively.

The resurfacing has been rapid and relatively recent as evidenced by the paucity of craters. Apparently, impacts have had a minimal role in shaping the current surface. In Figure 15.1, for example, only crater Pwyll, with its extensive white rays, is readily evident at this scale (lower left). The paucity of craters tells us that the current surface must be very young. It is continually reprocessed at such a great rate that most of the observable terrain, structures, and materials have probably been in place < 50 My (Zahnle *et al.*, 2003).

Even from the global appearance we see indications that the surface has been reworked by stress and heat, forming the tectonic and chaotic terrain, respectively. This chapter shows how these processes modified the surface and have provided, in diverse ways, access to the ocean. The relatively dark, orange-brown material that marks sites of these processes is indistinguishable whether it appears along a tectonic lineament or around a patch of chaotic terrain. It may represent concentration of impurities by thermal effects due to the near exposure of warm liquid or be the most recently exposed oceanic substances. All this activity is rapid, recent, and thus likely on-going.

15.2.2 Tectonics

15.2.2.1 Ridges and cracks

In regional-scale images (~1 km per pixel), many of the global-scale dark lines resolve into double lines (Figure 15.2). Counting the brighter zone between the dark lines, Voyager scientists named these features 'triple bands' (Lucchitta and Soderblom, 1982). Figure 15.2 shows an X-shaped intersection of global-scale triple bands that is located to the left hand side of Figure 15.1. In both figures, rays from crater Pwyll 1000 km to the south are visible across this region. Figure 15.2 has nearly vertical illumination aligned with the camera's point of view (i.e., low phase angle), so the visible features represent albedo and colour

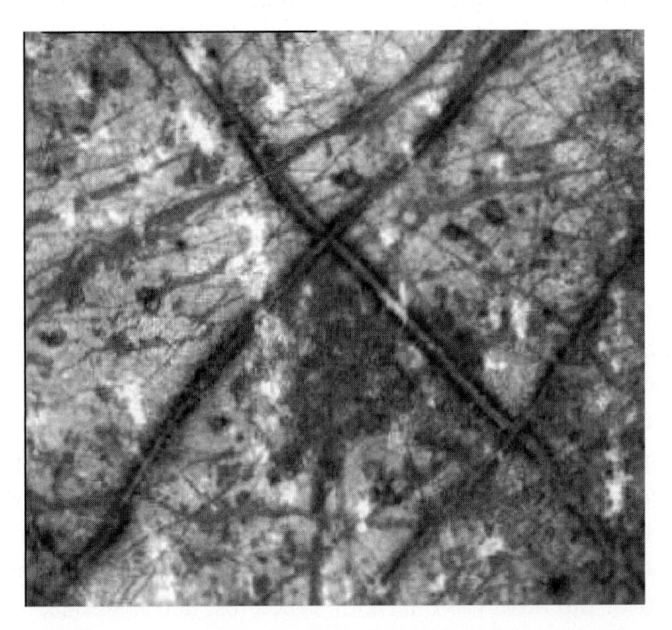

Fig. 15.2. A higher-resolution view of the prominent intersection of lineaments at the left of Figure 15.1 shows that the lines resolve into double dark lines known as 'triple bands'. The dark splotch just south of the intersection is Conamara Chaos, which is about 80 km across. The white streaks are rays from crater Pwyll, 1000 km to the south.

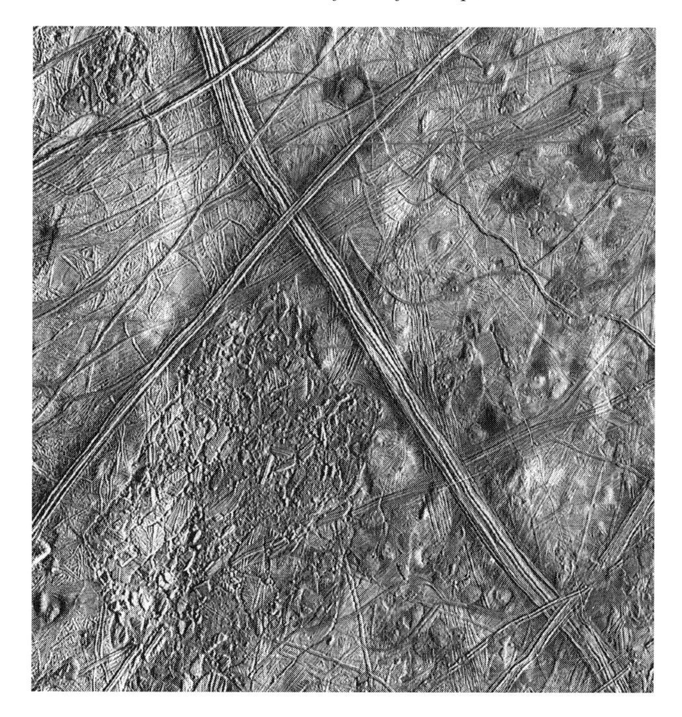

Fig. 15.3. The area in Figure 15.2 is shown here with oblique illumination and resolution of ∼200 m, showing more morphological detail. The lineaments are complexes of double ridges. The double dark lines in Figure 15.2 prove to be only the faint darkening alongside the ridge complexes. Conamara is the archetype of chaotic terrain: rafts of older surface, displaced in a lumpy matrix.

variations. The same area viewed at ∼200 m resolution and with more oblique illumination (Figure 15.3), reveals the morphology of the large-scale lineaments, showing them to be complexes of ridges. Each ridge complex consists of sets of ridges, roughly parallel, although crossing or intertwined in various places.

How does this structural form (in Figure 15.3) relate to the appearance at lower resolution and vertical illumination in Figure 15.2? As shown in Figure 15.3, along either side of the major ridge complexes the adjacent terrain is slightly darker than average. This subtle, diffuse darkening is what shows as the thin dark lineaments on global scale images (e.g., Figure 15.1) or as the double dark components of the 'triple bands' at kilometre resolution (e.g., Figure 15.2). The ridge systems, which are structurally the most significant characteristic of the global-scale lineaments, lie between the dark margins and at lower resolution can only be seen as the relatively bright centre line of the triple bands.

The dark coloration on the margins, which on global and regional scale images was most prominent, proves to be quite subtle in the higher resolution images

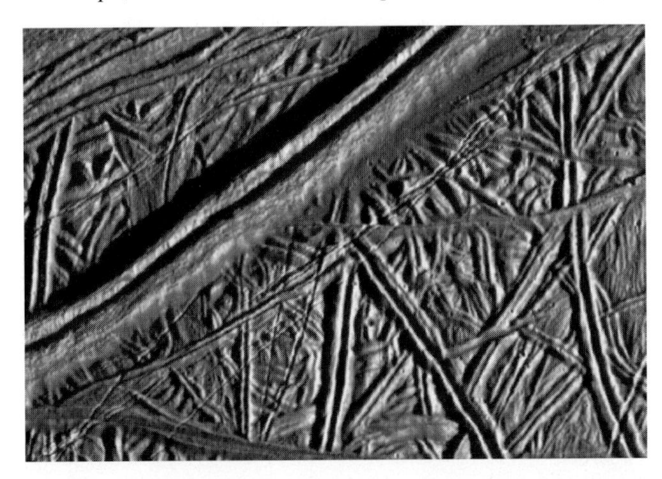

Fig. 15.4. Densely ridged terrain at the top of Figure 15.3, here at much higher resolution. The large double ridge is about 2 km across.

with oblique illumination (e.g., Figure 15.3). In general, the dark areas do not correlate with any single type of morphological structure, although they do tend to be somewhat smoother than average. The association with ridge complexes is very important from the perspective of planetary geochemistry: the ridges probably mark cracks along which oceanic substances have been able to reach the surface; thus, these margins are the first of several examples of darkening associated with ocean water near or at the surface.

Examination of the ridge complexes that make up the global lineaments (e.g., in Figure 15.3) shows them to be composed of multiple sets of double ridges. In fact double ridges are ubiquitous on Europan tectonic terrain. While ridges appear in several varieties (Greenberg *et al.*, 1998), the common denominator seems to be the simple double-ridge system. Numerous examples, besides those that make up the global-scale lineaments, can be seen in Figure 15.3. Figure 15.4 shows a close up of terrain (just north of that shown in Figure 15.3) densely packed with double ridges. Figure 15.5 shows double ridges in the highest-resolution image ever taken of Europa, with a road-cut-like cross-section that offers a view of the interiors of some ridges.

As ridges formed across the surface, they covered over the preexisting terrain. A single ridge pair can often cover over 1000 km^2. In some places, ridges have crossed over ridges repeatedly in the same area creating densely ridged terrain like that shown in Figure 15.4.

Such areas often appear smooth in images with resolution inadequate to resolve the individual ridges. Geologists have mapped such smooth-seeming terrains as 'background plains' (e.g., Prockter *et al.* (1999)), promoting an assumption that

Fig. 15.5. The highest-resolution image taken by Galileo (6 m per pixel). The view is oblique (not straight down) as if looking sideways at the ground from an aeroplane window. In this frame, a couple of ridge pairs come down from the upper left and are cut off by chaos in the foreground, giving a road-cut-style cross-section.

this is the oldest type of terrain on Europa (Greeley *et al.*, 2000). That terminology might suggest an initial slate-cleaning event, but such 'plains' are really only an artefact of resolution. There is no evidence that resurfacing has ever been other than a continual process nor that there has been any change in styles of resurfacing over time. Looking back in time, there is a geological horizon, but that limit probably represents the oldest features that are still recognizable from a continual history of resurfacing that started even earlier. There is no evidence that the time horizon represents a single slate-cleaning event.

There is considerable evidence regarding the mechanism that creates the double ridges. The global scale lineaments correlate reasonably well with theoretical tidal stress patterns (Section 15.3.2), indicating that they are associated with tensile cracks in the crust. Because these lineaments seem to comprise complexes of double ridges, it is reasonably assumed that simple double ridges are similarly associated with cracks. Moreover, it has been generally assumed that the double nature is due to ridges running along each side of a crack.

Identification of ridges with cracks has been reinforced by investigations of the locations, orientations, geometries, appearance at higher resolution, displacement

Fig. 15.6. An image from the Voyager spacecraft, showing cycloidal ridges: chains of arcs, with each arcuate segment ~100 km long. Running from near the lower left corner up to a dark parallelogram is Astypalaea Linea.

of adjacent crust, and sequences of formation over time. The studies have shown that these various characteristics can be explained in terms of tidal stress. For example, distinctive and ubiquitous cycloid-shaped lineament patterns (e.g., Figure 15.6) follow from the periodic variation of tidal stress during crack propagation (see Section 15.3.2). Cycloidal crack patterns also provide strong evidence for a liquid water ocean under the ice crust, because an ocean is required in order to give adequate tidal amplitude for these distinctive patterns to form (Hoppa *et al.*, 1999a).

15.2.2.2 Tectonic displacement

Strike-slip displacement (where the surface on one side of a crack has sheared past the surface on the other side, as along California's San Andreas fault) is common on Europa and has important implications (Tufts *et al.*, 1999; Hoppa *et al.*, 1999b; 2000; Sarid *et al.*, 2002). Observed examples of displacement fit a theoretical model driven by tides and requiring that cracks penetrate to a low-viscosity 'decoupling layer' such as a liquid ocean. Strike-slip generally occurs along ridges (e.g., Figure 15.7), rarely along any of the observed simple, ridgeless cracks. Thus, strike-slip displacements suggest that ridges may form along cracks that penetrate entirely through the solid ice crust to the liquid ocean. If so, then under much of Europa's tectonic terrain, the crust must have been thin enough (probably <10 km) to allow such penetration (Hoppa *et al.*, 1999a). Strike-slip displacement also has several other crucial implications, including evidence for non-synchronous rotation and for recent polar wander, as discussed in Section 15.3.2.

Surface displacement also includes opening of cracks, filled by broad bands on the surface (Figure 15.8). Recognizable in Voyager and Galileo low-resolution images (e.g., Schenk and McKinnon (1989), Pappalardo and Sullivan (1996)), at higher resolution most of these dilation bands exhibit similar characteristics, especially fine

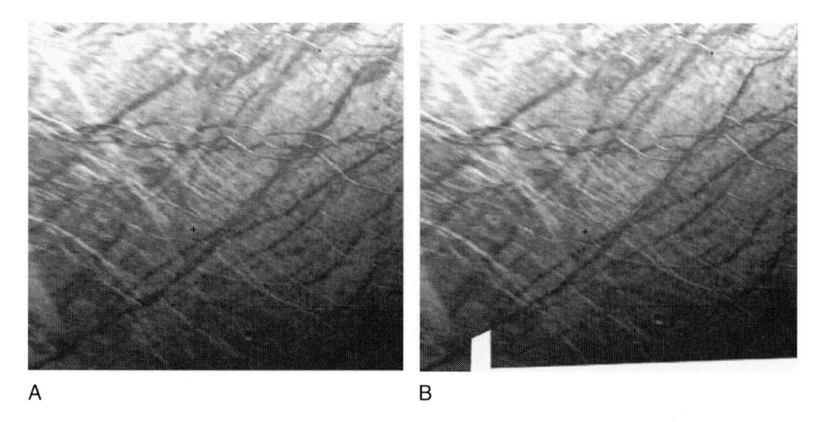

Fig. 15.7. A. A normal projection of Astypalaea Linea (cf. Figure 15.6), showing the wispy bright lines that end abruptly at Astypalaea. Also note the dark parallelogram at one end. B. A reconstruction shows that Astypalaea is a strike-slip (shear) fault. Note (going back in time) the realignment of the wispy white lines and the closing of the parallelogram.

parallel ridges within most bands, usually symmetrical about the centreline. Most bands are reconstructable (e.g., Figure 15.8), demonstrating their dilational nature and they compose a structural continuum with the forms of ridge systems (Tufts *et al.*, 2000). The mobility represented by bands is consistent with the underlying liquid ocean inferred from the mobility of the strike-slip displacement mechanism and from the tidal amplitude of the cycloidal cracks.

The large amounts of new surface created during dilation of the bands, which has been going on as far back in time as we can recognize in the tectonic record, raised the question of how the surface-area budget has been balanced. Evidence for the processes that remove surface on Earth (plate subduction or Himalaya-like mountain building) is not found on Europa. Nevertheless, sites of surface convergence have been found by reconstruction of surface plate motions (Sarid *et al.*, 2002). These sites display bands with internal striations, rounded boundaries, shallow lips at their edges, and only low topography. Unlike on Earth, it appears that there was not much solid material to resist the convergence of surface plates.

15.2.3 Chaotic terrain

The dark splotches in the low-resolution global views (e.g., Figure 15.1) represent the other major type of terrain on Europa, *chaotic*, and correspondingly the other major type of resurfacing, thermal. Resolution of $\sim$200 m per pixel with appropriate illumination reveals the character of chaotic terrain. The archetypal example is Conamara Chaos, just south of the intersection of the crossing global lineaments shown in Figures 15.2 and 15.3. Typical of chaotic terrain, this region has been

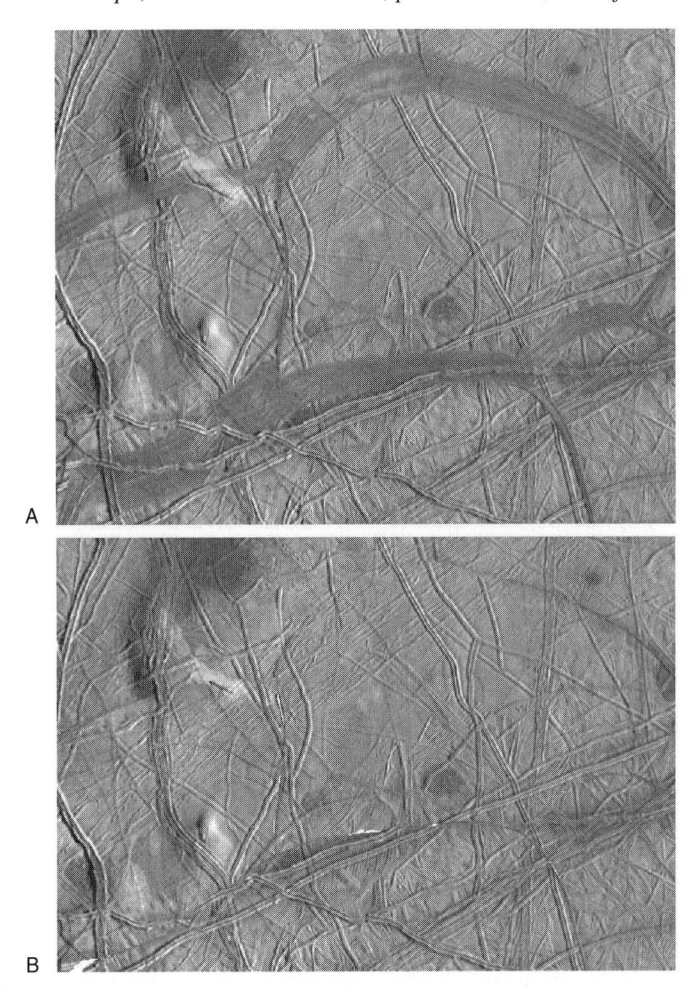

Fig. 15.8. A. Examples of dark dilational bands (also visible in Figure 15.1). B. Reconstruction of the bands shows them to be dilational, as plates of the surface have separated, allowing in-filling of fresh material.

thermally disrupted, leaving a lumpy matrix with somewhat displaced rafts, on whose surfaces fragments of the previous surface are clearly visible. Conamara, like other chaotic areas, has the appearance of a site at which the crust had melted, allowing blocks of surface ice to float to slightly displaced locations before refreezing back into place (Carr *et al.*, 1998). Similar features are common in Arctic sea ice and even frozen lakes in terrestrial temperate regions, where the underlying liquid has been exposed. Thus chaotic terrain, like tectonics, appears to represent significant exposure of the liquid ocean at the surface (Greenberg *et al.*, 1999).

While Conamara is a good example, some of the most important results about chaotic terrain were only revealed by identifying and characterizing all chaos regions as completely as possible with the Galileo data set (Greenberg *et al.*, 1999; Riley *et al.*, 2000), while quantitatively accounting for observational selection effects.

Chaotic terrain is very common, covering nearly half of Europa's surface (Greenberg *et al.*, 1999; Riley *et al.*, 2000). The other half is covered by tectonic terrain (the densely packed and overlaid ridges, cracks, and bands). Chaotic terrain is not necessarily young, and it probably formed at various times and places throughout Europa's geological history. Examples of chaotic terrain are found in all degrees of degradation, as cracking, ridge formation, and thermal effects destroy the older terrain. In general, it is more difficult to recognize older or smaller patches of chaotic terrain, which can introduce an observational selection bias favouring younger and larger examples, as demonstrated by Hoppa *et al.* (2001a). Thus, for example, impressions that chaos is systematically younger than tectonic terrain (e.g., Prockter *et al.* (1999), Greeley *et al.* (2000), Figueredo and Greeley (2004)) are probably artefacts of this observational bias. No evidence has yet been developed for any systematic change over time in frequency or style of chaos formation. In fact, where high-resolution images are available, some chaotic terrain is older than most of the tectonics nearby (e.g., Riley *et al.* (2006)).

15.3 Tides

The activity evidenced on Europa's surface is driven by tides. Tides generate periodic global stresses that crack and displace surface crust, as the thin shell of ice must accommodate to the changing global shape of the ocean below it. Moreover, tidal friction provides the dominant internal heat source, which keeps most of the thick water layer in the liquid state, as well as allowing occasional local or regional melting of the ice crust, either all the way through or just varying the ice thickness. Tides also generate rotational torque that (as discussed below) tends to maintain non-synchronous rotation, which is especially plausible for Europa given that tidal heating makes it unlikely that there are frozen-in asymmetries capable of locking onto the direction of Jupiter. Non-synchronous rotation (though very nearly synchronous) also may add important components to the tidal stress as the direction of Jupiter varies all the way around relative to the body of the satellite. Over the long term, tidal deformation of the resonantly coupled Galilean satellites contributes to orbital change, which in turn modifies orbital eccentricities, which changes the amplitudes of periodic (orbit driven) tides.

All of these tidal effects, especially rotation (Section 15.3.1), stress (Section 15.3.2), and heat (Section 15.3.3), not only explain much of what we observe on the surface, but they also provide conditions and rates of change that may be able to support survival and evolution of life.

If Europa were in a circular orbit, rotating synchronously (with one face always toward Jupiter), there would be no variable tide. In general, even if a satellite's orbit were eccentric, tides would quickly circularize it, and then, if it rotated non-synchronously, tides would quickly torque it to synchroneity. If that were the case for Europa, it would be a heavily cratered, geologically inactive body.

However, the orbit of Europa is forced to remain eccentric by an orbital resonance with other satellites, with enormous consequences for tidal effects and the geology and geophysics. The resonance involves the orbits of Io, Europa, and Ganymede (Laplace, 1805). Europa's orbital period (and also its day) is about 3.6 days long. The orbital periods of these three satellites are locked into an orbital ratio (or 'commensurability') close to 1:2:4. Thus conjunction of Io and Europa (where they line up on one side of Jupiter) is locked to the direction of Europa's orbital apocentre (furthest point from Jupiter in its eccentric orbit). Moreover, conjunction of Europa and Ganymede is always aligned with Europa's pericentre (as reviewed by Greenberg (1982, 1989)).

Because of the periodicity of these geometrical configurations, these relationships enhance the mutual gravitational perturbations among the satellites. The effect is to maintain the alignments of conjunction, and to pump the eccentricities of the three satellites. The forced eccentricity has a value that depends on exactly how close the ratio of orbital periods is to the whole number ratio. The closer the system is to the exact commensurability, the greater are the forced eccentricities of the satellites.

The origin of the resonance itself may be a result of tidal effects that modified orbital periods until the system became locked in resonance. More likely, the resonance formed as orbits evolved during the formation of the Jovian system, and then evolved significantly under the influence of tides. For a complete discussion, see Greenberg (2005). However the resonance formed, it currently drives the significant eccentricity of Europa.

The orbital eccentricity causes tides on Europa to change periodically over the orbital period of 3.6 days. Over the course of each orbit, Jupiter not only approaches and recedes from Europa, but it also advances and regresses in Europa-centric longitude. The tidal elongation of the figure of Europa thus changes both in magnitude and in direction. The change in orientation of the elongation of Europa does not represent rotation of the body, but rather a 'remolding' of the figure of Europa in response to the changing direction of Jupiter. Those periodic tidal variations are responsible for the character of Europa's surface. They stress the crust causing cracks, they provide frictional heat for melting the ice, they induce rotational torque yielding slightly non-synchronous rotation and modifying stress patterns, and they drive long-term orbital change. These effects are only operative because the resonance forces the orbit of Europa to remain eccentric.

15.3.1 Tidal torque: rotation effects

The lag in the tidal response to Jupiter creates a torque on Europa's rotation which, averaged over each orbit, tends to drive the rotation to a rate slightly different from synchronous. That rate depends on details of the tidal response, but according to theory is expected to be only slightly faster than synchronous (Greenberg and Weidenschilling, 1984).

Even given Europa's orbital e, the rotation might be synchronous if a non-spherically symmetric, frozen-in density distribution (like that of the Earth's Moon) were locked to the direction of Jupiter. Given that Europa is substantially heated by tidal friction (see below), it may not be able to support such a frozen-in asymmetry. It is also conceivable that the silicate interior is locked to the direction of Jupiter by a mass asymmetry, while the ice crust, uncoupled from the silicate by an intervening liquid water layer, rotates non-synchronously due to the tidal torque.

Even a completely solid Europa would rotate non-synchronously as long as mass asymmetries are small enough. Non-synchronous rotation would not in itself imply the existence of an ocean. However, both the existence of an ocean and non-synchronous rotation are made possible by the substantial tidal heating.

Observational evidence places some constraints on the actual rotation rate. A comparison of Europa's orientation during the Voyager 2 encounter with that observed 18 y later by Galileo showed no detectable deviation from synchroneity. Hoppa *et al.* (1999c) found that any deviation must be small (as predicted), with a period $>$12 000 y, relative to the direction of Jupiter.

In principle, further evidence could come from changes in the orientations of cracking over time, as terrain moved west-to-east through the theoretical tidal stress field. The possibility that the observed tectonic patterns might contain some record of such reorientation was noted by Helfenstein and Parmentier (1985) and McEwen (1986). Studies of cross-cutting sequences of lineaments from early Galileo spacecraft images suggested just such a variation in the azimuths over time (Geissler *et al.*, 1998).

However, more recent studies using higher-resolution images and exploring several regions suggest that the systematic result was an artefact of an incomplete data set (Sarid *et al.*, 2004, 2005). The sequence of azimuth formation is not continuous and can only be consistent with non-synchronous rotation if one or two cracks (at most) form during each cycle of rotation. That result is reasonable, because once one crack forms in a given region, it would relieve further tidal stress until it has rotated into a different stress regime. This change over time is important in the evolution of the hypothetical ecosystem discussed in Section 15.4.

Other lines of evidence do suggest non-synchronous rotation. As the tectonic effects of tidal stress began to be better understood, various features were found to have likely formed further west than their current positions, including strike-slip faults (Hoppa *et al.*, 1999b, 2000) and cycloidal crack patterns. From the latter, Hoppa *et al.* (2001b) inferred a non-synchronous period of $<$250 000 y, but $>$12 000 y based on the Voyager–Galileo comparison discussed above. There is also similar evidence from strike-slip faults for substantial 'polar wander' within the past few million years, in which the ice shell, uncoupled from the interior, was reoriented relative to the spin axis (Sarid *et al.*, 2002). Supporting evidence for polar wander comes from asymmetries in distributions of small pits, uplifts, and chaotic terrain (Greenberg, 2005).

15.3.2 Tidal stress: tectonic effects

15.3.2.1 Crack patterns

Global-scale lineament patterns (e.g., those in Figure 15.1) correlate roughly with tidal stress patterns (i.e., major global- and regional-scale lineaments are generally orthogonal to directions of maximum tension), showing that the lineaments result from cracking (e.g., Greenberg *et al.* (1998)). Because the morphology of these lineaments involves large-scale ridge systems, we infer that ridges result from cracks.

A distinct and ubiquitous shape for cracks and ridges on Europa is a chain of arcs, known as cycloids, first discovered by Voyager (Figure 15.6) (Smith *et al.*, 1979; Lucchitta and Soderblom, 1982). Many of these are marked by double ridges; cycloidal cracks that have not yet developed ridges are also common; many dilational bands appear to have initiated as arcuate cracks; even the major strike-slip fault Astypalaea appears to have begun as a chain of arcuate cracks. Typically these features have arcuate segments $\sim$100 km long and the chains run for $\sim$1000 km.

This distinctive style of cracking probably occurs as a result of the periodic changes in tidal stress due to the orbital eccentricity (Hoppa *et al.*, 1999a). Suppose a given crack initiates at a time when the tidal tension exceeds the local strength of the ice. It will then propagate across the surface perpendicular to the local tidal tension. On a timescale of hours, as Europa moves in its orbit, the orientation of the stress varies. Hence the crack propagates in an arc until the stress falls below a critical value necessary to continue cracking. Crack propagation comes to a halt. A few hours later, the tension returns to a high enough value, now in a different direction, so crack propagation resumes at an angle to the direction at which it had stopped. Thus a series of arcs is created (each corresponding to one orbit's worth of propagation), with cusps between them.

The characteristics of observed arcuate features can be reproduced theoretically, using strength values plausible for bulk ice ($\sim10^5$ Pa) and a speed of propagation of a few kilometres per hour. Because each arc segment represents propagation during one Europan orbit, a typical cycloidal crack must have formed in about one (terrestrial) month. In general, the orientation and curvature of the cycloidal chain as a whole is determined by the location at which the crack formed. Thus one can determine the formation location by fitting the model to those characteristics; such modelling (Hoppa *et al.*, 2001b; Hurford *et al.*, 2007) constrains Europa's rotation (Section 15.3.1).

In all cases a subsurface ocean was required in order to have adequate tidal amplitude to form the observed cycloidal geometries. If there were no ocean, and the tidal amplitude were correspondingly small, cracking would require the ice to be even weaker. In that case, the shapes of cycloids would be distinctly different from what is seen on Europa. Thus the presence of cycloidal lineaments, and their correlation with theoretical characteristics, provided the first widely convincing evidence for the hypothesized liquid water ocean (Hoppa *et al.*, 1999a). Because many cycloids formed during the past few million years, they suggest that the ocean still exists. More recently, studies of Europa's effect on the Jovian magnetosphere provided corroborating evidence for a current liquid ocean (Khurana *et al.*, 1998; Zimmer *et al.*, 2000).

15.3.2.2 Tidally active cracks

Once a crack forms, it relieves the tidal stress in its vicinity. Further cracking is unlikely as long as the crack remains 'active', that is until it anneals so that it can support transverse tension again. During the time that the crack is active and unannealed, the periodic tidal distortion of the satellite's icy shell results in a working of the crack. That is, depending on time, location, and crack orientation, there may be a regular opening and closing of the crack and/or a periodic shearing along the crack. The repeated opening and closing would pump ice and slush from the ocean to the surface, creating the ubiquitous double ridges and their modified forms. Moreover, such working appears to have also driven strike-slip displacement along many cracks on Europa, which in turn indicates that cracks penetrate all the way down to the liquid ocean, a strong constraint on ice thickness. The next two subsections describe how the working of active cracks may build double ridges and drive strike-slip displacement.

Ridge formation The regular opening and closing of a crack can build a ridge by a process of tidal extrusion (Greenberg *et al.*, 1998), which operates as illustrated in Figure 15.9: as the tides open cracks (Figure 15.9A), water flows to the float line, where it boils in the vacuum and freezes owing to the cold. As the walls of the crack close a few hours later (Figure 15.9B), a slurry of crushed ice, slush, and water is squeezed to the surface and deposited on both sides of the crack (Figure 15.9C, D).

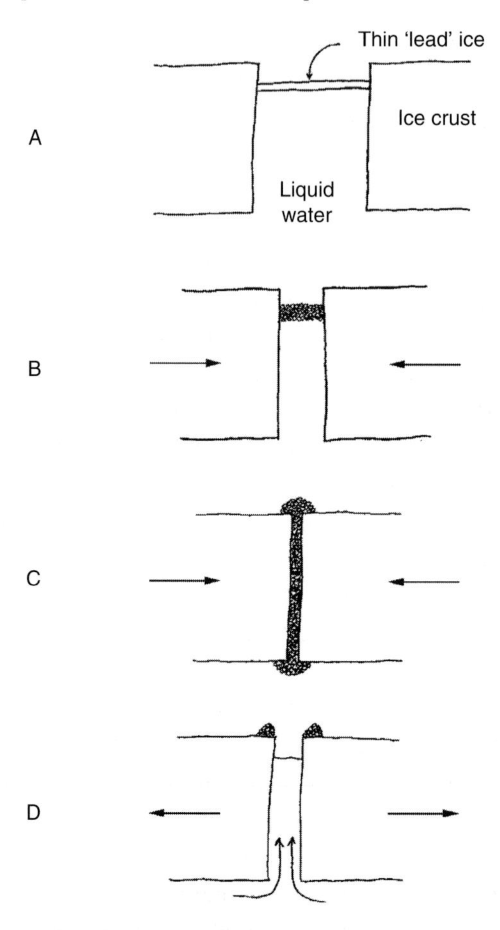

Fig. 15.9. A schematic of the diurnal steps of ridge-building as described in the text.

Given the frequency of the process, ridges of typical size (100 m high and 1 km wide) can grow quickly, in as little as 20 000 y. Identification of a mechanism for ridge formation that is so fast is important, because many generations of Europan ridges have formed over the past 10^8 y (the maximum age of the surface on the basis of the paucity of craters); also, each ridge probably had to form well within one rotational period.

Thus ridges appear to form by a process in which active cracks are bathed in liquid water (including whatever impurities the ocean contains) on a daily basis. Any localized heating due to friction at these cracks would further maintain this process. Fagents *et al.* (2000) have proposed that the darkening that flanks major ridge systems may be related to heat along the lineament, allowing impurities in the ice to be concentrated at the surface. Given the exposure of liquid water during the ridge-formation process described above, the dark material flanking major ridge systems is plausibly associated with oceanic substances. This material might have been emplaced at the margins of ridges as oceanic fluid spread through porous ridges (Greenberg *et al.*, 1998), or geyser-like plumes might have sprayed the surface during the daily exposure of liquid water to the vacuum (Kadel *et al.*, 1998).

Another model of ridge formation assumes that long linear diapirs (low-density blobs in the viscous crust) have risen from below the surface, tilting it upward along the sides of cracks (Head *et al.*, 1999). That model was predicated on hypothetical solid-state diapirism in a tens-of-kilometres thick layer of ice, and assumes uniform upwelling over hundreds of kilometres for each major ridge pair, which seems implausible. Moreover, the linear diapir model is inconsistent with the observed properties of ridges in a variety of ways (e.g., Greenberg (2005)).

Variations on the theme of double ridge formation are common (Greenberg *et al.*, 1998; Greenberg, 2005). The multiple-ridge complexes that compose the centre stripes of many of the long triple bands probably formed as parallel, lateral cracks due to the weight of earlier ridges on the thin ice. These lateral cracks then become activated by tidal working and build their own ridges. Other ridges are wider and symmetrically lineated along their length, and are found where the adjacent terrain has separated somewhat. The dilation may have been due to incomplete extrusion of solid ice during daily crack closure (e.g., Figure 15.9C), which left jammed ice to gradually pry the cracks open.

Strike-slip displacement Strike-slip displacement (e.g., Figure 15.7) is common and widely distributed, often with long extent and large offset displacement (Schenk and McKinnon, 1989; Tufts, 1998; Tufts *et al.*, 1999; Hoppa *et al.*, 1999b, 2000; Sarid *et al.*, 2002). Examples include a 170-km-long fault in the far north with >80 km of shear offset, and a long, bent cycloidal crack whose shear offset indicates that a cohesive plate >400 km across rotated by about 1°. The 40 km shear offset fault along 800-km-long Astypalaea Linea near the south pole (Tufts *et al.*, 1999; see Figures 15.6 and 15.7) was originally cycloid-shaped with double ridges. Under shear, ridges on opposite sides of the cracks moved in opposite directions, like trains passing on separate tracks. Those parts of the original crack that were oblique to the shear direction were pulled apart, yielding parallelograms of in-filled material, presumably slush from below. These sites show in detail the structure at a location where we know the crust was pulled apart, an important point of reference for interpretation of dilational bands (Section 15.2.2).

Tidal stress in the Astypalaea region drives the shear displacement by a mechanism analogous to walking (Tufts, 1998; Hoppa *et al.*, 1999b). Over each orbital period, the tide goes through a sequence starting with tensile stress across the fault, followed 21 hours later by right-lateral shear, followed 21 hours later by compression across the fault, followed 21 hours later by left-lateral shear. Because the left-lateral shear stress occurs after the fault is compressed, friction at the crack may resist displacement, while right-lateral stress occurs immediately after the crack is opened by tension. Thus tides drive shear in the right-lateral sense in a ratcheting process. The mechanism is similar to walking, where an animal repeatedly separates a foot from the ground (analogous to a crack opening), moves it forward (analogous to shear displacement), presses it to the ground (analogous to compressive closing of a crack), and pushes it backward (analogous to the reverse shear phase), resulting in forward motion. On Europa this process moves plates of crust.

Surveys of strike-slip offsets over much of Europa show that they fit the predictions of the tidal-walking theory quite well (Hoppa *et al.*, 2000; Sarid *et al.*, 2002). Furthermore, the fit is even better if one takes into account some non-synchronous rotation and at least one polar wander event. Another result of the surveys of strike-slip was identification of large-plate displacements that showed where surface convergence had taken place (Sarid *et al.*, 2002). The latter result is important because it shows how the surface-area budget may be balanced, even given the large amount of new surface created where cracks have dilated (Section 15.2.2).

The success of tidal walking in explaining the observed faulting argues strongly that the decoupling layer over which the surface plates slide is a fluid. Penetration to such a layer is required for the daily steps in the 'tidal-walking' model, because a fluid can deform as necessary on the short timescale of the orbit-driven tides. A thick, ductile, warm ice layer may, in principle, allow lateral displacement above it (Schenk and McKinnon, 1989; Golombek and Banerdt, 1990), but no studies have yet shown that tidal walking could be viable in that case.

Thus strike-slip requires that cracks penetrate through most of the crust, which in turn requires that the ice be quite thin, because the tidal tensile stress available for creation of the cracks is not high, only about 40 kPa. Such low tensile stresses are overwhelmed by the compressive hydrostatic overburden pressure at a depth of only $\sim$100 m. These cracks could be driven to greater depths by the insertion of liquid water or in-falling solid material. Cracks may also go to greater depth because additional stress is concentrated at the tip of the crack. It is unlikely that cracks could penetrate more than a few kilometres on Europa. Thus our model of strike-slip displacement implies that Europa's ice crust overlies liquid water, that cracks readily penetrate from the surface to the liquid ocean, and therefore the ice must be fairly thin.

Furthermore, Hoppa *et al.* (1999b) noted that strike-slip offsets are along ridge pairs, not simple cracks, implying that ridges form along cracks that penetrate all the way down to liquid water. This result is consistent with the evidence that ridges form as a result of the working of cracks that link the ocean to the surface (Section 15.3.2).

The observations and theory of strike-slip displacement have helped constrain the physical character of Europa in ways that are critical to the potential setting for life, including evidence for non-synchronous rotation, for polar wander, for penetration of cracks to the ocean, for significant crust displacement driven by the tidal walking process, and for zones of surface convergence.

15.3.3 Tidal heating

15.3.3.1 Liquid water ocean with an ice crust

As the figure of a satellite is distorted periodically due to tides, friction may generate substantial heat. Even with assumptions about unknown parameter values (especially the dissipation parameter Q) that would tend to minimize heating, tidal friction could maintain a liquid water ocean (e.g., Cassen *et al.*, 1979; Squyres *et al.*, 1983), but only if one already exists: without a global ocean, the tidal amplitude would be inadequate to generate the heat to create one. However, the range of plausible parameters could admit much higher heating rates: With plausible material parameters, the tidal dissipation rate could be $>6 \times 10^{12}$ W, or >0.2 W m^{-2} at the surface (O'Brien *et al.*, 2002). In these estimates, the bulk of the heating is in the rocky mantle under the ocean. Geissler *et al.* (2001) noted that with partial melting of silicate, dissipation could be as much as 9×10^{12} W, or >0.3 W m^{-2} at the surface.

Even though such a total heat flux is very high from the point of view of geophysical implications, it is below the current limit of detectability, swamped out by the reradiated solar energy (Spencer *et al.*, 1999).

A heat flux >0.2 W m^{-2} would imply a steady-state conductive ice layer thinner than $\sim$3 km. Convection is probably not possible in such thin ice. On the other hand, the ice could be much thicker and still transport the same amount of heat, if and only if it were convective. Thus either a thin-ice model (thinner than 10 km) with conductive heat transport or a thick ice model (thicker than $\sim$20 km) with convective heat transport could be consistent with the plausible heat flux value estimated above. The thickness required for convection depends on unknown material properties. It must be greater than $\sim$20 km (McKinnon, 1999; Wang and Stevenson, 2000), but convection may require ice so thick that a liquid ocean would be precluded (Spohn and Schubert, 2001), a result that favours the thin conducting ice layer.

Thermal transport models may still be too dependent on uncertain parameters for them to definitively discriminate between thin conductive or thick convective ice. Even if the relevant parameters were known, both possibilities might provide a stable steady state. In that case, Europa's actual current physical condition may depend on its thermal history. If the heating rate has increased toward its current value, the ice would probably be thick and convecting. If the heating rate has decreased from significantly higher rates, then the current ice would more likely be thin and conducting. Because tidal heating is driven by orbital eccentricity, the history of Europa's orbit may be crucial.

15.3.3.2 Formation of chaotic terrain

Chaotic terrain (Section 15.2.3) has the appearance of sites of melt-through from below, consistent with the thin crust that we infer from the tidal-tectonic model (Greenberg *et al.*, 1998, 1999). The characteristics of chaos, surveyed by Riley *et al.* (2000), including the rafts, the relations with preexisting ridges, the types of shoreline, the apparent topography, and the association with dark material are all explained by melting from below, so that a lake of liquid water is exposed for some finite time before refreezing (Greenberg *et al.*, 1999). Moreover, morphological similarities between chaotic terrain and craters that appear to have punctured the crust reinforce the argument that both represent penetration to liquid.

Dark halos seen around chaotic terrain are also consistent with our interpretations of both ridge formation and chaos formation as processes involving exposure of liquid water. The dark halos are analogous to the diffuse dark margins along major ridge systems (yielding the 'triple band' appearance at low resolution). The dark material is spectrally indistinguishable in the two cases (e.g., Fanale *et al.* (1999)). It is consistent in both cases for the dark material to have been brought up from the ocean and deposited around sites of temporary oceanic exposure, especially given that it includes hydrated salts (McCord *et al.*, 1998a). (The darkening agents are unknown, but may include various organic and sulphur compounds.)

Only modest, temporary concentrations of tidal heat are required for substantial melt-through. For example, if a relatively modest 1% of the total internal heat flux were concentrated over an area about 100 km across, the crust would melt through in only a few thousand years, with broad exposure over tens of kilometres wide in only $\sim 10^4$ y (O'Brien *et al.*, 2002). The modest concentrations needed for such melt-through might originate at volcanic vents. Thermal plumes within the ocean could keep the heat localized as it rises to the ice (Thomson and Delaney, 2001; Goodman *et al.*, 2004).

Study of chaotic terrain implies that melt-through occurs at various times and places. It follows that the thickness of ice is quite variable with time and place. Chaos formation has been common, currently covering nearly half the surface (Riley *et al.*, 2000), and continual, displaying a wide range of degrees of degradation by subsequent cracking and ridge formation (Greenberg *et al.*, 1999). The sizes of chaos patches range from over 1000 km across down to at least as small as recognizable on available images, ~1 km across. Persistent claims that chaos is relatively young are based on an observational artefact: old or small examples are harder to see. In fact, over the surface age, formation of areas of chaotic terrain has been frequent and continually interleaved with tectonics (Riley *et al.*, 2006), so at any given location melt-through has occurred at least every few million years.

15.4 The permeable crust: conditions for a Europan biosphere

The two major terrain-forming processes on Europa are: (1) melt-through, creating chaos, and (2) tectonic processes of cracking and subsequent ridge formation, dilation, and strike-slip. These two major processes have continually destroyed preexisting terrain, depending on whether local or regional heat concentration was adequate for large-scale melt-through or small enough for refreezing and continuation of tectonism. The processes that create both chaotic and tectonic terrains (and probably cratering as well) all include transport or exposure of oceanic water through the crust to the surface.

Whether life was able to begin on Europa, or exists there now, remains unknown, but the physical conditions seem propitious (e.g., Greenberg (2005)). Change probably occurs over various timescales, which may provide reasonable stability for life, while also driving adaptation and evolution.

Cracks are formed in the crust due to tidal stress and many penetrate from the surface down to the liquid water ocean. The cracks are subsequently opened and closed with the orbit-driven tidal working of the body. Thus, on a timescale of days, water flows up to the float line during the hours of opening, and is squeezed out during the hours of closing. Slush and crushed ice are forced to the surface, while most of the water flows back into the ocean. The regular, periodic tidal flow transports substances and heat between the surface and the ocean.

At the surface, oxidants are continually produced by disequilibrium processes such as photolysis by solar ultraviolet radiation, and especially radiolysis by charged particles. Significant reservoirs of oxygen have been spectrally detected on Europa's surface in the form of H_2O_2, H_2SO_4, and CO_2 (Carlson *et al.*, 1999; McCord *et al.*, 1998b), while molecular oxygen and ozone are inferred on the basis of Europa's oxygen atmosphere (Hall *et al.*, 1995) and the detection of these compounds on other icy satellites (Calvin *et al.*, 1996; Noll *et al.*, 1996). Moreover, impact of cometary material should also provide a source of organic materials and other fuels at the surface, such as those detected on the other icy satellites (McCord *et al.*, 1998c). In addition, significant quantities of sulphur and other materials may be continually ejected and transported from Io to Europa.

The ocean probably contains endogenic substances such as salt, sulphur compounds, and organics (e.g., Kargel *et al.* (2000)), as well as surface materials that may be transported through the ice. Oceanic substances most likely have been exposed as the orange-brown darkening along major ridge systems and around chaotic terrain. While the coloration displayed in images taken at visible-to-near-infrared wavelengths is not diagnostic of composition, near-infrared spectra are indicative of frozen brines (McCord *et al.*, 1998a), as well as sulphuric acid and related compounds (Carlson *et al.*, 1999). The orangish brown appearance at visible wavelengths may be consistent with organics, sulphur compounds, or other unknowns. The ocean probably contains a wide range of biologically important substances.

The ice at any location may contain layers of oceanic substances, which are deposited at the top during ridge formation, and then work their way deeper as the ice maintains thickness by melting at the bottom (until a melt-through event resets the entire thickness as a single layer of refrozen ocean). This process can bury surface materials, eventually feeding (or recycling) them into the ocean. For oxidants, burial is especially important to prevent their destruction at the surface, and impact gardening may help with the initial burial (Phillips and Chyba, 2001).

Chemical disequilibrium among materials at various levels in a crack is maintained by production at the top and the oceanic reservoir at the bottom, while the ebb and flow of water continually transports and mixes these substances vertically during the tidal cycle. Transport is through an ambient temperature gradient $\sim$0.10 °C m^{-1}, from 0 °C at the base of the crust to about 170 °C colder at the surface.

The physical conditions in such an opening in the ice might well support life as illustrated schematically in Figure 15.10. No organisms could survive near the surface, where bombardment by energetic charged particles in the Jovian magnetosphere would disrupt organic molecules (Varnes and Jakosky, 1999) within $\sim$1 cm of the surface. Nevertheless, sunlight adequate for photosynthesis could penetrate a few metres, farther than necessary to protect organisms from radiation damage (Reynolds *et al.*, 1983; Lunine and Lorenz, 1997; Chyba, 2000). Thus, as long as some part of the ecosystem of the crack occupies the appropriate depth, it may be able to exploit photosynthesis. Such organisms might benefit from anchoring themselves at an appropriate depth where they might photosynthesize, although they would also need to survive the part of the day when the tide drains away and temperatures drop. Other non-photosynthesizing organisms might anchor themselves at other depths, and exploit the passing daily flow. Their hold would be precarious, as the liquid water could melt their anchorage away. Alternatively, some might be plated over by newly frozen water, and frozen into the wall. The individuals that are not anchored, or that lose their anchorage, would go with the tidal flow. Organisms adapted to holding onto the walls might try to reattach their anchors. Others might be adapted to exploiting movement along with the mixing

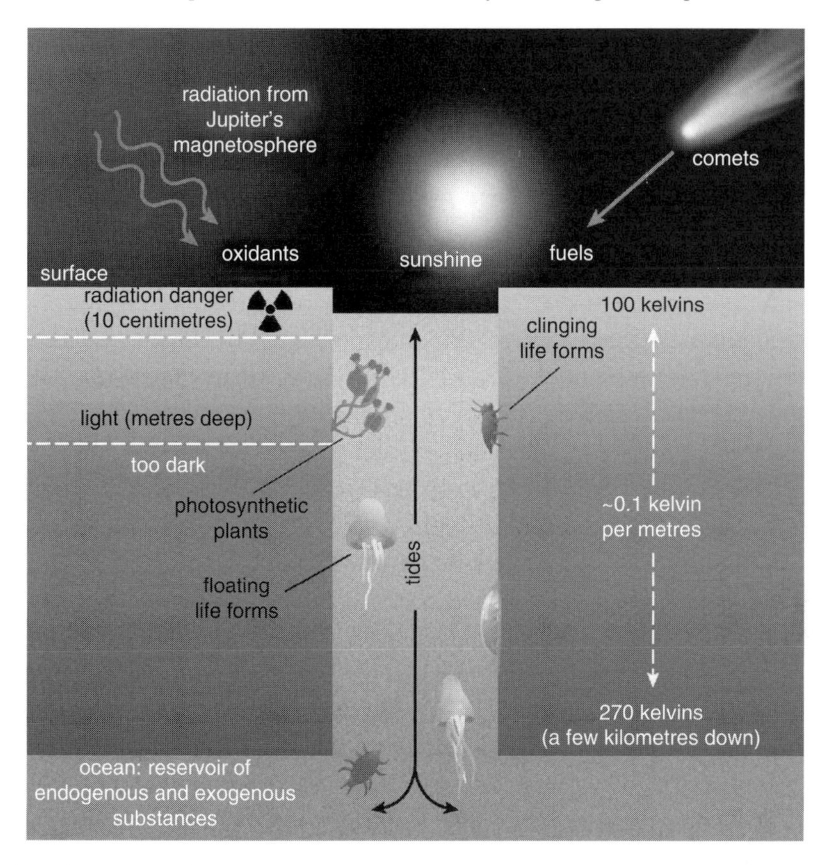

Fig. 15.10. Tidal flow through a working crack provides a potentially habitable setting, linking the surface (with its low temperature, radiation-produced oxidants, cometary organic fuels, and sunlight) with the ocean (with its brew of endo- and exogenic substances and relative warmth). Photosynthetic organisms (represented here by the tulip icon) might anchor themselves to exploit the zone between the surface radiation danger and the deeper darkness. Other organisms (the tick icon) might hold on to the side to exploit the flow of water and the disequilibrium chemistry. The hold would be difficult, with melting releasing some of these creatures into the flow, and with freezing plating others into the wall of the crack. Other organisms (jellyfish icon) might exploit the tides by riding with the flow. This setting would become hostile after a few thousand years as Europa rotates relative to Jupiter, the local tidal stress changes, and the crack freezes shut. Organisms would need to have evolved strategies for survival by hibernating in the ice or moving elsewhere through the ocean. (Artwork by Barbara Aulicino/*American Scientist*.)

flow. A substantial fraction of that population would be squeezed into the ocean each day, and then flow up again with the next tide.

A given crack is probably active over thousands of years, because rotation is nearly synchronous and the crack remains in the same tidal-strain regime, allowing for a degree of stability for any ecosystem or organisms within it. Over longer times, with non-synchronous rotation a given site moves to a substantially different tidal-strain regime in 10^3–10^5 y. Then the tidal working of any particular crack is likely to cease. The crack would seal closed, freezing any immobile organisms within it, while some portion of its organism population might be locked out of the crack in the ocean below.

For the population of a deactivated crack to survive, (a) it must have adequate mobility to find its way to a still active (generally more recently created) crack, or else or in addition (b) the portion of the population that is frozen into the ice must be able to survive until subsequently released by a thaw. At any given location, a melt event probably has occurred every few million years, liberating frozen organisms to float free and perhaps find their way into a habitable niche. Alternatively, in the timescale for non-synchronous rotation (less than a million years), fresh cracks through the region would cross the paths of the older refrozen cracks, liberating organisms into a niche similar to where they had lived before. Survival in a frozen state for the requisite few million years seems plausible, given evidence for similar survival in Antarctic ice (Priscu *et al.*, 1999). The need to survive change may provide a driver for adaptation and mobility, as well as opportunity for evolution.

We have shown that this model creates environments in the crust that may be suitable for life. Moreover, it provides a way for life to exist and prosper in the ocean as well, by providing access to necessary oxidants (Gaidos *et al.*, 1999; Chyba and Phillips, 2001) and linkage between oceanic and intracrust ecosystems. Oceanic life would be part of the same ecosystem as organisms in the crust. Components of the ecosystem might adapt to exploit suboceanic conditions, such as possible sites of volcanism. If there is an inhabited biosphere on Europa, it most likely extends from within the ocean up to the surface. While we can only speculate on conditions within the ocean, we have observational evidence for conditions in the crust, and the evidence points toward a potentially habitable setting.

This model is based on the surface manifestations of tectonic processing and chaotic terrain formation. The entire surface is very young, <2% of the age of the Solar System according to the paucity of impact craters. Because the tectonic and melting processes described above were recent, they may well continue today. Whether any organisms exist on Europa to exploit this setting is unknown.

Spacecraft exploration of Europa is likely to continue in the future. With the likelihood that the liquid ocean is linked to the surface in multiple ways, Europa's biosphere may be exposed at the surface, facilitating exploration and also contamination. Unless exploration is planned very carefully, we may discover life on Europa that we had inadvertently planted there ourselves. The advantage from the point of view of exploration is that, if landing sites are chosen wisely, it may not be necessary to drill down to the ocean in order to sample the deep. Oceanic materials, possibly including organisms, may be readily accessible at or near the surface. Thus, the search for life in that habitable zone may be less daunting than has been assumed in the past.

Acknowledgements

This chapter reviews a body of evidence produced by many people. Most of the data come from the Galileo mission. For help developing the interpretation presented here, I thank Paul Geissler, Greg Hoppa, Terry Hurford, Dave O'Brien, Jeannie Riley, Alyssa Sarid, and the late Randy Tufts. The images were processed at the Planetary Image Research Laboratory of the University of Arizona. This work was supported by a grant from NASA's Outer Planets Research program.

References

Anderson, J. D., *et al.* (1998). Europa's differentiated internal structure. *Science*, **281**, 2019–2022.

Calvin, W. M., Johnson, R. E., and Spencer, J. R. (1996). O_2 on Ganymede: spectral characteristics and plasma formation mechanisms. *Geophys. Res. Lett.*, **23**, 673–676.

Carlson, R. W., Johnson, R. E., and Anderson, M. S. (1999). Sulfuric acid on Europa and the radiolytic sulfur cycle. *Science*, **286**, 97–99.

Carr, M. H., *et al.*, (1998). Evidence for a subsurface ocean on Europa. *Nature*, **391**, 363–365.

Cassen, P., Reynolds, R. T., and Peale, S. J. (1979). Is there liquid water on Europa? *Geophys. Res. Lett.*, **6**, 731–734.

Chyba, C. F. (2000). Energy for microbial life on Europa. *Nature*, **403**, 381–382.

Chyba, C. F., and Phillips, C. B. (2001). Surface-subsurface exchange and the prospects for life on Europa. *Lunar and Planetary Science XXXII* (abstract 2140), CD-ROM.

Fagents, S. A., Greeley, R., Sullivan, R. J., Pappalardo, R. T., and Prockter, L. M. (2000). Cryomagmatic mechanisms for the formation of Rhadamanthys Linea, triple band margins, and other low-albedo features on Europa. *Icarus*, **144**, 54–88.

Fanale, F. P., *et al.* (1999). Galileo's multiinstrument spectral view of Europa's surface composition. *Icarus*, **139**, 179–188.

Figueredo, P., and Greeley, R. (2004). Resurfacing history of Europa from pole-to-pole geological mapping. *Icarus*, **167**, 287.

Gaidos, E. J., Nealson, K. H., and Kirschvink, J. L. (1999). Life in ice-covered oceans. *Science*, **284**, 1631–1633.

Geissler, P., *et al.* (1998). Evidence for non-synchronous rotation of Europa. *Nature*, **391**, 368–370.

Geissler, P., O'Brien, D. P., and Greenberg, R. (2001). Silicate volcanism on Europa. *Lunar and Planetary Science Conference* ***XXXII*** (abstract 2068).

Goodman, J. C., *et al.* (2004). Hydrothermal plume dynamics on Europa: implications for chaos formation. *J. Geophys. Res. – Planets*, **109**, E03008.

Greeley, R., *et al.* (2000). Geologic mapping of Europa. *J. Geophys. Res. Planets*, **105**, 22559–22578.

Greenberg, R. (1982). Orbital evolution of the galilean satellites, in *The Satellites of Jupiter*, ed. D. Morrison. Tuscon: University of Arizona Press, pp. 65–92.

Greenberg, R. (1989). Time-varying orbits and tidal heating of the Galilean satellites, in *Time-Variable Phenomena in the Jovian System*, eds. M. J. S. Belton, R. A. West, and J. Rahe. NASA Special Publication No. 494.

Greenberg, R. (2005). *Europa, the Ocean Moon*. New York: Springer-Praxis Books.

Greenberg, R., and Weidenschilling, S. J. (1984). How fast do Galilean satellites spin? *Icarus*, **58**, 186–196.

Greenberg, R., *et al.* (1998). Tectonic processes on Europa: tidal stresses, mechanical response, and visible features. *Icarus*, **135**, 64–78.

Greenberg, R., *et al.* (1999). Chaos on Europa. *Icarus*, **141**, 263–286.

Golombek, M. P., and Banerdt, W. B. (1990). Constraints on the subsurface structure of Europa. *Icarus*, **83**, 441–452.

Hall, D. T., *et al.* (1995). Detection of an oxygen atmosphere on Jupiter's moon Europa. *Nature,* **373,** 677–679.

Head, J. W., Pappalardo, R. T., and Sullivan, R. J. (1999). Europa: morphologic characteristics of ridges and triple bands from Galileo data (E4 and E6) and assessment of a linear diapirism model. *J. Geophys. Res.*, **104**, 24223–24236.

Helfenstein, P., and Parmentier, E. M. (1985). Patterns of fracture and tidal stresses due to nonsynchronous rotation: implications for fracturing on Europa. *Icarus*, **61**, 175–184.

Hoppa, G. V., *et al.* (1999a). Formation of cycloidal features on Europa. *Science*, **285**, 1899–1902.

Hoppa, G. V., *et al.* (1999b). Strike-slip faults on Europa: global shear patterns driven by tidal stress. *Icarus*, **141**, 287–298.

Hoppa, G. V., *et al.* (1999c). Rotation of Europa: constraints from terminator and limb positions. *Icarus*, **137**, 341–347.

Hoppa, G. V., *et al.* (2000). Distribution of strike-slip faults on Europa. *J. Geophys. Res. Planets*, **105**, E9, 22617–22627.

Hoppa, G. V., *et al.* (2001a). Observational selection effects in Europa image data: Identification of chaotic terrain. *Icarus*, **151**, 181–189.

Hoppa, G. V., *et al.* (2001b). Europa's rate of rotation derived from the tectonic sequence in the Astypalaea region. *Icarus*, **153**, 208–213.

Hubbard, G. S. (2005). Astrobiology: The First Decade. IAF International Astronautical Congress, Fukuoka, Japan.

Hurford, T. A., Sarid, A., and Greenberg, R. (2007). Cycloidal cracks on Europa: Improved modeling and rotation effects. *Icarus*, **186**, 218–233.

Kadel, S. D., *et al.* (1998). Trough-bounding ridge pairs on Europa – considerations for an endogenic model of formation. *Lunar and Planetary Science Conference XXIX*, (abstract 1070).

Kargel, J. S., *et al.* (2000). Europa's crust and ocean: origin, composition, and the prospects for life. *Icarus*, **148**, 226–265.

Kasting, J. F., *et al.* (1993). Habitable zones around main sequence stars. *Icarus*, **101,** 108.

Khurana, K. K., *et al.* (1998). Induced magnetic fields as evidence for sub-surface oceans in Europa and Callisto. *Nature*, **395**, 777–780.

Laplace, P. S. (1805). *Mecanique Celeste*, Vol. 4. Paris: Courcier. Translation by N. Bowditch, reprinted 1966. New York: Chelsea.

Lipps, J. H., and Rieboldt, S. (2005). Habitats and taphonomy of Europa. *Icarus*, **177**, 515–527.

Lucchitta, B. K., and Soderblom, L. A. (1982). The geology of Europa, in *The Satellites of Jupiter*, ed. D. Morrison. Tuscon: University of Arizona Press, pp. 521–555.

Lunine, J. I., and Lorenz, R. D. (1997). Light and heat in cracks on Europa: implications for prebiotic synthesis. *Lunar Planet. Sci. Conference XXVIII*, 855–856.

McEwen, A. S. (1986). Tidal reorientation and the fracturing of Jupiter's moon Europa. *Nature*, **321**, 49–51.

McCord, T. B., *et al.* (1998a). Salts on Europa's surface detected by Galileo's Near-Infrared Mapping Spectrometer. *Science*, **280**, 1242–1245.

McCord, T. B., *et al.* (1998b). Non-water-ice constituents in the surface material of the icy galilean satellites from the Galileo near-infrared mapping spectrometer investigation. *J. Geophys. Res. - Planets*, **103**, 8603–8623.

McCord, T. B., *et al.* (1998c). Organics and other molecules in the surfaces of Callisto and Ganymede. *Science*, **278**, 271–275.

McKinnon, W. B., (1999). Convective instability in Europa's floating ice shell. *Geophys. Res. Lett.*, **26,** 951–954.

Menou, K. and Tabachnik, S. (2003). Dynamical habitability of known extrasolar planetary systems. *Astrophys. J.*, **583**, 473.

Noll, K. S., Johnson, R. E., Lane, A. L., Domingue, D. L., and Weaver, H. A. (1996). Detection of ozone on Ganymede. *Science*, **273**, 341–343.

O'Brien, D. P., Geissler, P., and Greenberg, R. (2002). A melt-through model for chaos formation on Europa. *Icarus*, **156**, 152–161.

Pappalardo, R. T., and Sullivan, R. J. (1996). Evidence for separation across a gray band on Europa. *Icarus*, **123**, 557–567.

Phillips, C. B., and Chyba, C. F. (2001). Impact gardening rates on Europa. *Lunar and Planetary Science XXXII* (abstract 2111), CD-ROM.

Pilcher, C. B., Ridgeway, S. T., and McCord, T. B. (1972). Galilean satellites: identification of water frost. *Science*, **178**, 1087–1089.

Priscu, J. C. *et al.* (1999). Geomicrobiology of subglacial ice above Lake Vostok, Antarctica. *Science*, **286**, 2141–2144.

Prockter, L. M., *et al.* (1999). Europa: stratigraphy and geological history of the anti-Jovian region. *J. Geophys. Res. Planets*, **104**, 16531–16540.

Reynolds, R. T., *et al.* (1983). On the habitability of Europa. *Icarus*, **56**, 246–254.

Raymond, S. N., and Barnes, R. (2005). Predicting planets in known planetary systems II: Testing for Saturn mass planets. *Astrophys. J.*, **619**, 549.

Riley, J., *et al.* (2000). Distribution of chaos on Europa. *J. Geophys. Res. Planets*, **105**, E9, 22599–22615.

Riley, J., Greenberg, R., and Sarid, A. (2006). Europa's south pole region: A sequential reconstruction of surface modification processes. *Earth and Planetary Science Letters*, **248**, 808–821.

Sarid, A. R., *et al.* (2002). Polar wander and surface convergence on Europa: evidence from a survey of strike-slip displacement, *Icarus*, **158**, 24–41.

Sarid, A. R., *et al.* (2004). Crack azimuths on Europa: time sequence in the southern leading face. *Icarus*, **168**, 144–157.

Sarid, A. R., *et al.* (2005). Crack azimuths on Europa: the G1 lineament sequence revisited. *Icarus*, **173**, 469–479.

Schenk, P., and McKinnon, W. B. (1989). Fault offsets and lateral crustal movement on Europa: evidence for a mobile ice shell. *Icarus*, **79**, 75–100.

Smith, B. A., *et al.* (1979) The Galilean satellites and Jupiter: Voyager 2 imaging science results. *Science*, **206**, 927–950.

Spencer, J. R., Tamppari, L. K., Martin, T. Z., and Travis, L. D. (1999). Temperatures on Europa from Galileo PPR: nighttime thermal anomalies, *Science*. **284**, 1514–1516.

Spohn, T., and Schubert, G. (2001). Internal oceans of the galilean satellites of Jupiter. *Jupiter Conference*, Boulder, Colorado.

Squyres, S. W., *et al.* (1983). Liquid water and active resurfacing on Europa. *Nature*, **301**, 225–226.

Thomson, R. E., and Delaney, J. R. (2001). Evidence for a weakly stratified Europan ocean sustained by seafloor heat flux. *J. Geophys. Res.*, **106**, 12355–12365.

Tufts, B. R. (1998). Lithospheric displacement features on Europa and their interpretation, Ph.D. thesis, University of Arizona, Tucson.

Tufts, B. R., *et al.* (1999). Astypalaea Linea: a San Andreas-sized strike-slip fault on Europa. *Icarus*, **141**, 53–64.

Tufts, B. R., *et al.* (2000). Lithospheric dilation on Europa. *Icarus*, **146**, 75–97.

Varnes, E. S., and Jakosky, B. M. (1999). Lifetime of organic molecules at the surface of Europa. *Lunar and Planetary Science XXX* (abstract No. 1082), CD-ROM.

Wang, H., and Stevenson, D. J. (2000). Convection and internal melting of Europa's ice shell. *Lunar and Planetary Science XXXI* (abstract 1293).

Ward, P. D., and Brownlee, D. (2000). *Rare Earth.* New York: Copernicus Books.

Zahnle, K. L., *et al.* (2003). Cratering rates in the outer solar system. *Icarus*, **163**, 263–289.

Zimmer, C., Khurana, K. K., and Kivelson, M. G. (2000). Subsurface oceans on Europa and Callisto: Constraints from Galileo magnetometer observations. *Icarus*, **147**, 329–347.

Index